High Dilution Effects on Cells and Integrated Systems

Istituto Italiano per gli Studi Filosofici

Series on Biophysics and Biocybernetics

Coordinating Editor: C. Taddei-Ferretti

Vol. 1: Biophysics of Photoreception: Molecular and Phototransductive Events
edited by: C. Taddei-Ferretti

Vol. 2: Biocybernetics of Vision: Integrative Mechanisms and Cognitive Processes
edited by: C. Taddei-Ferretti

Vol. 4: Macromolecular Interplay in Brain Associative Mechanisms
edited by: A. Neugebauer

ISTITUTO ITALIANO PER GLI STUDI FILOSOFICI
SERIES ON BIOPHYSICS AND BIOCYBERNETICS
Vol. 3 – Biophysics

Istituto Italiano per gli Studi Filosofici

High Dilution Effects on Cells and Integrated Systems

Proceedings of the International School of Biophysics
Casamicciola, Napoli, Italy, 23–28 October 1995

Edited by

C. Taddei-Ferretti
Istituto di Cibernetica, CNR, Arco Felice, Napoli, Italy

P. Marotta
International Center for the Study of the History of Medicine, IRFEST, Université Louis Pasteur, Strasbourg, France and Istituto Italiano per gli Studi Filosofici, Napoli, Italy

World Scientific
Singapore • New Jersey • London • Hong Kong

Published by

World Scientific Publishing Co. Pte. Ltd.

P O Box 128, Farrer Road, Singapore 912805

USA office: Suite 1B, 1060 Main Street, River Edge, NJ 07661

UK office: 57 Shelton Street, Covent Garden, London WC2H 9HE

British Library Cataloguing-in-Publication Data
A catalogue record for this book is available from the British Library.

HIGH DILUTION EFFECTS ON CELLS AND INTEGRATED SYSTEMS

ISBN 981-02-3216-0

This book is printed on acid-free paper.

Printed in Singapore by Uto-Print

PREFACE

"... a meeting ... promising to seriously discuss a question which interests many but is often not dared to touch by scientists because they are afraid to loose their face ...": it is worthwhile to begin this volume with the words of Hennig Stieve. He was a member of the Advisory Board and an invited lecturer of the 1994 International School of Biophysics, as well as a participant in 1995 of the same International School, the Proceedings of which are in the present volume.

This volume is the third one of the Series on Biophysics and Biocybernetics promoted by the Istituto Italiano per gli Studi Flosofici.

The words of Hennig Stieve are related to the fact that the problem of the effects of highly diluted and ultra-highly diluted substances became a frontier between those who have the evidence of these effects in their therapeutical practice, but give explanations of a rather mystical type, and those who, due to the Avogadro-number obstacle, deny the possibility that the high-dilution effects be a subject of scientific investigation, although they may have a personal interest in it.

However, the scientific inconsistency of the motivation of such a refusal, which goes against the ideal of a search that is proper for the scientific enterprise, as well as the existence at present of several scientific institutions in the world in which the high-dilution problem is seriously investigated, induced the Istituto Italiano per gli Studi Filosofici to promote and to support the "International School of Biophysics on High Dilution Effects on Cells and Integrated Systems", which was inaugurated at the site of the Istituto, Palazzo Serra di Cassano, Naples, Italy, and was held at Casamicciola on the isle of Ischia, Italy, on October 23-28, 1995, under the direction of one of us (CTF).

The organization of the School was carried forward by the Istituto di Cibernetica of the Italian Consiglio Nazionale delle Ricerche (CNR).

The topics covered ranged from "Biophysical Aspects", to "Clinical Evidences" and to "Structure and Communication".

The reported effects of high dilutions on simple and complex systems (cells, organs and organisms of the vegetable and animal kingdom) included those on *in vitro* and *in vivo* models, cellular metabolism regulation, immune system, nervous system, intoxicated organs and organisms and developmental models.

The physical properties of high dilutions have been considered, such as the organization properties of water molecules in the presence and after the presence of solute molecules, the energy characteristics of empty and full water clusters and their dynamical interactions with proteins.

Among the mechanisms responsible for the high-dilution effects a non-molecular transfer of information has been hypothesized by several researchers.

Others have stressed the impact which the accumulated knowledge on high-dilution effects may have on a change of the scientific paradigm applied to medicine, toward the emergence of a new paradigm concerned with different levels of reciprocal information transfer, including environment, patient, symptoms, remedy, physician.

Also subjects only indirectly related to the high-dilution effect problem have received space among those of the International School of Biophysics and in the present volume. Thus the papers on biocommunication may be fruitfully considered; one paper deals with the organism sensitivity, a phenomenon that is met during the investigations on the highly diluted substances, also in some investigations reported in other contributions to the present volume; another paper is an example of the philosophical considerations that the high-dilution problem may trigger.

We would like to thank the members of the School Advisory Board for their fruitful advice and suggestions, as well as for their helpful cooperation also as reviewers of the participants' communications: M. Bastide (F), A.A. Berezin (CND), J. Jacobs (USA), M. Oberbaum (IL), N.C. Sukul (IN), R. Van Wijk (NL).

We would also acknowledge the partial financial support of the Gruppo Nazionale di Cibernetica e Biofisica, CNR, the Istituto di Cibernetica, CNR, and the Delchimica Scientific Glassware, as well as the auspices given by the Società Italiana di Biofisica Pura ed Applicata.

Furthermore, only due to the help of the Istituto Italiano per gli Studi Filosofici it has been possible to award grants to several deserving participants needing financial help.

We wish to thank all the scientists who agreed to lecture, who contributed to the School with their discussions and who favoured, toghether with all the participants, a charming atmosphere in a highly stimulating scientific milieu.

Finally, we thank the members of the local organizing committee: A. Cotugno, S. Piantedosi, S. Santillo, of the Istituto di Cibernetica, CNR; without their work the School could not have been realized, nor its cordial atmosphere obtained.

The beauty of the isle of Ischia and the courtesy of the staff of the Hotel Gran Paradiso at Casamicciola, where the School was held, completed the pleasantness of the environment.

Cloe Taddei-Ferretti

Paolo Marotta

CONTENTS

STRUCTURE AND COMMUNICATION

CLOSING DISCUSSION

PARTICIPANTS

INTRODUCTORY LECTURE

BASIC RESEARCH ON HIGH DILUTION EFFECTS

MADELEINE BASTIDE
Laboratory of Immunology,Faculty of Pharmacy,
University of Montpellier 1, France

ABSTRACT

Basic research on high dilution effects started with homeopathic therapy. So many models have been tested that we have tried to classify them according to the general concept of regulation. Firstly, succussed dilutions must be separated from unsuccussed very low doses from a physical point of view. This leads us to discuss the validity of the controls in high dilution research. Then, following a classification according to regulating effects, one can consider that some research in the field of "homeopathic research" can be relevant to cybernetic regulation and, in some cases, very low dose effects can be described as cybernetic regulatory signals. Hormetic models and application of the Arndt-Schultz law are based on the identity principle and are related to variations of concentration. They are presented and differentiated from the self recovery process which exists as a function of time. By using unsuccussed molecular and succussed non-molecular dilutions, the hormetic model supports a learning process which must be related to informative concepts. Starting from this primary level of informative process and by comparison with the phylogenic evolution of the immune system as an informative system, we can elaborate a progressive information organisation of the high dilution effects. Endogenous molecules have a specific regulatory function while highly diluted exogenous molecules will only be informative in the framework of the similia principle.

1. Introduction

Studies on the effects of high dilutions started with homeopathic therapy. Without such an example, no scientist could imagine to test such diluted substances that theoretically contain no molecules. It has been proved by therapeutical use that extremely diluted solutions even beyond Avogadro number could have biological effects. This is a very controversial affirmation which has been considered by the scientific community as a scientific error. It looks like an error because nobody is able to explain why and how these high dilutions can work However, experiments investigating this phenomenon have been performed by various scientists, most of them attempting to prove the effects of homeopathy.

Since many papers have been published, this work does not claim to list exhaustively all published research. The proposed models are so different and the results so various that it is necessary to clarify the subject. Many classifications can be proposed: we prefer trying to classify them without any chronological approach but by ordering them according to their relationship with the general concept of regulation. After describing rapidly the physical properties of such high dilutions, we propose a classification according to cybernetic regulation, hormesis, the Arndt-Schultz law and a new paradigme which implies the communication between living bodies by receiving signifying information.

2. Physical studies of high dilutions

The physical properties of high dilutions have been studied. A high dilution is a solution of a material (pure chemical or biological substance, vegetal or biological extract etc...) serially diluted with succussion between each dilution in a solvent, classically water or water+alcohol. This is also called potentised dilutions. The physical modifications of the potentised solutions have been tested by NMR by Demangeat *et al* [1, 2]. It seems that the successed diluted solution presents significant differences compared to the successed water control allowing the hypothesis of a remaining modification related to the original molecule. Apart from these two works, no scientific repeatable study has really been performed [3]. Two approaches may be proposed: the structuralist one, derived from the mechanistic paradigm, demonstrating a conservative structure of the original molecule printed in the solvent [4, 5, 6]. The second one is based on the diffusion of an informative carrier from the successed solution. This information has been transferred by Endler *et al* [7] or Van Wijk *et al* [8] either through glass from a sealed phial , or as an electromagnetic signal by mean of an input coil linked to a filter and an amplifier by Endler and Smith [9]. The high dilution signal of histamin transmitted to an isolated guinea-pig heart has been inhibited by a magnetic field [10]. Our personal observations of pharmacological activities of the successed solvent (compared to the unsuccessed solvent) in some biological models prompt us to think that the successed solvent is a modified structure avid to trap information from inside or outside the body of the receiver.

We must ask ourselves now whether the true control of a successed high dilution is the successed solvent. We will demonstrate in an other chapter [11] that the solvent plus a substance X, once diluted and successed, becomes a new substance X (the potentized substance X) which is different from successed solvent + successed substance. This observation is important considering the controls to be used to test pharmacologically successed remedies, the true control being the unsuccessed remedy or the unsuccessed solvent.

3. Pharmacological experimentations related to cybernetic regulation

Firstly, we understand now that the dilution-succussion is a fundamental modality to obtain pharmacologically active high dilutions. We will distinguish this kind of preparation even though they contain still molecules i.e. diluted less than 10C (10^{-20}) from the unsuccussed very low concentrations. We can find in classical research either very low concentrations of pheromons which act at the level of 10^{-16} M [12], or neuropeptides introduced directly into the brain at 10^{-6} pg[13], or kinds of catalytic effects of metal at picomolar concentrations[14] or immunomediators at 10^{-14} M[15] or neuropeptide activity on lymphocytes [16] as examples. No comparison with the effects of succussed high dilutions is available.

The pharmacological effect may be related to the presence of molecules. Some models published on "very low dose effects" are only related to a cybernetic regulation as a signal function. Two kinds of work presented as "very low dose effect" studies for the demonstration of the activity of low succussed dilutions have been published. Bonavida *et al* [17, 18, 19] demonstrates the synergistic activity of a cytotoxic cytokine, Tumor Necrosis factor α (TNF α), and cytotoxic drugs such as adriamycine or cis-platinum , or microbial toxins. To summarize, they demonstrate the increase of the cytotoxic effects by association of very low doses of TNF α (in a range of 6 to 600 picomols when associated with 0.1 to 10 nM of diphteria toxin). These concentrations are normally inactive. Bellavite *et al* [20, 21] report the reverse effect of a formylpeptide chemotactic substance (fMLP) on various phases of activation of human blood neutrophil granulocytes in vitro. The two concentrations giving opposite effects were 1-5 nM (low) and 100-500 nM (high). These two kinds of research are an excellent example of the confusion between cybernetic regulation as a function of a signal-molecule allowing modifications of the regulation (maybe by enhancing the cAMP formation in the cells by using the toxin in the TNF model). We are strictly in the cybernetic information network with a signal effect. In this case, only low potencies are active. The cybernetic regulation requests signals given at molecular level. Moreover, these are two "in vitro" models and no examples of this kind exist by using pharmacologically "in vivo" such *unsuccussed* low concentrations.

4. The self recovery, or rebound effect

There exists a phenomenon of inversion of effects as a function of time in the same individual which has received the stressor. We observed that immunosuppressed BALB/c mice by 100mg of cyclophosphamide show a very low lymphoblast transformation test (LTT) induced by mitogenic substances 2 days after the challenge, but an opposite effect is observed 10 days later when the LTT of the treated mice becomes statistically superior to the control mice. Arinaga demonstrates that a single dose of cisplatin in cancer patients stimulates the cytokine production of peripheral blood monocytes [27]. This self-recovery is also called the rebound effect and is the consequence of the immunosuppressive effect. It is a biological phenomenon which exists as *a function of time* after a strong pharmacological or toxic effect as shown in figure 1: the organism presents the opposite manifestations

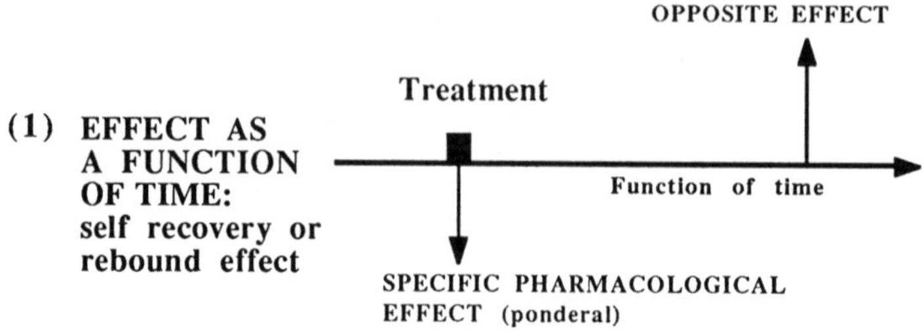

as a dynamic reaction against poisoning. It is not related to a general immunological mechanism (except in an immunosuppressive effect) and has no relationship with the placebo. This self-recovery is the reaction of the living body to aggression and self-recovery uses the appropriate and specific tools to reach a new equilibrium after the aggression . The forces it uses in order to find a new balance are so strong that they go beyond the final equilibrium.

5. Experimentations in the hormetic model

5.1. General principle of hormesis

The modern notion of hormesis originated with the observations of Southam and Erlich[22] and was developped by Stebbing [23]. Hormesis has no relationship with the similia law which is one of the characteristics of homeopathy. Hormesis is a very well known phenomenon which demonstrates the reverse toxic action of a substance which becomes a stimulating agent at a lower concentration (figure 2).

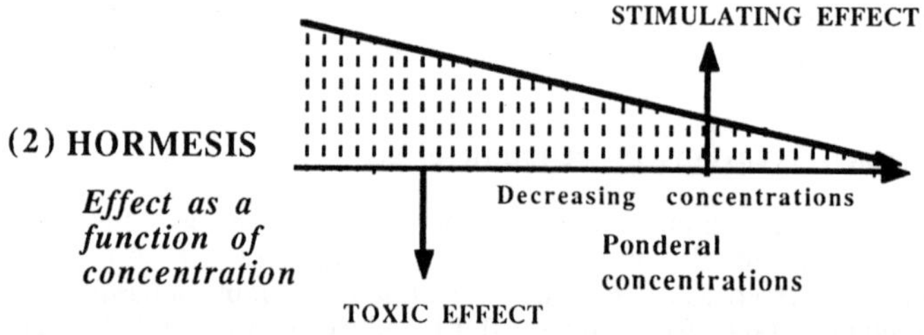

This was demonstrated with ponderal doses, the reversion of the effect being obtained only by a small decrease of the concentration (10-100 times lower). Models have been created with various poisons and on all possible organisms whatever their level in evolution, from procaryotic cells to plants, from eucaryotic cells to mammals (see the review by Oberbaum *et al*, [24]). This reverse effect was demonstrated even with very low doses by Wagner *et al* [25] who showed that the concentration of 25 μg of vincristine, a cytostatic remedy, inhibited the growth of lymphocytes while 1pg stimulated the cold treated cells.

5-2. The Arndt-Schultz law and the generalisation of hormesis

The demonstration of the reverse effect as a function of concentration is an ancient notion proposed in 1877 by Schultz [26]. Arndt at about the same time wrote that "weak stimuli slightly accelerate the vital activity, middle strong stimuli raise it and very strong ones halt it "[24].

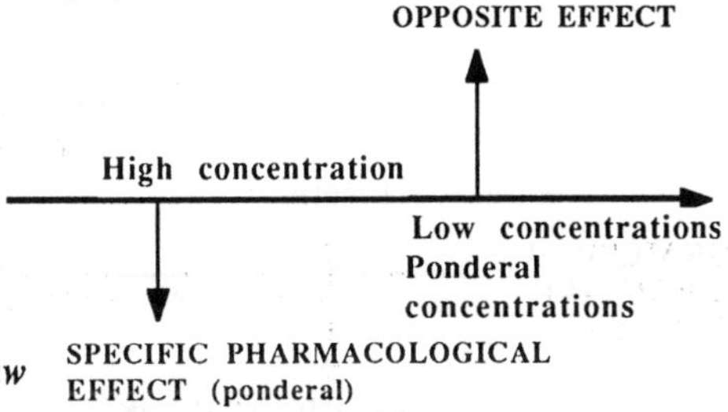

(3) HORMESIS applied to any agent
Effect as a function of concentration: the Arndt-Schultz law

We can propose a kind of generalisation which seems to be a general phenomenon whatever the substances (figure 3). For example, high concentrations of caffeine both stimulates and depresses respiratory centers; chemicals, alcaloids, metals etc.. are listed in manuals of Pharmacology with indications of the opposite effects observed as a function of concentration.

5.3. The application of hormesis: the principle of identity

The administration of low doses of a poison to an organism teaches it how to fight against the same poison. The famous king Mithridate, using daily small doses of various poisons, could not be poisoned and died by a sword-thrust. The protection is strictly given by identity between the pretreatment and the poison and is strictly specific (figure 4 and 5).

Pretreatment by low concentrations of poison have been used in cell models, in various living organisms. Heavy metals have been used, or arsenite, or many other products. Weis *et al* [28] showed that, after cadmium intoxication of fish, fin regeneration occurred more rapidly if a pretreatment with a lower dose of cadmium had been applied. Van Wijk et al [29, 30, 31] tried to analyze the general cellular adaptation syndrome by using cellular models. All the events described concern the adaptative process during the cell cycle after toxic (arsenite) treatment related to the action of heat shock proteins. All the mechanisms described are of interest and the action on the cell cycle is very well done. Another recent example is given by Conforti et al [32] in an immunological induction of arthritis in rats by injection of Freund adjuvant (*Mycobacterium butyricum* associated to paraffin oil) into the hind paw. Lower doses (10 to 100 times lower) of adjuvant injected intraperitoneally starting 6 days before the arthritogenic injection or 10 days after significantly decreased the arthritic process. The general observation of a stimulation of life and longevity has been observed after hormetic stimulation by toxic substances[33].

We propose that pretreatment allows a learning process aimed to establish a protection against the "waited danger". Specific appropriate tools are conceived and performed by the living organism in order to fight against the aggression. When the low dose is given after the toxic challenge, it allows an amplification of the organized defence, helping its achievement.

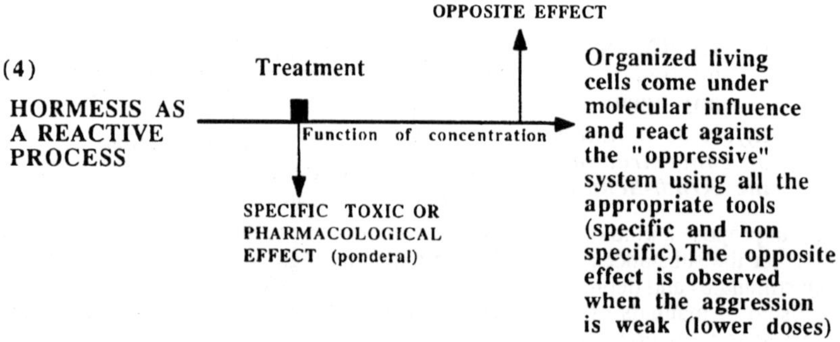

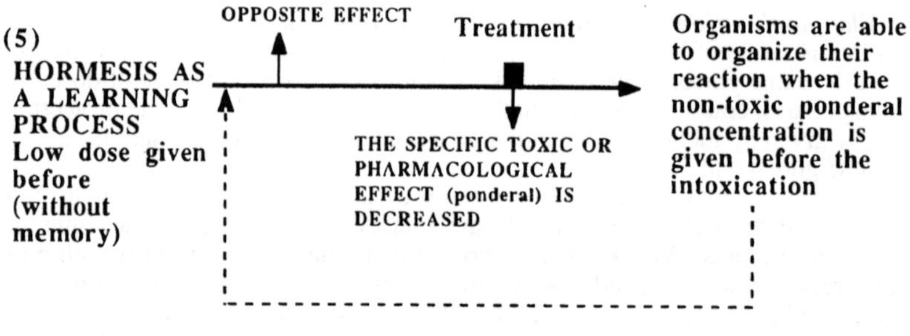

5.4. The principle of identity applied to high dilution effects

The same observations have been made by researchers using succussed high dilutions of the poison itself (figure 6).

This may be observed either with pretreatment or after poisoning: in the first case, the living organism is, as before, in the situation of a learning process; in the second case, the defence is increased by an addition of information. Succussed high dilutions are supposed to be an informing structure. Examples of protection by succussed high dilutions of heavy metal are given [34, 35] by using 10^{-30} or 10^{-40}M of mercury chloride. Taddei *et al* [36] study the effect of high dilutions of different teratogenic substances administered before and after the teratogenic challenge. The results demonstrate the specificity of the principle of identity as the best results preventing caffeine teratogenic effects were obtained with coffee; however the adenine did not protect against the adenine itself.

Doutremepuich *et al* [37, 38] have shown in vivo and in vitro that high dilutions of aspirin had a thrombogenic effect whereas classical ponderal doses have an antithrombogenic effect. The same principle of experimentation is presented by Pongratz on plants [39] by using.silver nitrate to stimulate the growth of wheat.

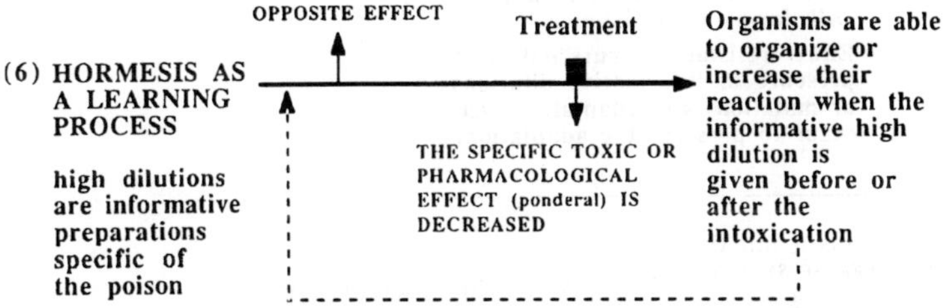

6. From the hormetic learning process to high level informative learning processes: model of the immune system phylogenesis

All forms of resistance seem to start with hormesis, this kind of primitive learning process which functions in all living organisms, from procaryotic cells to the most achieved mammals as Hominidies. We can propose that the resistance of bacteria to antibiotics, the resistance of parasites, the resistance of malignant cells to antimitotic substances (multi drug resistance) could originate in an hormetic model, a real learning process depending on the informative recognition of these poisons. The "memory" could be either a permanent presence of informative structures in water, for instance, and no true memory would exist or an induction of genetic mutations which would preserve the momentaneous adaptation. This kind of evolution does not belong to strict darwinism.

Starting from the hormetic model as a general law of defence of living organisms, it is interesting to analyse the evolution of the immune system which represents the achieved and specialized protective organisation of living organisms against pathogenic aggressions. The first organisation which appeared and which is preserved in the most ancient evolutionary branch of arthropods represented by insects belongs to the hormetic model (figure 7) [40]. The first observation of Metalnikoff [41] in 1920 demonstrated that caterpillars infected by bacteria were protected if they had received just before infection a small injection of the same bacteria. No memory was observed. The defense was non-specific. We now know that antibacterial peptides are induced by the first contact. No true immunological memory occurs (the true memory-lymphocytes only start with Chordates apparition when true lymphocytes with recognizing molecules appeared).

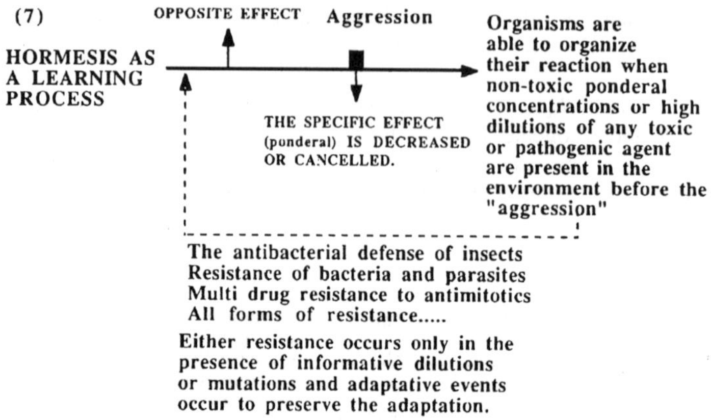

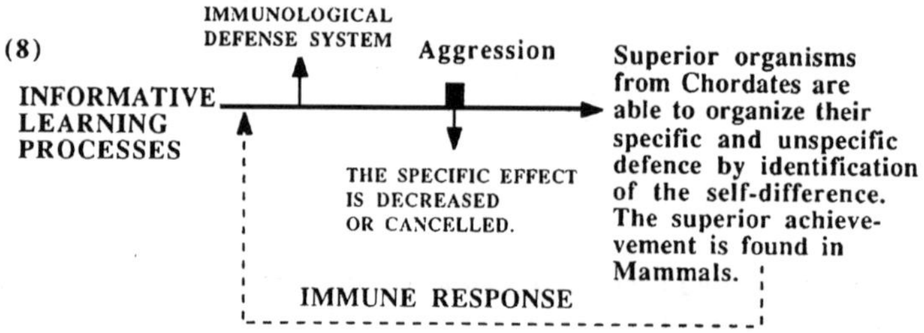

It is interesting to compare the immune defence of insects to the immune defence of mammals (figure 8). Mammals possess a recognizing structure allowing identification of the self represented by the Major Histocompatibility Complex (MHC).

The receptors of thymic lymphocytes are able to identify the "self difference": this phenomenon describes the fact that as long as T cell receptor is faced with a self structure alone such as MHC associated with a self antigenic peptide, the lymphocyte remain neutral. The moment it identifies a difference i.e.

the presence of an external antigenic peptide different from the self associated to the MHC, the immune system is activated.

If we compare these two extreme systems of defence, the archaic one as hormesis with a very simple learning process, no memory, and the achieved one, analysing the self, recognizing the "difference", with a strong physical memory, we can identify the evolution of organization structures able to perceive and treat information." According to the Bateson's definition, information is "a difference which makes a difference" [41].

7. Experimental models using high dilutions of endogenous molecules

Following the model of evolution in the informative sytems allowing the preservation of the species, we can now evoke other informative models using high dilutions belonging either to the self such as endogenous molecules, or not belonging to the self as exogenous molecules. It is easy now to understand that a framework for understanding is necessary in order to observe an "effect" of this information. Many experimental models have been carried out with endogenous molecules, most of them relevant to the immune system. Descriptions of an immunomodulatory activity of succussed dilutions of bursin or thymulin are described in this book [43, 44, 45, 46]; higly diluted histamin even placed in sealed vials or the highly diluted antigen itself was administered in isolated hearts of guinea pigs immunized with ovalbumin induced significant coronary flow variations [47, 48, 49]. Thyroxine was administered in highly diluted succussed solution (10^{-30}) to frogs at the end of the metamorphosis. Significant modifications of the motility were observed by Endler *et al* [40, 51]. In these models, high dilutions bring information of endogenous molecules which are automatically recognized by the living organism (whole body or organ). The effects of this endogenous molecule information belong to various mechanisms according to the nature and the level of information and the state of the receiver. A dilutiondependent effect characterizes a high level of information.

(9) INFORMATIVE PROCESSES BY ENDOGENOUS MOLECULES

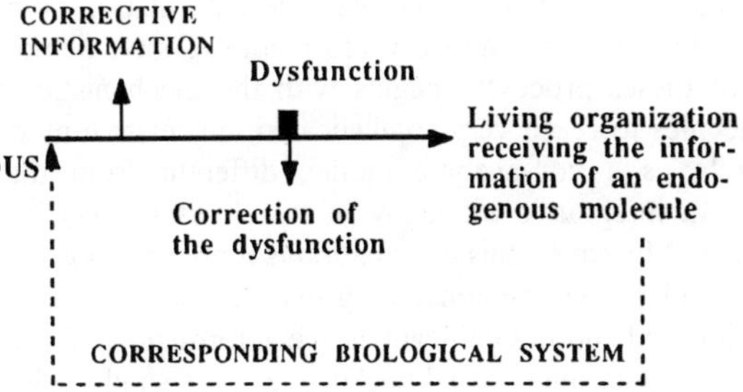

8. Experimental models using high dilutions of exogenous molecules

These models are based on the similia principle. Experimental models of this type are rare: they require the similarity between the symptoms of the remedy and the pathology as in the Oberbaum model [52, 53]. Results could be compared to the therapeutic effect in homeopathic therapies. The information of the remedy is able to modify the subject itself by switching to a"new" normal state·[11]. In these models, we only use the image of the remedy's symptoms.We need neither identity between poison and high dilution of the same poison nor pretreatment as in hormesis. These models clearly belong to a different approach which relies on signifying communication at the level of the living body [1 l].This is exactly what we call the homeopathic model.

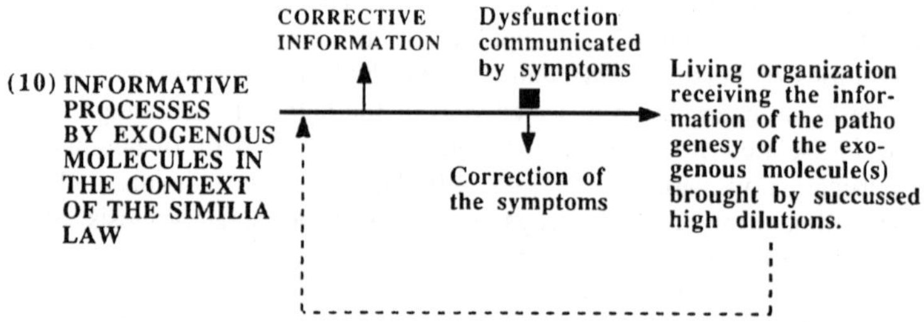

9. Conclusion

Basic research on high dilution effects leads us to consider many possibilities of regulatory mechanisms because a successed high dilution is the carrier of information corresponding to the original molecule. These regulation modalities are specific to living organisms able to receive and treat information. As we demonstrated in this chapter, the hierarchy of theses processes begins with the mechanistic cybernetic regulation using signal-molecules to reach very sophisticated informative modalities. The particularity of this hierarchy is a non-analytic model, different from the fractal model: we have established an integrative hierarchy in which each level is included in the following superior level. Moreover, this information is more and more significant for the body from the first to the last step, demonstrating that the more complex the biological system, the more specific and more significant is the information. We started from the molecular signal in cybernetics and reached tine' image in the similia principle.

References

[1] J.L.Demangeat, C.Demangeat, P.Gries, B.Poitevin, A.Constantinesco, "Modifications des remps de relaxation RMN à 4 MHz des protons du solvant dans les très hautes dilutions salines de silice/lactose", *J.Med.Nucl.Biophys*, 16, 2,pp 135-145, 1992.

[2] J.L.Demangeat, P.Gries, B.Poitevin, "Modification of 4 MHz N.M.R. water proton relaxation times in very high diluted aqueous solutions", in *Signal and Images*, M.Bastide Ed,Kluwer Academic Publisher, Dordrecht, in press.

[3] J.Schulte, P.C.Endler, "Outline of experimental physical methods to investigate specific structures of ultra high diluted solvents", pp 99-104, in *Ultra High Dilution, Physiology and Physics*, Endler and Schulte Eds, Kluwer Academic Publisher, Dordrecht, 1994.

[4] J.Schulte, "Conservation of structures in aqueous ultra high dilutions", pp 105-115, in *Ultra High Dilution, Physiology and Physics*, Endler and Schulte Eds, Kluwer Academic Publisher, Dordrecht, 1994.

[5] E. Del Giudice, "Is the memory of water a physical impossibility", pp 117-119,pp 5-19, in *Ultra High Dilution, Physiology and Physics*, Endler and Schulte Eds, Kluwer Academic Publisher, Dordrecht, 1994.

[6] G.S.Agnastostatos, "Small water clusters (clathrates) in the preparation process of homeopathy", pp 121-128, pp 5-19, in *Ultra High Dilution, Physiology and Physics*, Endler and Schulte Eds, Kluwer Academic Publisher, Dordrecht, 1994.

[7] P.C.Endler, W.Pongratz, C.W.Smith, J.Schulte, "Non-molecular information transfer from thyroxine to frogs with regard to homeopathic toxicology", *Vet.Human Tox.*, 37, pp 259-260, 1995.

[8] R.Van Wijk and F.A.C.Wiegant, "Physiological effects of homeopathic medicines in closed phials; a critical evaluation", pp 81-95, in *Ultra High Dilution, Physiology and Physics*, Endler and Schulte Eds, Kluwer Academic Publisher, Dordrecht, 1994.

[9] P.C.Endler, W.Pongratz, C.W.Smith, J.Schulte, F.Senekowitsch, M.Citro, "Non molecular information transfer from thyroxine to frogs", in *Signal and Images*, M.Bastide Ed, Kluwer Academic Publisher, Dordrecht, in press.

[10] L.Hadji, B.Arnoux and J.Benveniste, "Effect of dilute histamine on coronary flow of guinea-pig isolated heart. Inhibition by a magnetic field". *FASEB*, n°7040, 1992.

[11] M.Bastide, A.Lagache, "A new paradigm applied to high dilution effects on the living body", see the corresponding chapter in this book.

[12] O.Andersen, K.B.Doving, "Gonadotropin releasing hormone, a novel olfactory stimulant in fish", *Neuroreport*, 2, pp 458-460, 1991.

[13] S.Pavel, D.Psatta, R.Goldstein, "Slow-wave sleep induced in cats by extremely small amounts of synthetic and pineal vasotocin injected into the third ventricle of the brain", *Brain Res. Bull.*, 2, pp 251-254, 1977.

[14] J.Markovac, G.W.Goldstein "Picomolar concentrations of lead stimulate brain protein kinase C", *Nature*, 334, pp 71-73, 1988.

[15] J. Leung-Tack, J.Martinez, J.L.Sansot, Y.Manuel, A.Coll, "Inhibition of phagocyte functions by a synthetic peptide lys-pro-pro-arg (postin)", *Protides Biol.Fluids Proc.Colloq.*, 34, pp 205-208, 1986.

[16] F.Boudard, M.Bastide, "Inhibition of mouse T-cell proliferation by CGRP and VIP: effects of these neuropeptides on IL-2 production and cAMP synthesis", *J.Neurosc.Res.*, 29, pp 29-41, 1991.

[17] T.Tsuchitani, J.Zighelboim, J.Berek, B.Bonavida, "Potentiation of cytotoxicity against human ovarian cell-lines with combinations of subtoxic concentrations of tumor necrosis factor and adrianamycin or cisplatinum", J.Cell.Pharmacol., 2, pp 1-11, 1991.

[18] H.Morimoto, J.S.Safrit, B.Bonvida, "Synergistic effect of tumor necrosis factor α and diphteria toxin-mediated cytotoxicity in sensitive and resistant human ovarian tumor cell lines", J.Immunol, 147, pp 2609-2616, 1991.

[19] J.T.Safrit, B.Bonavida, "Hierarchy of in vitro sensitivity and resistance of tumor cells to cytotoxic effector cells, cytokines, drugs and toxins", Cancer Immunol. Immunother., 8 pages, Springer-Verlag Publisher, Heidelberg, 1992.

[20] P.Bellavite, S.Chirumbolo, G.Lippi, G.Andrioli, L.Bonazzi, L.Ferro, "Dual effcts of formylpeptides on the adhesion of endotoxin-primed human neutrophils", Cell.Biochem.Funct, 11, pp 231-239, 1993.

[21] P.Bellavite, S.Chirumbolo, C.Santonastaso, D.Biasi, S.Lussignoli, G.Andrioli, "Dose-dependence of the various functional responses of neutrophils to formylpeptides", in *Signal and Images*, M.Bastide Ed, Kluwer Academic Publisher, Dordrecht, in press.

[22] C.M.Southam, J.Erlich, "Effects of extracts of western red-cedar heartwood on certain wood-decaying fungi in culture", *Phytopathology*, 33, pp 515-524, 1948.

[23] A.R.D.Stebbing, "Hormesis- Stimulation of colony growth in Campanularia flexuosa, (hydrozoa) by copper, cadmium and other toxicants", *Aquatic Tox.*, 1, pp 227-238, 1981.

[24] M.Oberbaum, J.Cambar, "Hormesis: dose-dependent reverse effects of low and very low doses", pp 5-18, in*Ultra High Dilution, Physiology and Physics,* Endler and Schulte Eds, Kluwer Academic Publisher, Dordrecht, 1994.

[25] H.Wagner, B.Kreher, K.Jurcic, "In vitro stimulation of human granulocytes and lymphocytes by pico- and fentogram quantities of cytostatic agents", *Arzneim.Forsch./Drug Res.*, 38, pp 273-275, 1988.

[26] H.Schultz, "Uber die theorie der arzneimittelwirkung", *Virchows Archiv*, 108, pp 423-434, 1877.

[27] S.Arinaga, M.Adashi, N.Karimine, H.Inoue, T.Asoh, H.Ueo, T.Akiyoshi, "Enhanced induction of lymphokine-activated killer activity follwing a single dose of cisplatin in cancer patients", *Int.J.Immunpharmac.*, 16, pp 519-524, 1994.

[28] P.Weis, J.S.Weis, "Cadmium acclimation and hormesis in Fundulus heteroclitus during fin regeneration", *Environ.Res.*, 39, pp 356-363, 1986.

[29] R.van Wijk, H.Ooms, F.A.C. Wiegant, J.E.M.Souren, J.H.Ovelgönne, J.M.van Aken, A.W.J.M.Bol, "A molecular basis for understanding the benefits from subharmful doses of toxicants; an experimental approach to the concepts of hormesis and the homeopathic similia law", *Environ.Manag. Health,* 5, pp13-25, 1994.

[30] R.Van Wijk, M.Welters, J.A.Souren, H.Ovelgonne, F.A.Wiegant. "Serum-stimulated cell cycle progression and stress protein synthesis in C3H10T1/2 fibroblasts treated with sodium arsenite", *J.Cell.Physiology*, 155, pp 265-272, 1993.

[31] R.van Wijk, F.A.C. Wiegant, in *Cultured mammalian cells in homeopathy research- the similia principle in self recovery*, University Utrecht Publisher, 1994.

[32] A.Conforti, S.Lussignoli, S.Bertrani, R.Ortolani, G.Verlato, P.Bellavite, "Intraperitoneal administration of adjuvant inhibits the development of adjuvant arthritis in rats", Int.J.Immunopathol.Immunopharmacol., 8, pp 113-121, 1995.

[33] P.J.Neafsey, "Longevity hormesis, a review", *Mechanims Ageing Develop*, 54, pp 1-13, 1990.

[34] J.C.Cal, F.Larue, J.Guillemain, J.Cambar, "Chronobiological approach of protective effects of Mercurius corrosivus against mercury-induced nephrotoxicity", *Ann. Rev. Chrnonopharmacol.*, 3, pp 99 102, 1986.

[35] A.Delbancut, "Contribution à l'étude des effets de hautes dilutions de métaux vis-à-vis de la cytotoxicité du Cadmium sur des cultures de cellules tubulaires rénales", *Thèse Université Bordeaux II, France*, Juillet 1994.

[36] C. Taddei Ferreti and A Cotugno, "Treatment of the teratogenicity induced in mice by caffeine or adenine", *Progr.Biochem.Biotechnol.*, 3, pp.*in press*, 1995.

[37] C.Doutremepuich, O. De Seze, D.Le Roy, M.C.Lalanne, M.C.Anne, "Aspirin at very ultra low dosage in healthy volunteers effects on bleeding time, platelet aggregation and coagulation", *Haemostasis*, 20, pp 99-105, 1990.

[38] C.Doutremepuich, O.Aguejouf, D.Pintigny, M.N.Sertillanges and O. De Seze, "Thrombogenic properties of ultra-low-doses of acetylsalicylic acid in a vessel model of laser-induced thrombus formation", *Thrombosis Reseach*, 76, pp 225-229, 1994.

[39] W.Pongratz and P.C.Endler, " Reappraisal of a classical botanical experiment in ultra high dilution research. Energetic coupling in a wheat model", pp 19-26, in *Ultra High Dilution, Physiology and Physics*, Endler and Schulte Eds, Kluwer Academic Publisher, Dordrecht, 1994.

[40] J.A.Hoffman, J.L.Dimarc, P.Bulet, ,"Les peptides antibactériens inductibles des insectes", Med.Sci., 8, pp 432-439, 1992.

[41] S.Metalnikoff, "Immunité naturelle ou acquise des chenilles de Galleria mellonella, C.R.Soc.Biol., 83, pp 278-280, 1920.

[42] G.Bateson, *Steps to an ecology of mind*, Chandler Publisher, New-York, 1972.

[43] M.Bastide, F.Boudard, "High dilutions as a tool of immunomodulation", see the corresponding chapter in this book.

[44] M.Bastide, "Immunological examples on ultra high dilution research", pp 27-33, in *Ultra High Dilution, Physiology and Physics*, Endler and Schulte Eds, Kluwer Academic Publisher, Dordrecht, 1994.

[45] M.Bastide, M.Doucet-Jaboeuf, V.Daurat, " Activity and chronopharmacology of very low doses of physiological immune inducers",*Immunol. Today*, **6**, pp 234-235, 1985.

[46] B.J.Youbicier-Simo, F.Boudard., M.Mekaouche, M.Bastide., J.D.Baylé. "Effects of embryonic bursectomy and *in ovo* administration of highly diluted bursin on adrenocorticotropic and immune response of chickens", *Int.J.Immunotherapy*, 9, pp 169-180, 1993.

[47] L.Hadji, B.Arnoux, J.Benveniste, "Effect of dilute histamine on coronary flow of guinea pig isolated heart. Inhibition by a magnetic field", FASEB J., 5, pp 1583, 1991.

[48] J.Benveniste, B.Arnoux and L.Hadji, " Highly dilute antigen increases coronary flow of isolated heart from immunized guinea-pigs", *FASEB* n° 3900, 1992.

[49] M.H.Litime, J.Afssa and J.Benveniste,"Antigen signalling at high dilution", *FASEB*, n° 3488, 1992.

[50] P.C.Endler, W.Pongratz, R.Van Wijk, F.A.C.Wiegant, K.Waltl, M.Gehrer, H.Hilgers, "A zoological example on ultra high dilution research. Energetic coupling between the dilution and the organism in a model of amphibia", pp 39-68, in *Ultra High Dilution, Physiology and Physics*, Endler and Schulte Eds, Kluwer Academic Publisher, Dordrecht, 1994.

[51] P.C.Endler, Corresponding chapter in this book.

[52] M.Oberbaum, R.Markovits, Z.Weisman, A.Kalinkevits, Z.Bentwich," Wound healing by Homeopathic Silica Dilutions in Mice", *Harefuah*, 123, pp 79-82, 1992.

[53] M.Oberbaum, see the corresponding chapter in this book.

BIOPHYSICAL ASPECTS

HORMESIS

MENACHEM OBERBAUM
Ruth Ben-Ari Institute for Clinical Immunology, Kaplan Hospital, Rehovot, Israel

1. Introduction

Hormesis, also known by its synonyms Arndt-Schulz Law, Hueppe's rule, or hormoligosis, is a discipline that exists on the periphery of conventional science. The rules of hormesis use catain aspects of the *Law of Similars* in homeopathy, although the fields are not the same. The term hormesis, first proposed in 1943 by Southam and Ehrlich (1) and lata expanded by Luckey (2), describes the "stimulatory effect of subinhibitory concentrations of any toxic substance on any organism." The view that high and low concentrations of potentially toxic substances respectively suppress and stimulate biological systems developed from the observations of Southam and Ehrlich (1) that growth of fungal plant pathogens was enhanced by low concentrations of a western red cedar (*Thuja plicata*) extract that normally shows fungistatic or fungicidal properties at 10- to 100-fold higher concentrations. The authors attributed the growth-enhancement effect to an initial response that was followed by progressive desensitization to subinhibitory concentrations of a toxic constituent of the extract.

Even if not recognized as such, adaptations of the Law of Similarity abound in the practice of conventional medicine and pharmacology. For example, derivatives of mercury, whose toxic effects include oliguria and anuria, are used in many classic diuretic agents, such as mersalyl and chlormerodrin. Although toxic quantities of digitalis, a drug prepared from the seeds and leaves of the foxglove plant, can produce a clinical picture of arrhythmic tachycardia, treatment of atrial fibrillation is one of the major attributes of this drug. Emetine is a drug that at high doses causes ulcers of the colon and diarrhea, symptoms that resemble amebiasis. But nevertheless, conventional pharmacologists have long advocated using Emetine as an anti-amebic agent (3). Thus, even the most conservative scientist cannot denounce the Law of Similars.

The biological definition of hormesis was amplified by Neafsey (4), who called it "a term which has been applied to a variety of stimulatory responses to low doses of otherwise toxic substances which improve health and enhance longevity." Hormetic effects include "improved fecundity, hatching, viability, growth, weight gain, development, reproductive life span, and wound healing in exposed animals," as well as "P-450 enzyme induction, decreased tumor incidence, increased resistance to infection and disease, radiation resistance, and longevity enhancement" (30, 33-36) (cited from Ref. 4). The McGraw-Hill *Dictionary of Scientific and Technical Terms* defines hormesis as "providing stimulus by nontoxic amounts of a toxic agent." Despite reservations about many elements of the concept, the term has been accepted for phenomena in various areas where substances are suppressive in high concentrations and stimulatory at low concentrations.

Despite the wealth of papers that have been published on the subject during the past century, the category of *hormesis* has been added to *Biological Abstracts* only during the past decade. Interestingly, an on-line search of *Biological Abstracts* reveals that many of the articles listed to date relate to radiation hormesis in Japan, whose population was exposed to the fallout of two atomic bombs in 1945. Other information retrieval systems do not include this term or any of its synonyms. Although accepted as a part of conventional science, hormesis has long been the target of scorn and suspicion. Indeed, Hugo Schulz, a founder of the field, was ridiculed by conventional scientists as "the Griefswald homeopath" (5).

2. The Arndt-Schultz Law

In 1877 Hugo Schulz claimed that the effect of a stimulus on a living cell is indirect and proportional to its intensity or quantity (6). Concomitantly, the psychiatrist Rudolf Arndt proposed his *Basic Law of Biology,* which states that "weak stimuli accelerate slightly the vital activity, middle strong raise it, strong suppress, and very strong stimuli halt it" (7). After meeting with Arndt, Schulz experienced "a sudden enlightenment" that he had experimental proof of Arndt s basic postulate. He then set out to prove this law (8) by studying in yeast the effect of subtoxic concentrations of such poisons as iodine, bromine, mercuric chloride, arsenious acid, chromic acid, salicylic acid, and formic acid. Administration of low doses of the poisons to yeast cells increased their carbon dioxide production by three-to-six orders of magnitude, which Schulz (7) interpreted as stimulation of the yeast's metabolism.

From the results of their studies, Schulz conceived the Arndt-Schulz law stating that "every stimulus on a living cell elicits an activity, which is inversely proportional to the intensity of the stimulus" (7). Schulz postulated that if this biological law is correct, "it must be possible to use a sufficiently small quantity of a correctly chosen, organ-specific drug in order to raise the depressed vital energies to an extent that the physiological norm will be restored completely or as much as possible" (6). Both scientists concluded that the phenomenon of stimulation-depression was a fundamental characteristic of the excitability of protoplasm (8). Subsequently, Hueppe described similar phenomena in bacteria (9).

Early reactions to the work of Arndt and Schulz ranged from total rejection to reservations about the nonspecific character of the law. Clark (cited in Ref 4) claimed that polyphasic responses to drugs are often seen in complex systems and biphasic actions on tissues are not unusual. Because phasic phenomena are also observed in preparations of purified enzymes, there is no need to attribute "mysterious" properties to living tissues. Clark also stated that in many cases, high and low doses of drugs produce a single, graded response.

Acetylcholine, for example, exerts an inhibitory effect on frog heart muscle and a contractile effect on *rectus abdominis* fibers. The enhancing effect of low concentrations of cyanide on organisms or isolated organs is attributed to inactivation of inhibiktory trace metals.

Growth stimulation, in yeast for example, may occur because the destruction of one part of a population provides nutrition for the survivors.

Interest in hormesis flourished in the 1920s, stimulating many scientists not only to repeat the yeast experiments of Schulz but also to investigate very low concentrations of other toxic substances (10-25). Low doses of heavy metals (10), pesticides (11), antibiotics (12-14), trace elements (15), chlorinated hydrocarbons (16), and such physical agents as heat, cold, oxygen, and X-rays (17-20) were administered to such diverse organisms as fungi (1), bacteria (12-15), mammals (17), fish (18), insects (19), yeast (21), protozoa (22) algae (23), and plants (24). Although the protocols of the experiments would not meet the rigorous statistical standards demanded today by modern science, the findings provided evidence that the phenomenon of hormesis may be ubiquitous.

Holz *et al.* (25) repeated the experiments of Schulz, but had a different interpretation of the results. In response to the administration of either high or low concentrations of "yeast poisons," the yeast cells were shown to mobilize glycogen, which is metabolized via glucose to carbon dioxide and water. High concentrations of the poisons suppress glucose metabolism, whereas low concentrations do not. Thus, Holz believed that the so-called stimulation of carbon dioxide production reported by Schulz was simply due to the mobilization of glycogen and to the failure of low concentrations of "yeast poisons" to suppress glucose metabolism (cited from Ref 26).

A decade later Wilder (27) proposed to expand the concept of the Arndt-Schulz law as follows: "The direction of a response of a body function to any agent depends to a large degree on the initial level of that function at the start of the experiment. The higher this initial level, the smaller is the response to function-raising, and the greater is the response to function-depressing agents. At more extreme initial levels, there is a progressive tendency to no response, and to paradoxic reactions, i.e., a reversal of the usual direction of response."

Conversely, at lower initial levels, the response to function-raising agents is greater and the response to function depressing agents is decreased, which, according to Wilder, contradicts the conventional wisdom in medicine. He noted that "most investigators do not take the initial levels into account at all. If they do, they operate usually with the tacit assumption that the opposite of our law is true: e.g., that a hypertensive individual will respond to adrenalin with a higher rise in pressure than one who is normotensive."

Whereas the Arndt-Schulz law was based on pharmacological dosage, Wilder (27) perceived the varying sensitivity of different organisms to a given dose level as the basis for his rule: a small dose for one individual may be medium for another and large for a third, thereby affecting each individual in a different manner. In either case, "the reversal of the usual direction of response" is the phenomenon of interest (cited from Ref. 28).

3. Hormoligosis

In 1956 Luckey (29) proposed the term "hormoligosis" (Greek *hormos* - enhance; *oligas* - little, few) to describe a phenomenon that is induced by a substance that is stimulatory in small quantities, harmful in higher quantities (therapeutic), and toxic in high quantities. Luckey's idea was based on the observation that adding minute amounts of antibiotics to bird food accelerated the growth rate of germ-free birds. Because the growth-accelerating effect could not be due to changes in bacterial flora as the birds were germ-free, he postulated that the antibiotics directly stimulated growth in such birds. Using three standard pharmacology textbooks as references (namely, Goodman & Gilman; Sollman; Drill), Townsend and Luckey (30) published in 1960 a list of 100 substances that were known to inhibit at high concentrations to stimulate at low concentrations; they suggested adding to the list all known stimulatory substances because such materials become toxic at suff~ciently high concentrations. The authors characterized the following basic forms of dose-response curves (Fig. 1):

1. α-Curve: This is the most well-known pattern, which describes a situation in which low to subtoxic concentrations of a toxic substance produce no effect, whereas concentrations above a borderline progressively suppress. The

substances described by this curve do not appear on the Townsend-Luckey list because the curve describes the behavior of substances or drugs that do not evoke a stimulatory response.

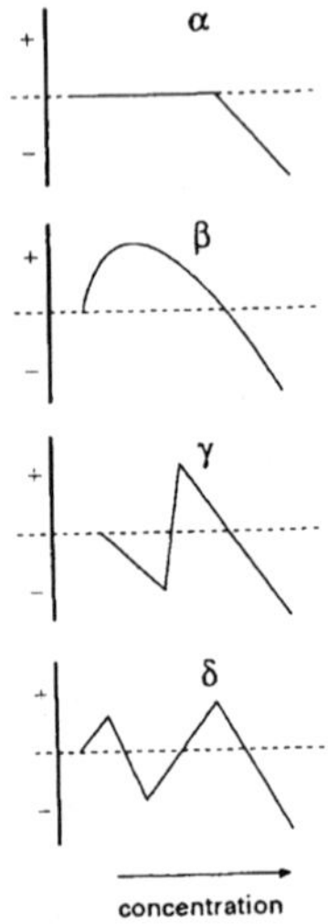

Figure 1 - The four basic types of doses-response curve:
α-curve, describing the simple augmentation of toxicity, when increasing the dose concentration.
β-curve, describing an initial stimulation followed by toxicity augmentation (hormesis).
γ-curve, where an initial depression is followed by a stimulation and an uncompensated toxicity, when increasing the dose (rare).
δ-curve, where a stimulation is followed by an inhibition, which in turn is followed by a stimulation, ending with an uncompensated toxicity (rare). (Modified from Ref. 30)

2. β-Curve: This pattern, which is observed most frequently, describes the behavior of most drugs that were examined by Townsend and Luckey. One peak of typical stimulation is observed at subtoxic concentrations, which is followed by progressive inhibition as the concentration increases to a toxic level. Any substance for which precise and comprehensive data are available will show a β-curve pattern.
3. γ- and δ-curves: The validity of these curves depicting low concentrations is questionable because the data are sparse.

Luckey called the phenomenon of hormoligosis *hormesis* and labeled all stimulatory agents *hormetica* (2), which later included vitamins and trace elements as well (32). Like Schulz, Luckey deemed hormesis as a biological law.

Developing the concept of hormesis further, Luckey stated that "any agent, be it physical, chemical, or biologic, may be expected to be stimulatory when it is given in much smaller doses that those found to be harmful. The recipient organism will respond more readily if it exists in suboptimum conditions of nurture and nature" (ref. 2, p. 88). Figure 2 shows what Luckey calls the *zero-equivalent point,* the concentration of any toxin at which no effect occurs. This concentration of a toxic material depends on diet, environment, condition, and animal species or strain.

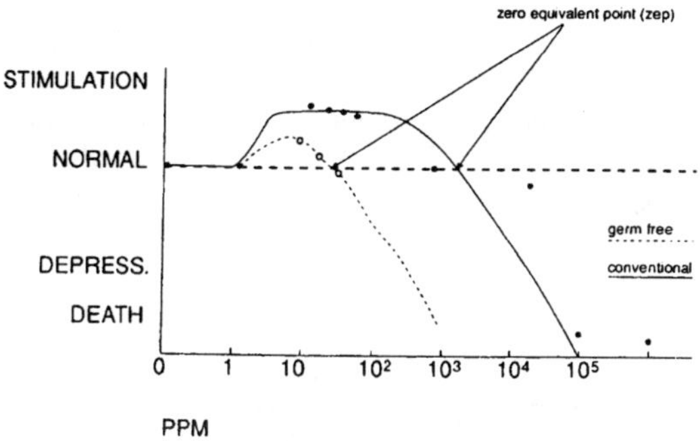

Figure 2 - Dose-response curve of oxytetracycline as a hormetic substance. Chicks received the oxytetracycline continuously for 4 weeks. The solid line represents the growth response of "classic chicks"; the broken line represents that of germfree chicks. The difference between the two groups suggests that microorganisms could have a particular influence on hormetic effects (Modified from Ref. 31)

At present, neither taxonomic limitations to growth hormesis nor limitations about the substances causing this phenomenon have been found. Because the same pattern of hormesis occurs in different organisms that have been exposed to low concentrations of various toxic materials (7,9,21,36-38), the control mechanism of this phenomenon must be aimed at low organizational steps and is not lost at high organizational steps (31,39).

Sagan (40-41) believed that because hormesis and toxicity must exist as two parallel phenomena, the stimulating, hormetic effect of a toxin does not exclude a toxic effect on other parameters (Fig. 3). Whether one effect overwhelms the other depends on the dose, environmental conditions, age, and class of the organism.

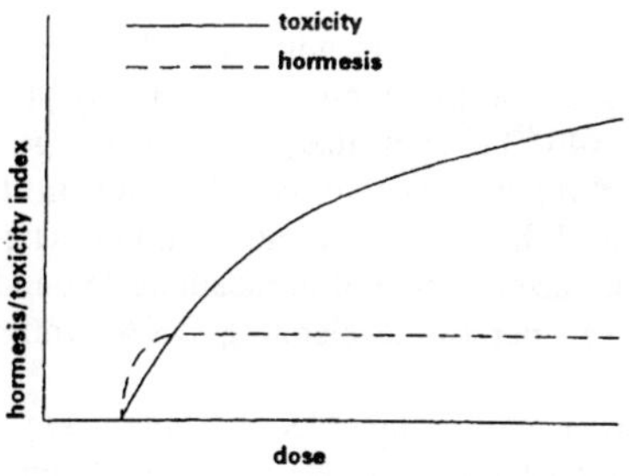

Figure 3 - Hormesis and toxicity as phenomena occurring in parallel. The overwhelming action of one on the other depends on dose (in addition to environmental conditions, age and class of organism) (Modified from Ref. 41)

In the 1980s, the following empirical formula was proposed to explain and predict the hormetic phenomenon (42-44).

$$E(D)=\alpha 0 - \alpha_1 D + \alpha_2 D_2$$
$$\alpha i > 0$$

where *E(D) is* the effect (cancer, for example) at dose *D,* and αi values are constants. Because *E(D) is* negative at low doses, the benefit outweighs the harmful effect, whereas the damaging effect prevails at high concentrations (Fig. 4) (cited from 4).

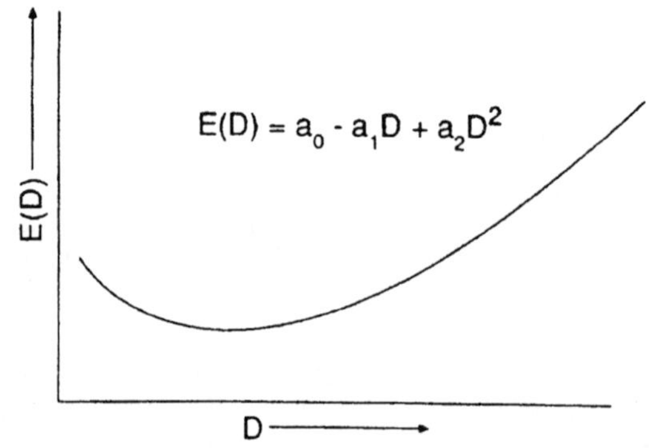

Figure 4 - Linear quadratic model of hormesis and toxicity. E(D) versus D where E(D) is effect (e.g. of cancer) at dose D. At low doses, E(D) is negative and the hormetic, beneficial effect overshadows the harmful one. At high doses the harmful, toxic effect predominates (Modified from Ref. 43)

4. Possible Mechanisms

The biochemical mechanism of the phenomenon of hormesis has not yet been explained. Liu *et al.* (45) suggest that the immune system may play a role in radiation hormesis, whereas Congdon (46) believes that a rebound regenerative hyperplasia during the early stage of chronic exposure could result in radiation longevity. Feinendegen *et al.* (47) advanced the theory that radiation hormetic effects could result from the low-dose, radiation-induced release of intracellular free radicals, which in turn could inhibit thymidine kinase, thereby decelerating DNA synthesis and enabling a prolonged period of DNA repair.

Chelation or complexation may be involved in some cases, such as claims that copper augments the growth of bacteria by displacing micronutrients, which then allows growth (48,49). According to Nickel and Finlay (24), antibiotics form complexes with toxic metabolises in plants or increase the permeability of the plant cell wall, thereby increasing the uptake of water, which facilitates plant growth. Dunstan *et al.* (50) agreed that antibiotics increase the permeability of the cell wall, thus creating favorable growth conditions.

Fabricant (51) showed a compensatory cell proliferation response in mice that had been exposed to continuous low-dose radiation, manifested by an increased cell-birth rate, a decreased cell loss in certain tissues, and a shortened stem-cell cycle. The accelerated cell-birth rate was caused by the following elements: (a) regenerative proliferation of surviving cells; (b) enlargement of the proliferative compartment because of cell migration from nonirradiated areas of the tissue; (c) increased output from the stem-cell compartment; and (d) prolonged transit time through the proliferative compartment.

Kondo (52) has suggested that *altruistic cell suicide* might explain the phenomenon of radiation-induced hormesis. In this scenario, eliminating the injured cell would benefit the organism as a whole if the presence of a dead cell stimulated the proliferation of healthy cells. The author calls this event *cell-replacement repair* in contrast to DNA repair, which he thinks is incomplete in most cases. After reviewing a study by James and Makinodan (53), Kondo concluded that in normal mice, the suicide of radiosensitive T-cells stimulated the activation and proliferation of precursor T-cells, thereby augmenting the production of effector T-cells. The enhanced radiation hormetic effect seen in immunosuppressed mice was due to a larger population of radiosensitive cells. Kondo then suggested that continuously upregulating the immune response would augment antitumor responses, which could result in longevity hormesis if spontaneously occurring tumors were susceptible to the cytotoxicity of killer cells.

Another possible explanation for hormetic mechanisms is that stress-protein synthesis might cause the life-enhancing effect (4,54): Heidrick *et al.* (54) showed that the antioxidant activity of 2-mercaptoethanol retarded free-radical damage in spleen cells of exposed mice. As 2-mercaptoethanol is also known as an inducer of glucose-regulated stress protein in mammalian cells, it is possible that a connection exists between the enhanced longevity resulting from 2-mercaptoethanol administration and the production of stress protein.

Thorough investigations by Stebbing (31,38,39,54-56) revealed that the dose-response curve (39) is the only element that is common to all the hormetic phenomena studied.

In *in vitro* growth experiments on the colonial hydroid *Campanularia flexulosa,* Stebbing first showed that providing a solution containing low concentrations of copper ion elicited typical growth-enhancing effects, whereas high concentrations suppressed growth. Initially, the author assumed that the concentration of copper ion in sea water must be suboptimal for growth of the organism and that adding a low concentration of copper ion to the experiments enhanced growth. He rejected this idea, however, when he found that such other toxicants as cadmium, synthetic biocides, and a low concentration of salts evoked the same phenomenon (38).

In the following experiments, Stebbing *et al.* (56) showed that after exposure to a toxic agent, the growth rate oscillates with a decreasing amplitude until equilibrium is reached (Fig. 5A). After observing a similar curve in a different system, *Rhodotorula rubra,* (Fig. 5B), he postulated that the oscillating growth rate represents the response of the control system to an input that disturbs the equilibrium (58). Stebbing (39) noted that similar toxin-induced oscillatoric perturbation of the growth rate had been observed by other scientists. Thus he concluded that growth hormesis may be "the consequence of regulatory overcorrection by biosynthetic control mechanisms to low levels of inhibitory challenge resulting in growth that is greater than normal" (31). Thus, Stebbing believes that in itself, growth increase is not an adaptation to reduce the growth inhibition but rather a consequence of the adaptive behavior of ratesensitive control mechanisms of the

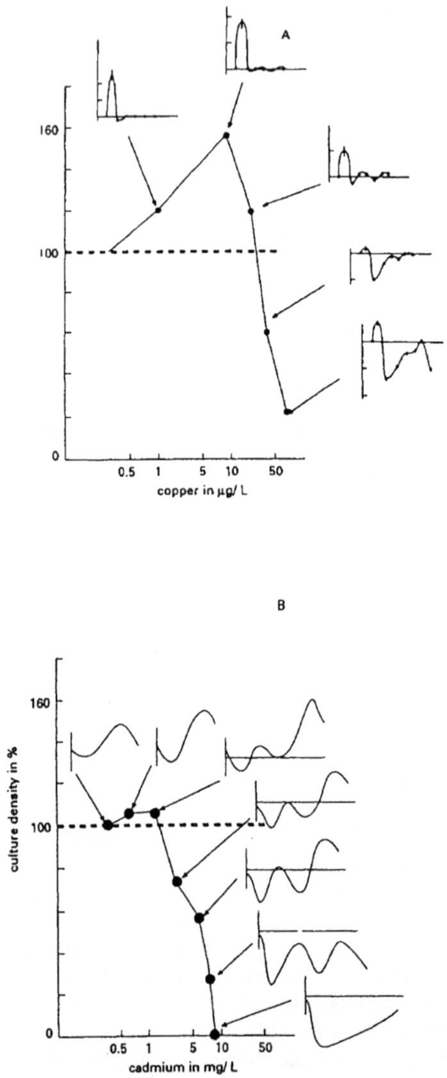

Figure 5 - Curve A - Mean size of the colonies of Laomedea flexuosa after exposure to different doses of copper. Curve B - Mean size of cultures of Rhodotorula rubra after exposure for 8 hours to different concentrations of cadmium. The small curves describe the oscillative perturbation of the growth rate after exposure to the toxicants (modified from Ref. 51)

organism. Whereas hormesis appears to be the consequence of regulatory overcorrection on the one hand, inhibition can be interpreted as the consequence of regulatory undercorrection on the other. When the concentration of toxic agents is high, the ability of the organism to contraregulate is rapidly exhausted, which leads to a fast, noncompensable decrease in the growth rate.

Stebbing perceives hormesis as a simple cybernetic system (31), in which oscillation depends on how the receptor, the integration system, and the effector react to the specific properties of the perturbation. He also considers that the amplitude of the initial overcorrection is not the only factor affecting the strength of the hormetic response. For example, Fig. 5B shows that the strongest hormetic effect is exerted by a concentration of 5 Cu^{++}/l rather than by a concentration of 1 Cu^{++}/l where the overcorrection is the strongest.

5. Longevity Hormesis

Until the 1960s, the main thrust of research on hormesis was the growth-enhancing aspect. Sacher and Trucco (59) were the first to connect longevity enhancement with the Arndt-Schulz law by demonstrating that in animals, low doses of toxicants increase and improve survival. The authors attributed this phenomenon to an induction of increased informational content of systems, which is equivalent to a decreased fluctuation variance of the system. During the next decade, more evidence accumulated about the longevity-enhancing effect of low doses of toxicants.

Hormetic effects had to be distinguished from other life-prolonging efforts, such as reduction of caloric intake (4,61-63).

In 1977 Sacher (60) published a book chapter called the "Life Table Modification and Life Prolongation," where he differentiated between *proper action* and *real hormetic action.* He defined proper action as reducing the accumulation of aging lesions by a specific biochemical mechanism that exerts a direct protection against age-dependent deterioration (such as vitamins, steroids, other hormones, and essential substances). By contrast, he characterized hormetic effects as nonspecific because they can be induced by a variety of chemically different organic and inorganic substances, radiation, and other stimuli (4). Sacher states that a hormetic response is a function of the state of the organism rather than that of the stimulant. He emphasizes that hormesis does not increase the longevity potential but rather enables the biological system to attain its inherent longevity. Thus, hormetic effects can be achieved in depressed or ill patients but not in healthy individuals.

6. Discussion

Although the mechanism underlying hormesis has not yet been explained, the observations that have accumulated to date indicate that any potential mechanism has the following four characteristics:

1. The hormetic effect occurs after exposure to a low concentration of an otherwise noxious agent.
2. The hormetic effect is reversible and does not accumulate after termination of exposure to the causal agent.
3. The hormetic effect can be induced by a variety of chemically unrelated substances.
4. The hormetic mechanism must be conserved throughout evolution.

Many reports on hormesis have been published, of which a small representative collection is cited in this chapter. Serendipitous hormetic phenomena that occur in conventional clinical and scientific research are traditionally ignored as artifacts or anecdotes. Such attitudes have emerged because hormetic effects are inconsistent with the basic principle of conventional science that a stronger stimulus must exert a stronger effect and, conversely, that a weaker stimulus exerts a weaker effect. Thus, an enhancing effect of lower doses of a noxious agent would not be accepted according to this paradigm, which implies that the toxic effect of a toxic agent can be reduced only by reducing the quantity, and that the reduced quantity cannot enhance or have a positive effect. That proof of a possible mode of action does not yet exist for this phenomenon further compounds these difficulties. Most research on hormesis has been conducted to demonstrate not only the existence of the phenomenon but also its universality. Very few investigators have undertaken to explore the underlying mechanisms of hormesis.

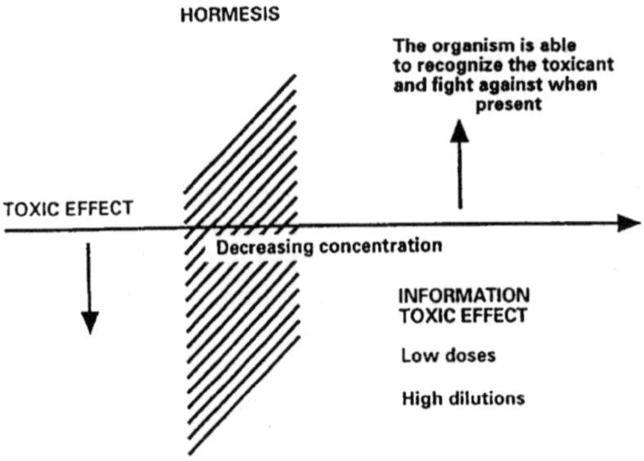

Figure 6 - Hormesis as an information process. The "toxic end-point concentration" is followed by atoxic concentrations that are unable to kill the living system but able to inform the living system about the properties of the toxicants (Modified from Bastide M., Lagache A., Essentials Elements I. Atelier Alpha Bleue, Paris p. 18)

Hormesis is not synonymous with homeopathy. Classic hormetic effects occur at dilutions that are much lower than those used in homeopathy, precluding the inevitable questions about how "such ultra-high dilutions, which do not contain even a single molecule, still retain biologic activity." Moreover, in contrast to homeopathy, no special techniques are required to prepare the active agent in hormesis. Nevertheless, certain similarities between homeopathy and hormesis that are beyond those of the basic laws cannot be ignored, especially where the common areas . contradict conventional standards. The following aspects of homeopathy are similar to the concept of hormesis:

1. Evidence has not yet emerged about whether homeopathic remedies can be prophylactic. Treatment by homeopathy requires that the individual display symptoms, which implies the presence of either a disease or a suppressed system. This principle agrees with Sacher's (58) idea that a depressed or ill system is a basic requirement of hormesis. Rather than being a real life prolongation, the hormetic effect abrogates certain deleterious environmental, life-shortening forces (4).

2. The principles for diluting and potentiating homeopathic remedies are based on the rationale that a toxic or suppressive substance is diluted to eliminate its noxious properties so that its enhancing properties emerge: thus, the higher the dilution, the stronger the enhancing activity of the potentially toxic substance.

Such similarities may suggest that homeopathy and hormesis may have a common basis, which could be an information process. Hormesis has been explained as a biochemical modification that is related to every toxic effect: although a biochemical mechanism that is common to all hormetic effects has not yet been discovered, each distinct mechanism is appropriate for enabling the biological system to adapt by neutralizing a particular toxic effect. Such adaptability implies that the living system can recognize the aggression and receive it as information. This approach would explain the variability of the defense mechanisms observed in various hormetic models.

Similarly, according to the hypothesis of Lagache (64,65) and Bastide and Lagache (66,67), the homeopathic remedy represents an informative template of the pathology of the subject that helps the organism to react. For example, high concentrations of a heavy metal can kill an organism, but the lethal effect ceases after diluting the substance to a critical point, called the toxic end-point concentration. Atoxic concentrations beyond this point are *unable* to kill the living system but are *able* to inform the system about the properties of the toxicant (Fig. 6), thus preparing the living system to recognize the toxicant and to attack it by developing an appropriate antagonistic mechanism that is adapted to the invading substance. Larue *et al.* (68) and Cal *et al.* (69) have shown that similar information can be obtained from high dilutions of toxicants. This new paradigm, designated as the *paradigm of signifiers,* presents hormesis as a general application of information processing.

References

[1] Southam C.M., Ehrlich J. Effects of western red cedar heartwood on certain wooddecaying fungi in culture. *Phytopathology* 1948; 33, 515-524.

[2] Luckey T.D. Hormology with inorganic compounds. In *Heavy metal-toxicity, safety and hormology*. Luckey TD, Venugopal B, Hutchenson (eds.) 1975. pp. 81-103. George Thieme Publ. Stuttgart-New York-San Francisco-London.

[3] Jouanny J., *The Essentials of Homeopathic Therapeutics*. pp. 14-15. Laboratoires Boiron,-Lyon-France: 1985.

[4] Neafsey P.J., Longevity Hormesis. A Review. *Mechanism of Ageing and Development*. 1990; 51, 1-31.

[5] Boxenbaum H., Neafsey P., Fournier D., Hormesis, Gomperz function and risk assessment. *Drug Met. Rev*. 1988; 19, 2: 195-229.

[6] Schulz H., Ueber die Theorie der Arzneimittelwirkung. *Virchows Archiv* 1877; 108: 423-434.

[7] Martius F., Das Arndt-Schulz Grundgesetz. Muench. *Med. Wschr.* 1923; 70, 31:

[8] Schulz H., Ueber hefegifte. *Pflueger Archiv fuer Physiologie* 1988; XLII: 517

[9] Hueppe F., *The Principles of Bacteriology*. 1896. Open Court. Chicago.

[10] Stare F.J., Arsenic as growth promotor. *Nutr. Rev.* 1956; 14, 206-209.

[11] Chapman R.K., Allen T.C., Stimulation and suppression of some vegetable plants by DDT. *J. Econ. Ent.* 1948; 41, 616-623.

[12] Finklestone-Sayliss, Sheff M.D., Paine C.G., Patrick L.B., The bacteriostatic action of p-amino-benzenesulphonamide upon haemolytic streptococci. *The Lancet* October 1937; 792-795.

[13] Randall W.A., Price C.W., Welch H., Demonstration of hormesis (increase in fatality rate) by penicillin. Am. *J. Public Health* 1947; 37, 421-425.

[14] Miller M.W., Green C.A., Kitchen H., Biphasic action of penicillin and other sulphonamide similarity. *Nature* 1945; 155, 210-211.

[15] Hochkiss M., Studies on salt action. VI. The stimulating and inhibitive effect of certain cations upon bacterial growth. *J. Bacteriol.* 1923; 8, 141-162.

[16] Hammett F.S., Reiman S.P., Effects of methylcholanthrene on developmental growth of obelia geniculata. *Am. J. Cancer* 1935; 25,807-808.

[17] Binger C.A.L., Faulkner J.M., Moor R.L., Oxygen poisoning in mammals. *J. Exp. Med.* 1927;45, 849-864.

[18] Faulkner J.M., Birger C.A.L., Oxygen poisoning in cold blooded animals. *J. Exp. Med.* 1927; 45, 865-871.

[19] Davey W.P., Prolongation of life of Tribolium confusum apparently due to small doses of X- Rays. *J. Exp. Zool.* 1919; 28, 447-458.

[20] Taylor N.B., Weld C.B., A study of the action of irradiated ergosterol and of its relationship to parathyroid function. *J. Cand. Med. Assoc.* 1931; 25,20-34.

[21] Branham S.E., The effects of certain chemical compounds upon course of gas production by Baker's yeast. *J. Bakteriol* 1929; 18, 247-264.

[22] Mottram J.C., An increase in the rate of growth of paramecium subjected to the blastogenic hydrocarbon 3:4 benzpyrene. *Nature* 1939; 144, 154.

[23] Dunstan W.M., Atkinson L.P., Nadi J. Stimulation and inhibition of phytoplankton growth by low molecular weight hydrocarbons. *Marine Biol.* 1975; 31, 305-310.

[24] Nickel L.G., Finlay A.C., Growth modifiers: antibiotics and their effects on plant growth. *Agric. Food Chem.* 1954; 2, 178-182.

[25] Holz P., Exner M., Schuman H.J., Zellreiz und Zellschaedigung: Untersuchungen ueber die "Reizgaerung" und "Reizatmung" der Hefezelle. *NaunynSchmiedeberg Archiv fuer Exptl. Pathol. und Pharmakol.* 1948; 205, 243-275.

[26] Linde K., *Dosisabhaengige Umkehreffekte eine differenzierende Literaturbetrachtung*. Dissertation presented to the Faculty of Medicine-LudwigMaximilian University, Munich 1991.

[27] Wilder J., The law of initial value in neurology and psychiatry: Facts and problems. *J. Nerv. Mental Dis.* 1957; 125, 73-86.

[28] Coulter H.L., Homeopathic Science and Modern Medicine, the Physics of Healing with Microdoses. *North Adantic Books*. 1981.

[29] Luckey T.D., in Proceedings of the First International Conference on the Use of Antibiotics in Agriculture, Publ. 397, *The National Academy of Sciences National Research Council*, Washington DC, 1956; pp. 135-145. [cit. from Ref 6, Boxenbaum H. et al.].

[30] Townsend J.F., Luckey T.D., Hormoligosis in Pharmacology. *J. Am. Med. Assoc.* 1960; 173, 44-48.

[31] Stebbing A.R.D., Hormesis-the stimulation of growth by low levels of inhibitor. *Science of the Total Environment* 1982; 22, 213-234.

[32] Luckey T.D., Activity spectrum of ingested toxicants. *Comp. Anim. Nutr.* 1977; 2, 144-178

[33] Calabrese E.J., McCarthy M.E., Kenyon E., The occurrence of chemically induced hormesis. *Health Phys.* 1987; 52, 531-541.

[34] Luckey T.D., *Hormesis with Ionizing Radiation.* CRC Press, Boca Raton, FL, 1980.

[35] Luckey, T.D., Physiological benefit from low levels of ionizing radiation. *Health Phys*. 1987; 43, 771-789.
[36] Furst A., Hormetic effects in pharmacology. *Health Phys*. 1987; 52, 527-531.
[37] Richet C., De ['action des doses minuscules de substances sur la fermentation lactique: Periodes d'acceleration et de relentissement. *Archiv. Intern. Phisiol.* 1906; 4, 18-50.
[38] Stebbing A.R.D., Hormesis- stimulation of growth in Campanularia flexuosa (Hydrozoa) by cadmium copper and other toxicants. *Aquatic Toxicol*. 1981; 1, 227-238.
[39] Stebbing A.R.D., Growth hormesis: a byproduct of control. *Health Phys*. 1987; 52, 543-547.
[40] Sagan L.A., What is hormesis and why haven't we heard about it before7 *Health Phys*. 1987; 52, 521-525.
[41] Sagan L.A., On radiation, paradigm, and hormesis. *Science*. 1989; 245.574.
[42] Hickey R.J., Bowers EJ, Clelland RC. Radiation hormesis, public help, and public policy: a commentary. *Health Phys*. 1983; 44, 207-219.
[43] Hickey R.J., More comments on radiation hormesis, epidemiology and public health. (letter). *Nuclear News. Mail* 1984; 46, 1159-1160.
[44] Hickey R.J., Clelland RC, Bowers EJ. More comments on radiation hormesis, epidemiology and public health. *Health Phys*. 1984; 46, 1159-1160.
[45] Liu Z.S., Liu W.H., Sun B.J., Radiation hormesis: its expression in the immune system. *Health Phys*. 1987; 52, 579-583.
[46] Congdon C.C., A review on certain low- level ionizing radiation studies in mice and guinea pigs. *Health Phys*. 1987; 52, 593- 597.
[47] Feinendegen L.E., Muehlensiepen, Bond V.P., Sondhaus CA. Intracellular stimulation of biochemical control mechanisms by low-dose, low-let irradiation. *Health Phys*. 1987; 52, 663-669.
[48] Jones G.E., in Hood D.W., (Ed.). University of Alaska Press, Alaska, *Organic Matter in Natural Waters* 1970.
[49] Davis H.C., Hidu H., Effects of pesticides on embryonic development of clams and oysters and on survival and growth of the larvae. *Fish Bull. Fish Wildl. Serv. US*. 1969; 67, 393- 404.
[50] Dunstan W.M., Atkinson L.P., Natoli J., Stimulation and inhibition of phytoplankton growth by low molecular weight hydrocarbons. *Mar. Biol.* 1975; 31 (4), 305-310.
[51] Fabricant J.I., Adaptation of cell renewal systems under continuous irradiation. *Health Phys*. 52:663-671, 1987.
[52] Kondo S., Altruistic cell suicide in relation to radiation hormesis *Int. J. Radiat. Biol*. 1988; 53, 95-102.
[53] James S.J., Makinodan T. T-cell potentiation in normal autoimmune-prone mice after extended exposure to low doses of ionizing radiation and/or caloric restriction. *Int. J. Radiat. Biol*. 1988; 53, 137-152.
[54] Heidrick M.L., Hendricks LC, Cook DE. Effect of dietary 2-mercaptoethanol on the life span, immune system, tumor incidence, and lipid peroxidation damage in spleen lymphocytes of aging BC3F1 mice. *Mech. Ageing Dev*. 1984; 27, 341358.
[55] Craig E.A., The Heat-Shock Response, *CRC Crit. Rev. Biochem.* 1985; 18, 239280.
[56] Stebbing A.R.D., The kinetics of growth control in a colonial hydroid. *J. Marine Biol. Assoc. United Kingdom* 1981; 61, 35-63.

[57] Stebbing A.R.D., Heath GW. Is growth controlled by a hierarchical system? *Zool. J. Linnean Soc.* 1984; 80, 345-367.
[58] Stebbing A.R.D., Norton J.P., Brinsley M.D., Dynamics of growth control in a marine yeast subjected to perturbation. *J. Gen. Microbiol.* 1984; 130, 1799-1808.
[59] Sacher G.A., Trucco E., *Biological Aspects of Aging*. (Ed. W. Shock). 244-251 Columbia University Press, New York: 1962.
[60] Sacher G.A., *Handbook of the Biology of Aging*. (Ed. Finch CE, Hayflick L). Chap. 24, 582-638. Van Nostrand. New-York: 1977.
[61] Goodrick C.L., Ingram D.K., Reynolds MA, Freeman JR, Cider LA. Effects of intermittent feeding on growth and life span in rats. *Gerontology*. 1982; 28 (4), 233-241.
[62] Young V.R., Diet as a modulator of aging and longevity. *Fed. Proc.* 1979; 38, 1994-2000.
[63] Cotte J., Bernard A., Effets de dilutions hahnemanniennes de Mercurius corrusivus sur la multiplication en culture de fibroblastes intoxiquespar le chlorure mercurique. In *Aspects de la recherche en homeopathie* (Eds. Boiron J., Abecassis J, Belon P). 1983; pp. 51-59. Editions Boiron. Lion.
[64] Lagache A., *Echos du Sensible*, 1988, Atelier Alpha Bleue Publishers, Paris.
[65] Lagache A., Evolution des modéles logiques dans la science contemporaine. In: M. Bastide ed. *Signals and Images*, Atelier Alpha Bleue Pablisher, Paris 1991, pp. 73-114.
[66] Bastide M., Lagache A., *The Paradigm of signifiers*, 1992 Atelier Alpha Bleue Pablisher, Paris.
[67] Bastide M., Lagache A., *Essential elements.* 1993, Atelier Alpha Bleue Pablisher, Paris.
[68] Larue F., Cal J.C., Dorian C., Cambar J., Influence du pretraitement par des dilutions infenitesimales de Mercurius corrosivus sur la mortalite induite par le chlorure mercurique. *Nephrologie*. 1985; 6: 86.
[69] Cal J.C., Larue F., Dorian C., Guillemain J., Dorfman P, Cambar J. Chronobiological approach of mercury-induced toxicity and of the protective effect of high dilutions of mercury against mercury-induced nephrotoxicity. *Lever Cells Drugs* 1988; 164: 481-485.

PHYSICO-CHEMICAL REGULATION SYSTEM OF LIPID PEROXIDATION AND ITS ROLE IN PATHOLOGICAL PROCESSES. ROLE OF THE LOW IRRADIATION EFFECT

E.B. BURLAKOVA
Institute of Chemical Physics, 117334, ul. Kosygina 4 Moscow, Russia

In 1956, the experiments on insertion of synthetic inhibitors of radical reactions to animals were pioneered. The model of peroxidation of membrane lipids was constructed. In 1957-1965, it was found that non-toxic synthetic inhibitors of radical reactions (antioxidants) are biologically active substances. It was shown that the antioxidants decrease the rate of free-radical peroxidation processes in membrane lipids and interact with natural antioxidants in lipids.

Chemiluminescence and oxidation models were used to determine the rate constants of reactions with peroxide radicals both of naturally occurring (tocopherols, ubiquinones, carotinoids, and the like) and synthetic (substituted phenols, oxipyridines, gallates, aminoalkylphenols, and the like) antioxidants and the criterion for selecting the most active ones was suggested.

In 1967, a hypothesis was proposed for the role of a free-radical mechanism in the cell reproduction. Later, the hypothesis was confirmed experimentally [1].
In 1976, a paper was published, in which the discovered system of regulation was described of cell metabolism by membranes based on correlation between the lipid peroxidation rate, on the one hand, and composition of lipids, membrane structure and functional activity, on the other [2].

The presence of such regulation system was established for nuclear, mitochondrial, microsomic, plasmatic (erythrocyte, leucocyte, photoreceptive) membranes, for bacteria, fish, birds, mammals including human beings (figure 1).

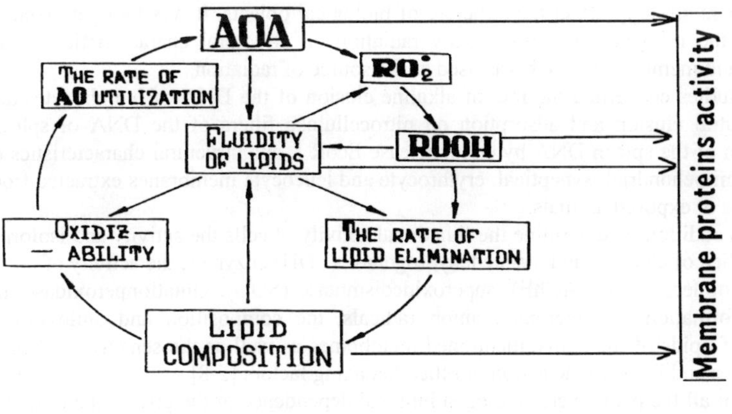

Figure 1. Scheme of the free radical mechanism of cell regulation.

The main components of this system are: the rate of generation of lipid peroxide radicals, antioxidants, lipid composition, structure (fluidity) of membrane lipid, membrane charge, activity of membrane-bound proteins, enzymes, receptors, channel-forming of proteins. All these parameters are in norm, both mutual-bounded structural and functional and the change of any parameter leads to the change of the rest. The presence of chain with negative feed-back bond (rising of the rate of lipids peroxidation resulting in the enrichment of lipids by the resistant to oxidation fractions and on the contrary) allows the system to carry out the function of adaptation to changes in the environment damaging factors and the transition of cellular metabolism on the other level. The period of relaxation of the regulation system depends on the nature of the objects (organism, organ, membranes species) and changes within 10^{-2} - 10^{-4} sec.

The system does not return to the norm in case of the bond break between the parameters, the change of bonds nature, or chronicle action of a damaging factor.

The physico-chemical system of membranes regulation is connected with the other systems - the phosphoinositol cycle, the regulation by cyclic nucleotides, etc.

As we know, the free-radical reactions are extraordinary sensitive to the action of physico-chemical factors at low doses. It was of interest to investigate the action of the low intensity irradiation on the structural and functional properties of membranes, its bonds with the changes in the cell genetical apparatus and in the cellular metabolism as a whole.

In recent years, the effects caused by chemical agents in ultra-low doses (ULDs) and by physical factors of low intensity has been a matter of great interest [3,4].

A peculiarity of the action of ULDs of the agents is a paradoxical dose-effect dependence that is, in most cases, of a bimodal (polymodal) nature. It is significant that the regions of active concentrations or doses of physical factors are separated by a so-called 'idle' or 'dead' zone: yet, the active regions exhibit similar effects for different concentrations, 10^6 to 10^8 times as high.

The effects of ULDs of agents have been studied for more than 10 years. For some time past, concentration has been on the investigation of the influence of low intensity radiation on the vital cellular components.

Numerous suggestions are made in the literature that damaging factors in the environment strongly affect the structure of biological objects. It was thought essential to study the influence of low-intensity radiation on structural characteristics of the genome and membranes. ^{137}Cs was used as the source of radiation.

Studies concerned the rate of alkaline elusion of the DNA of lymphocytes and liver, neutral elusion and adsorption on nitrocellulose filters of the DNA of spleen, restriction of the spleen DNA by endonuclease EcoR I, and structural characteristics of nuclear, mitochondrial, synaptical, erythrocyte and leucocyte membranes extracted from the organs of exposed animals.

In addition, to determine the functional activity of cells the activity and isoforms were studied of aldolase and lactatedehydrogenase (LDH) enzymes, the activity of acethylcholineestherase (AChE), superoxidedismutase (SOD), glutationperoxidase, the rate of formation of superoxide anion radicals, the composition and antioxidizing activity of lipids of the above-mentioned membranes as well as the sensitivity of cells, membranes and DNA to the action of other damaging factors [5-8].

For all the parameters studied, a bimodal dependence of the effect on a dose was found, i. e. the effect increased at low doses, reached its maximum (a low-dose maximum), then decreased (in some cases the sign of the effect changed to the opposite) and increased again as the dose increased. The value of the low dose maximum and the dose to reach it depended on the nature of the subjects, on the intensity of radiation and the time passed since irradiation.

A general rule was that the maximum was shifted to the region of lower doses as a result of a decrease in the radiation rate. Figures 2, 3, 4 show some of the results of the investigation.

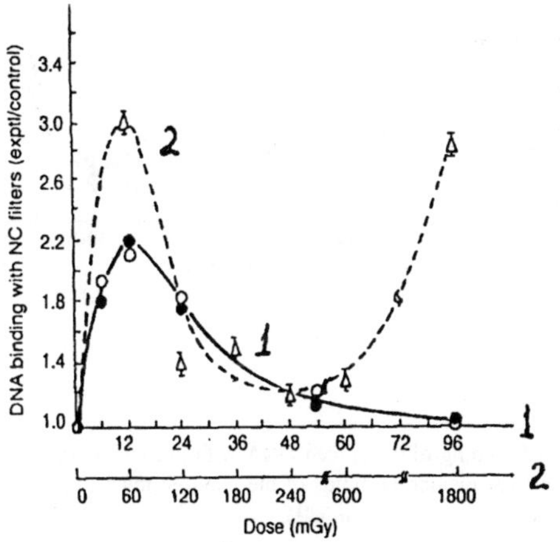

Figure 2. Per cent of binding of the exposed mice spleen DNA with nitrocellulose filters depending on the dose of irradiation (in relative units). Curves: (1) dose rate 4.1×10^{-3} mGy.min^{-1}; (2) dose rate 41×10^{-3} mGy.min^{-1}.

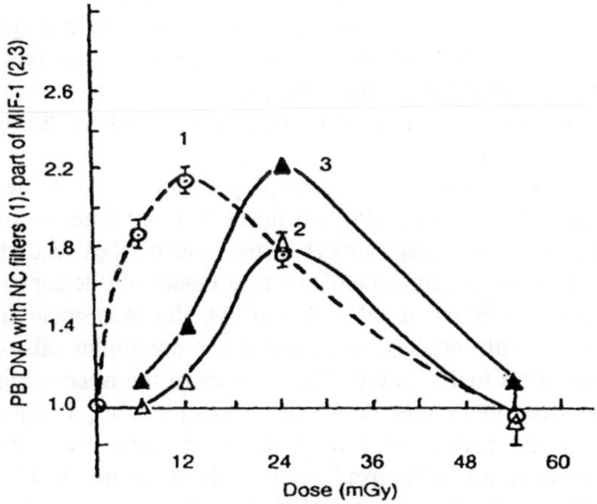

Figure 3. Per cent of binding of DNA with nitrocellulose filters: curve 1, relative units, and curves 2 and 3, the relative change of the MIF-1 intersperse reiterations share in the γ-irradiated mice spleen depending on the dose of irradiation (curve 2 was obtained by gel photometry, 3 photonegatives of the gels); dose rate 4.1×10^{-3} mGy.min^{-1} [7].

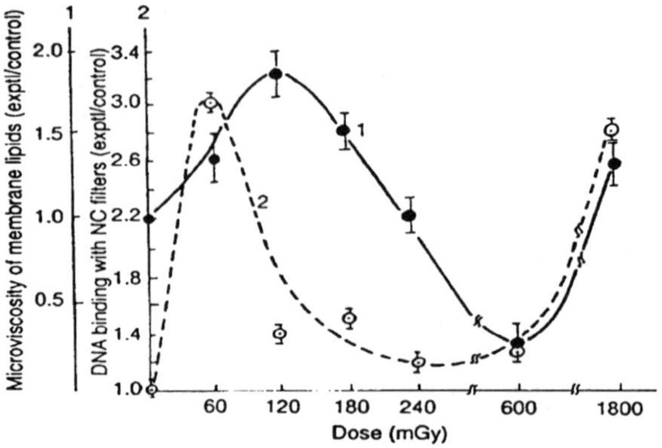

Figure 4. Curve 1, per cent of binding of the spleen DNA with nitrocellulose filters (relative units); and curve 2, microviscosity of nuclear membrane lipids of the exposed mice liver depending on the dose of irradiation (intensity of radiation 41 x 10^{-3} mGy.min^{-1}).

Figure 2 shows a change in the structural properties of DNA depending on dose rate. Figure 3 shows that the change in the per cent of binding (PB) the mice spleen DNA, depending on a radiation dose (curve 1), reaches its maximum at 12 mGy (the dose rate 4.1 x 10^{-3} mGy.min^{-1}). The PB values do not differ much from the control at a dose of 54 mGy. Curves 2 and 3 depict the results of the restriction analysis of the exposed mice spleen DNA; the analysis was carried out using endonuclease EcoR I. The data showed a change in the contribution of intersperse reiterations of MIF-1 in DNA that is indicative of the genome restructuring [7].

Note that the dose dependence corresponds to that for the per cent of binding DNA on nitrocellulose filters.

In addition, the changes in structural properties of the genome caused by lO to 54 mGy are comparable to those caused by the doses 20 to 40 times as high.

Figure 4 shows the change in PB of the spleen DNA (radiation rate 4.1 x 10^{-3} mGy.min^{-1}) and the dose dependence of microviscosity of nuclear membrane lipids. As with the structural properties of DNA (curve 1), the membrane lipid microviscosity changes dramatically with the dose of radiation; the maximum values were obtained for the radiation dose of 60 to 120 mGy. The changes in the microviscosity of erythrocyte membrane lipids were the same. As in the previous cases, it was observed that the maximum was in the region of low doses. The value was comparable with the microviscosity ones obtained after irradiadion by the doses of 1 to 1.2 Gy.

To evaluate the functional activity of the cells the data on the changes in kinetic parameters of membrane and cytosol enzymes in the cells of exposed animals were used.

The data are listed in Table 1.

Dose, mGy	Enzyme	K_M	V_M	V_{max}/K_M
6	AChE (Synaptosomes)	1.15	0.93	0.8
	Aldolase (Synaptosomes)	2.0	1.8	0.98
	LDH	1.25	1.5	1.2
12	AChE	1.5	1.1	0.75
	Aldolase	0.5	0.75	1.25
	LDH	1.3	2	1.55
24	AChE	1.6	1.4	0.87
	Aldolase	2.5	1.2	0.5
	LDH	0.88	0.88	1.0
54	AChE	0.8	0.94	1.2
	Aldolase	0.5	0.80	1.5
	LDH	2.0	1.0	2.0.

Table I. The changes in kinetic parameters of AChE, aldolase and LDH in organs of γ-irradiated mice (exptl/control): dose rate 4.1 X 10^{-3} $mGy.min^{-1}$.

As is seen from Table 1, the changes in the enzymes' kinetic proper;ties begin to develop at a dose of 6 mGy. Here, a prolonged retention was observed of the changed kinetic properties of enzymes caused by irradiation with 12 mGy. Also, a disturbance in the regulatory functions of enzymes was seen as a result of changes in the relation between the isozyme forms of lactatedehydrogenase and aldolase and between the activity of the enzyme and its substrate - for superoxidedismutase (Tables 2, 3). Thus, irradiation of low intensity results in a non-linear dose-dependent change in the functional activity of cells. It is important that sensitivity of separate macromolecules, cells and the organism as a whole to any other damaging factors also changed. To illustrate, an increase was observed in the sensitivity of DNA extracted from the organs of exposed mice (gamma-irradiation, 60 mGy) to a repeat irradiation as compared with the control. It was found that the low dose irradiation increased a degree of erythrocyte haemolysis of exposed mice; the CNS sensitivity to neuromediators, agonists and antagonists as well as the cell response to the action of regulators, to a repeat irradiation and to introduction of radiosensitizers also change. Similar results were obtained 20 years ago by the group headed by Yu.G.

Grigoryev in the experiments with dogs chronically exposed to radiation of low intensity to model the conditions of a long-term stay in space (Flight-to-Mars Project).

Such a dose dependence can be explained in terms of the concept that there is a scale of the doses causing damage in biological objects and pf those inducing the systems of restoration [8,9]. In fact, while the repair (adaptation) systems do not work, the effect increases with the dose: then it decreases as the repair develops, it may be eliminated or change its 'sign'; then it increases again with the dose as the damage prevails over reparation.

Dose, mGy	Isozymes				
	H_4	H_3M	H_2M_2	HM_3	M_4
6	1.21 ±0.03	1.27 ±0.06	0.95 ±0.04	1.04 ±0.10	1.07 ±0.09
12	0.83 ±0.10	0.74 ±0.02	1.1 ±0.1	0.78 ±0.02	0.67 ±0.01
24	1.15 ±0.06	0.81 ±0.09	0.75 ±0.09	0.88 ±0.11	0.71 ±0.11
54	1.35 ±0.19	0.84 ±0.06	0.84 ±0.05	0.91 ±0.08	1.09 ±0.05

Table II. The change in LDH isozymes activities in mice brain cytoplasma after irradiation: dose rate 4.1 x 10^{-3} $mGy.min^{-1}$.

Dose, mGy	$V_{O_2^{\cdot-}}$ / A_{SOD} (Microsomes)	$V_{O_2^{\cdot-}}$ / A_{SOD} (SMP)
0	1	1
6	1.8	1.6
12	3.4	0.8
24	1.7	1.8
54	2.0	1.2

Table III. The changes in the relation between the generation rate of superoxide radicals (VO2) and SOD activity in liver microsomes and liver submitochondrial particles (SMP) of mice after irradiation: dose rate 4.1 10 $mGy.min^{-1}$.

However, despite the numerous experimental findings in favour of this assumption, one cannot be quite positive about the cause of such dose-effect dependence in the region of low-dose irradiation. There are different viewpoints on the nature of this dependence, e.g., a concept of existence of a particular pool of cells that are sensitive to the low-intensity irradiation [10] and others [11].

Work focused on the investigation of similar biochemical and biophysical changes in the organisms of people exposed either to a chronic or to an acute low dose irradiation. Of special interest were the changes in the antioxidant status of the organism.

Numerous data obtained both at our laboratory and known from the literature permitted us to conclude that the antioxidant status of the organism is responsible for the frequency and severity of diseases, the probability of rehabilitation and sensitivity to the action of damaging factors [12].

Figure 5 shows the changes in the activity of superoxidedismutase in the blood plasma of people participating in the liquidation of the consequences of the Chernobyl catastrophe and exposed to radiation (the liquidators). Drastic changes were observed for doses less than 50 mSv. Similar dose dependences were observed while examining vitamin-antioxidants of the exposed people. In those cases the authors noted a decrease in the concentration of tocopherol and glutathione in the blood plasma; and a change in the A-E ratio at irradiation doses lower than 50 mSv [13].

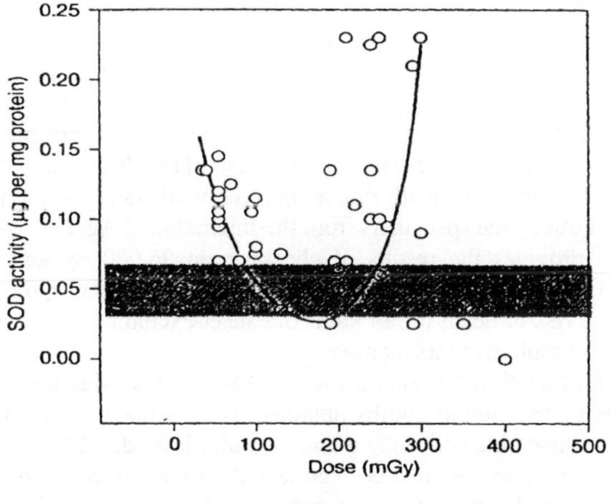

Figure 5. Superoxidedismutase activity in the blood plasma of the liquidators depending on the dose of radiation.

It was essential to find out whether the changes in the health of the exposed people show the similar regularities. However, the epidemiological data were not always obtained at the appropriate level to permit us to make definite qualitative conclusions.

Previously, it was noted that an increase in a number of disease cases among the liquidators shows a similar tendency to a non-monotonic, non-linear dose dependence [14]. Some biochemical and haematological parameters of the people exposed to radiation during the accidents at atomic power stations also exhibit a bimodal dose dependence [13,15]. However, it would be most interesting to study similar dependences for the long-term consequences of irradiation, i.e., malignant tumours. It is well known that ionizing radiation can play the role of a promoter as well as of an initiator of malignant tumours. An increase in the radiation intensity and doses (up to certain limits) results in a decrease in the promoting and in an increase in the initiating functions of irradiation. There is a wide discussion in the literature over the influence of low doses of radiation on the occurence of malignant radiogenic tumours and leucoses.

As a rule, the scientists who deny the fact of verification of a high probability of occurence of malignant tumours at low doses of irradiation put forward two arguments.

(1) It is almost impossible to reveal experimentally or epidemiologically an increase in a number of tumor-carriers caused by irradiation because this increase is within the limits of the difference in the percentage of cancers in different regions and the verification requires a great bulk of experimental data.

(2) Considering the cases of cancers and leucoses caused by low dose irradiation, which exceed significantly the control values (as was shown), and extrapolating the data to the high doses in accordance with the linear dependence, we can get that the number of expected tumors exceeds the real values by orders of magnitude. This is why the conclusions about the influence of low doses are doubtful [16].

The grounds for scepticism concerning the information on an increase in the number of deaths from tumours and leucoses caused by low dose irradiation were the results of the check when one and the same group of workers of an atomic power station yielded an increase in a number of deaths from cancer as compared with the control for doses of 20 or 50 mSv but a decrease for 100 mSv [17].

The analysis of the literature data on the leucoses death rate for different groups of exposed people working, as a rule, at atomic power stations showed the same tendency in the doseeffect dependence that was discovered in our experiments.

Recently, a fundamental paper has been published [18]. It contains the analysis of data on the cancer and leucosis death rate among the workers at atomic plants. The work had been carried out by the specialists from the International Agency for Cancer

Research and provides the results of checking on 96,000 workers of atomic plants in Canada, Great Britain and the USA. The authors concentrated on the estimation of a relative risk of death for all kinds of cancer excluding leukaemia, and for leukaemias excluding chronic lympholeucoses.

A relative risk of death from leucoses was 2.2 Sv^{-1}. This value lies between the value for the victims of the atomic bombs obtained using a linear model for the risk estimation, 3.7 Sv^{-1} and that obtained using a linear-quadratic model, 1.9 Sv^{-1}.

In 1994, the information that had been collected for the past 35 years on medical examinations of people in Japan exposed to radiation during the atomic bombing of Nagasaki and Hiroshima were published [19]. The authors of this investigation emphasize the fact of a decrease in the leucoses death rate in the group of people exposed to a dose of 80 mSv; they regard this fact as an evidence for a hormesis caused by low dose irradiation.

In Kossenko *et al.* [20], the data on the deaths from leucoses after the accident in the South Urals were presented.

An effort has been madeto pool the information in the literature [17-21] on the leucoses death rate for the workers of atomic plants and exposed people and to plot a corresponding dose rate dependence curve excluding chronic leucoses. This dependence is shown in Fig.6 and Table IV.

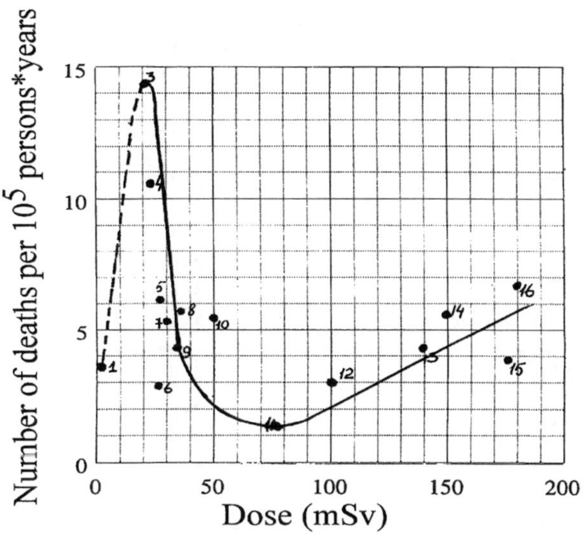

Figure 6. Leukemias deaths per 10^5 person.years depending on the dose of irradiation (the numbers of the points correspond to the numbers in Table IV).

It is evident that the results of emidemiological examinations of a great number of people can be represented as a bimodal dose dependence. The dependence amalgamates the cases when irradiation resulted in a decrease in the leucoses death rate below the control. The revealed tendency shows that the rate of deaths from leucoses caused by low-dose irradiation is comparable with that for the doses dozens of times higher. The analysis of the data obtained by the Oxford scientists [22] on the risk of malignant tumours other than the lung cancer among the miners in Northern Sweden showed the same regularities in the leucoses dose-dependence as those mentioned above. All these data can be explained in terms of the concept that small doses of the low intensity radiation promote but do not initiate carcinogenesis. The effect of the low intensity radiation can be attributed to the action of the radiation on the membranes.

It was shown [23] that the regularities in the action of low doses of radiation on the antioxidizing activity of the peroxide oxidation are similar to those exhibited by cocarcinogenic agents with reference to the same parameters [1].

Different dose-effect dependences for the regions of low and high doses permit us to suggest different mechanisms responsible for the increase in the leucoses death rate at low and high doses of irradiation. This is only a qualitative pattern that should be subjected to statistical processi~g with the resulting significantly 'smoothed' effects. However, it is believed that it is necessary to provide this plot in an effort to concentrate the investigators on searching the regularities and mechanisms of the effects at low doses, and to emphasize that it is not permissible to pool the data obtained for low and high doses because this levels the real effects.

Our findings and the data on experiments with exposed animals available in the literature count in favour of taking new approaches in studying the effects of low doses of radiation.

Irradiation locality	Dose, mSv	Leucoses death rate per 10^5 person.years	Ref.
1. Pilgrim 1983-1988	2	3.6*	21
2. UKAEA workers 1946-1979	20(20-50)	4.3*	17
3. Pilgrim 1979-1983	20	14.4*	21
4. Oakridge National Laboratory	21	10.4	18
5. Hanford	27	6	18
6. American Military Agency	27.6	2.5	18
7. Residents of Japan Group I	30	5.1	19
8. American Nuclear Agency	33.1	5.6	18
9. Rockyflats	35	4.0	18
10. UKAEA workers	50	5.22	17
11. Residents of Japan Group II	80	1.4	19
12. UKAEA workers	100	3.0*	17
13. Sellafield	139	4.2	18
14. Residents of Japan Group III	150	5.7	19
15. Residents at the Techa River, Group I	176	3.8	20
16. Residents at the Techa River, Group II	180	6.9	20
17. Residents at the Techa River, Group III	290	8.5	20
18. Residents of Japan Group IV	400	8.56	19
19. Residents at the Techa River, Group IV	610	6.5(14.3*)	20
20. Residents at the Techa River, Group V	780	7.9(17.3*)	20
21. Residents of Japan Group V	800	14.3	19
22. Residents at the Techa River, Group VI	1640	15.3	20
23. Residents of Japan Group VI	1800	28.6	19
24. Residents of Japan Group VII	2600	57	19
25. Residents of Japan Group VIII	3600	91	19

* The data are related to the control of 3.6 per 10^5 person.years

Table IV. The data on the leukaemias death rate for people after atomic bombing and nuclear accident and for workers in the nuclear industry (the numbers in the table correspond to the numbers of the points in Fig. 6).

References

[1] Burlakova, E.B., Alesenko, A.V., Molochkina, E.M., Palmina, N. P. and Khrapova, N.G. *Bioantioxidants in X-Ray Desease and Malignant Growth,* Moscow: Nauka, 1975.

[2] Aristarkhova, S.A., Arkhipova, G.V., Burlakova, E.B., Gvachariya,V.O., Glushchenko, N.N. and Khrapova, N.G. Dokl.Akad.Nauk SSSR, 227, 1342-1348 (1976).

[3] Burlakova, E.B., Konradov, A.A. and Khudyakov, I.V. *The Action of Chemical Agents at Superlor Doses on Biological Objects,* Izv.Akad.Nauk SSSR, Ser.Biol. 2, 184-193 (1990).

[4] Burlakova, E.B. *The Effect of ULD,* Vest.Akad.Nauk, 64(5), 425 431 (1994).

[5] Zhizhina, G.P., Skalatskaya, S.I. and Burlakova, E.B. *The Effect of Lor Dose Ionizing Radiation on Mice Spleen DNA,* Radiat.Biol. Radioekologiya 34(6) 759-762 (1994).

[6] Molochkina, E.M., Djaman, O.M., Ozerova, I.B., Burlakova, E.B. and Shishkina, L.N. *Biochemical Changes in Mice Brain Synaptosomes after Lor Dose Whole Body γ-Irradiation of Various Intensities* Radiat.Biol. Radioekologiya 35(6) (1995) (In press).

[7] Shugalii, A.V., Lichina, M.V., Zhizhina, G.P. and Skalatskaya, *S.I. Genome Effect of Lor Dose Prolonged γ-Irradiation of Mice and Its Consequences* Dokl.Ross.Akad.Nauk (1995) (In press).

[8] Smotryaeva, M.A., Kruglyakova, K.E., Shishkina, L.N., Blukhterova, N.V., Goloshapov, A.N. and Burlakova, E.B. *Structural and Biochemical Parameters of Mice Blood after Lor Dose γ-Irradiation of Different Intensities* Radiat.Biol. Radioekologiya (in press) (1995).

[9] Burlakova, E.B. *Consequences of Lor Level Radiation.* In: Low Level Radiation and Living State, Eds N.G.Huilgol, D.V. Gopmath and B.B.Singh (Narosa Publishing House, New Delhi, Madras, Bombay, Calcutta) pp. 57-62 1994).

[10] Spitkovskii, D.M. *The Concept of the Effect of Lor Doses of Ionizing Radiation on Cells and Its Possible Use for the Interpretation of Medicobiological Consequences of the Accident at the Chernobyl Nuclear Porer Station* Radiobiologiya, 32(3) 382 400 (1992).

[11] Eidus L.H. *About Mechanism of Lor Dose Effects Initiation,* Radiat.Biol. Radioekol. 34(6) 748-758 (1994).

[12] Pobedimskii, D.G. and Burlakova E.B. *Mechanisms of Antioxidant Action in Living Organisms in Atmospheric Oxidation and Antioxidants* Ed. G.Scott(Elsevier, (The Netherlands) 3(9) pp. 223-270 (1993).

[13] Neyfakh, E.A., Ivanenko, G.F. *Chernobyl Victims: Remote Consequences of Deficit of Essential Antioxidants and Lipoperoxide Pathway Disturbances.* In: Proc. 1st Int.Congress on Vitamins and Biofactors in Life Sciences, Kobe, Japan, Oct. 1991, I-V-6, 65 (1991).

[14] Chernobyl Catastrophe: Causes and Consequences (Expert conclusion) B2, *Medico-Biological and Genetic Consequences of Chernobyl Catastrophe* (Skaryna, Minsk) (1992).

[15] Czepeshi, T. and Fieldner T.M. *Reversible und Irreversible 13 Schadingung der Hamopoese nach Unerwarteter Ganzkörper Bestrahlung: Merkmale in Periferen Blut,* Wiener Klin. Wochenschr. 9 (1989).

[16] Klema, E., Shihab-Eldin, A. and Wilson, R. *Some Claims of Unusually Large Effects of Radiation,* 1989.

[17] Beral, V., Fraser, P., Booth, M. and Carpenter, L. *Epimemiological Studies of Workers in the Nuclear Industry.* In: Radiation and Health: The Biological Aspects of Low Level Exposure to Ionizing Radiation, Eds R.I. Russel and J.R.Southwood (Wiley and Sons, Chichester) pp. 97-106 (1987).

[18] IARC Study Group on Cancer Risk among Nuclear Industry Workers. *Direct Eestimates of Cancer Mortality due to* Low *Doses of Ionizing Radiation: an International Study.* Lancet, 344, 1039 1043 (1994).

[19] Sadao Hattori, *State of Research and Perspec*tive *on Radiation Hormesis in Japan.* BELLE Newsletter 3(1) 1-8 (1994).

[20] Kossenko, M.M., Degteva, *M.O.* and Petrushova, N.A. *Estimate of the Risk of Leukemia to Residents Exposed to Radiation as a Result of A Nuclear Accident im Southern Urals.* PSR Quart. 3, 187-197 (1992).
[21] Shulman, S. *Cancer around Nuclear Plant,* Nature, 347, 604 (1990).
[22] Darby, S.C., Radford, E.P. and Whitley, E. *Radon Exposure and Cancers Other Than Lung Cancer in Svedish Iron Miners.* Environ. Health Perspect. 103 (Suppl.2), 45-47 (1995).
[23] Polyakova, N.N., Shishkina, L.N. *Action of-Irradiation with Different Rate Dose on the Lipid Peroxidation in Tissues of Mi* ce. Rad.Biol.Radioecol., 35, 181-187 (1995).

EFFECTS OF METAL HIGH DILUTIONS ON CELLS AND INTEGRATED SYSTEMS

CAMBAR J., DELBANCUT A. and BARROUILLET M.P.
Groupe d'Etude de Physologie et Physiopathologie Rénales
Faculté de Pharmacie - 3, place de la Victoire - 33000, Bordeaux (France)

Chronic exposure to heavy metals, major environmental toxicants as mercury, cadmium or lead, can induce numerous pathologies at different organ levels as skin, kidneys, lungs, liver or nerves (for complements, see the very recent book "Toxicology of metals" by Goyer and Cherian, 1995).

This lecture will review the main papers already reported in the literature on the protective effects of very high dilutions of heavy metals in metal induced toxicity as well in vivo as in vitro.

If this review cannot be exhaustive, we have chosen the more significant results published in scientific publications with hard statistical considerations if possible.

Many reviews or thesis have reported for decades, with a critical attitude, the great amount of data showing, more or less significantly, clearly and undoubtfully the effects of high or very high dilutions (VHD) in quite all the domains of experimental biology, e.g. pharmacology, toxicology, endocrinology or immunology. (Poitevin, 1987 - Halm, 1993 - Cambar and coll., 1994 - Endler and Schulte, 1994 - Linde and coll., 1994).

Some of old described observations have been recently confirmed and ameliorated by new methodologies, Kolisko reviewed by Pelican and Unger and Pongrantz and coll. Likewise, the pioneer princeps studies of Lapp and Wurmser were later confirmed by Cazin and coll.

These studies, performed in vivo or in vitro, can be considered as toxicological studies because they intend to reduce, with different approaches, the heavy metal induced toxicity.

The use of high dilutions enters clearly this topics as one of the numerous agents or processes proposed to reduce more or less significantly the side effects of metals.

Indeed, many authors have already proposed a large reduction in metal toxicity by use of chelating agents or by the competition between different metals (see the **table 1** with numerous references).

The **table 1** reports some studies in classical experimental toxicology on *in vivo* and *in vitro* models on the protective effect of different metals towards toxic effects of the same metal (for example Cd vs Cd) or a different metal (for example Cd vs Hg or CP) showing, without any consideration of the use of very high dilutions, the effect of *identical* or *similar* metal. These data permit to open the discussion for advancing some explanation hypothesis about such observations.

Toxic agent	Protective agent	References
1 - In vivo animal models		
Cd	Cd	Terhar and coll., 1965
Cd	Cd	Ito and Sawauchi, 1966
Cd	Cd	Yoshikawa, 1973
Cd	Cd	Morita, 1984
Cd	Cd	Jin and Nordberg, 1986
Cd	Cd	Weis and Weis, 1986
Cd	Hg	Magos and coll., 1979
Cd	Hg	Webb and Magos, 1976
Mo	Hg	Yamane and Koizumi, 1982
Zn	Cd	Herkovitz and Perez-Col, 1990
Zn	Cd	Suzuki and coll., 1990
Zn	Cd	Liu and coll., 1992
Se	Cd	Flora and coll., 1982
Cd	CisPt	Bompart and coll., 1991
CisPt	CisPt	Litterst and Schweitzer, 1984
2 - In vitro animal models		
Cd	Cd	Jin and Norberg, 1986
Zn	Cd	Kaji and coll., 1992
Cu	Cd	Meshitsuka and coll., 1982

Table 1: Classical toxicological studies using metal as toxic and protective agents.

Toxic agent	Protective Agent	Material for the study	References
	Ag	Wheat coleoptile	Jones and Jenkins, 1981
	Hg	Wheat coleoptile	Boiron, 1965
	As	Wheat coleoptile	Boiron and Zervudachi, 1963
	Cu	Wheat coleoptile	Netien, 1962
	Cu	Wheat coleoptile	Netien and Girardet, 1963
	Cu	Chlorella vulgaris (alga)	Graviou and Boiron, 1971
	Cu	Chlorella vulgaris (alga)	Moss and coll., 1977
Cu	Cu	Wheat coleoptile	Netien and coll., 1966
Cu	Cu	Mustard beans	Auquière et coll., 1981 - 1982
Cu	Cu	Lens roots	Projetti et coll., 1985
As	As	Wheat coleoptile	Cier and coll., 1963

Table 2: Studies on metal ultra low doses in vegetal models.

It is very important to note that the problematic and methodological approaches are quite similar whatever the concentration of the substances used.

This review reports the main studies demonstrating the protective role of metal (or toxic agent) low and even very high dilutions (VHD) towards intoxication with metals.

Indeed, some authors have reported very exciting studies where very high dilutions (10^{-10} to 10^{-40} M and even more) are able to significantly reduce the metal toxicity whatever the metals or the biological models used. These VHD, obtained by successive centesimal deconcentrations can be considered as theoretical concentrations because of the famous number of Avogadro (10^{-24} M).

Figure 1 represents a part of the well-known element periodic classification of Mendeleiev where are colored, more or less deeply, the elements (metal or metalloid) particularly studied as highly diluted.

After a bibliographic survey on the metal high dilution use, we can consider 3 classes of "interesting" metals: the very largely used, cadmium, mercury, copper and arsenic, the middle used, lead, phosphor and silicium and some others used by only a little quantity of authors, bismuth, cobalt, iron, nickel, platin, silver or zinc.

In parallel with toxicological experiments that we will review with metal intoxications by ponderal doses and where metal high dilutions are given as preventive or curative agents, it can be noted that metal dilutions administrated alone act directly in normal tissues; for example, they can inhibit lysosome or mitochondria enzyme activities, the histamine release or change heart rate or cardiac rhythmicity.

Three parts with their own table (**tables 2, 3 and 4**) will report the main results obtained for decades in experimental toxicology on the protective effects of different metal high or very high dilutions towards toxic effects of the same (or not) metal, as well in vivo in experimental animals (**table 2**) or in plants (**table 3**), as in in vitro models (**table 4**). Some studies are only listed in the table and referenced in the bibliography without any description in text because of lack of place.

Effects of very high dilutions in vegetal models (table 2)

Many studies have been reported in different vegetal models, e.g. beans, plants or even alga.

Previous studies performed before 1970 have been criticized because of lack of satisfactory statistics and dosage methods, but have nevertheless the interest to already very early show that high dilutions act well on biological structures without any nervous system.

The principle of these previous studies is very similar. They compare the germination ability and the growth in different metal intoxicated vegetal models with or without very high dilutions of the same metal.

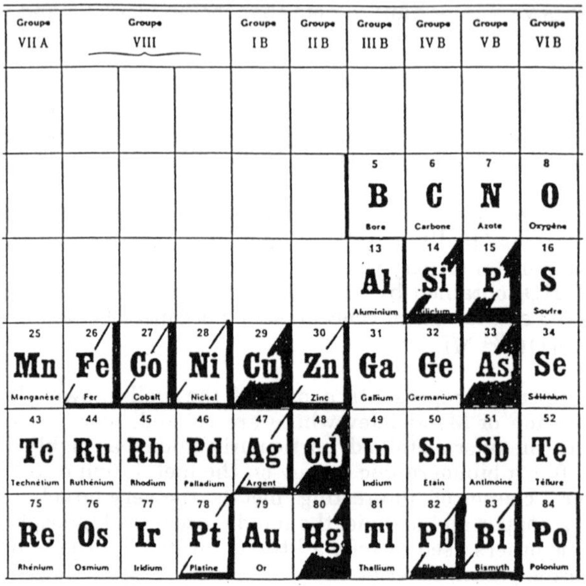

Figure 1 - Element periodic classification of Mendeleiev with the most studied metals (or metalloïd) in very high dilution experimentations.

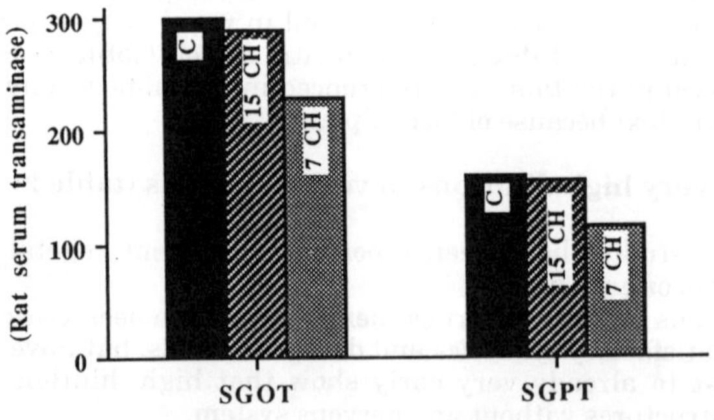

Figure 2 - Rat serum transaminases (SGOT, SGPT) in CCl_4 induced hepatitis with control (C), 7 and 15 CH phosphor dilutions (Bildet, 1975).

The historical studies of Kolisko during 30 years, reviewed in Netien and Girardet (1963) and confirmed by Pelikan and Unger (1971), with nice statistical precautions, have evidenced rhythmic alternating responses of VHD with stimulation or inhibition.

Likewise, in wheat coleoptile, Jones and Jenkins (1981) have observed that $AgNO_3$ had an inhibitory effect for 8 CH and a stimulating one for 14 CH.

Another type of experiment was led in plants (Boiron, 1965 - Boiron and Zervudachi, 1963 - Netien, 1962) on changes in respiratory functions, assessed by oxygen consumption in wheat coleoptile; mercury, cobalt and arsenic present a dilution height dependent rhythmic response (stimulation or inhibition).

Similar studies have considered intoxicated plants in the classical model of "identical metal" also used in animals. Effect of a pretreatment by Cu SO_4 (more generally 15 CH) is researched in intoxicated wheat or peas beans as compared with untreated intoxicated lots. High dilutions increase Cu elimination, growth and oxygen consumption.

More recently, Auquiere et coll. (1981), in mustard beans, and Projetti et coll. (1985), in lens roots, have confirmed the protective effect of Cu in Cu-intoxicated plants.

Effects of metal very high dilutions in animal models (table 3)

In animals, the use of metal VHD in experimental toxicology has been largely developed for decades.

The first one, historically, demonstrated their activity with a common methodology: the administration of a toxic in animal and the administration of the same "identical" or the analog "similar" toxic in VHD, either before intoxication (preventive-pretreatment) or after intoxication (curative treatment).

This protective effect is assessed either by mobilisation of the metal out of the organism or by the changes in the lethality or the organ toxicity of the metal.

First studies were reported by Lise Wurmser (see review in Wurmser, 1984) and are still noteworthy by their methodological precision and their historical interest. Lapp and Wurmser contributed fairly for the development of the VHD effect studies.

Shortly, a subacute toxic arseniate dose (1000γ) is given in guinea-pigs. The urinary arsenic elimination is compared immediatly after intoxication, 3, 7 weeks and 3 months in control (untreated) and in pretreated 4, 5 or 7 CH animals.

Immediatly after, As dilutions increase urinary As excretion in pretreated as in untreated animals, with an excretion advantage for high dilution pretreated guinea-pigs (42% vs 34-39%).

Although arsenic excretion has totally ceased, a new administration of 4, 5 or 7 CH dilution induced a significant metal elimination at the 3th or the 7th week, about 15 to 65γ, corresponding to 2 to 10% of the remaining As in the organism after the first intoxication (see details below).

Toxic agent	Protective agent	Aim of the study	References
Hg	Hg	Mice lethality	Cambar and coll., 1983
Hg	Hg	Mice lethality	Larue and coll., 1985 - 1986
Hg	Hg	Mice lethality variability	Cal and coll., 1988
Pb	Pb	Rat Pb elimination kinetics	Vischniac, 1965
Pb	Pb	Rat Pb elimination kinetics	Fisher and coll., 1987
Cd	Cd	Amphibian embryo lethality	Herkovitz and Perez-Col, 1991
As	As	Rat As elimination kinetics	Cazin and coll., 1987
As	As	Guinea-pig As elimination kinetics	Lapp and coll., 1958
As	As	Pigeon As elimination kinetics	Mouriquand and coll., 1961
Sb	Sb	Pigeon Sb elimination kinetics	Mouriquand and coll., 1961
Bi	Bi	Guinea-pig Bi elimination kinetics	Lapp and coll., 1958
P	P	Rat P-induced hepatitis severity	Bildet, 1975
CCl4	P	Rat CCl4-induced hepatitis severity	Bildet and coll., 1978

Table 3: Studies on metal ultra low doses in animal models.

Toxic agent	Protective agent	Material for the study	References
	Hg	Mice lymphocytes	Mansvelt and van Amons, 1975
	Si	Mice macrophages	Davenas and coll., 1987
Hg	Hg	Fibroblasts in culture	Boiron and coll., 1981
Cd	Cd	Renal cells in culture	Delbancut, 1994
Cd	Cd	Human lymphocyte T line	Pellegrini and coll., 1994
Cd	Cd	Hepatoma cultures	Van Wijk and Wiegant, 1994

Table 4: Studies on metal ultra low doses in in vitro cellular models.

	After 3 weeks		After 7 weeks	
	Patterns excreted	% As excreted	Patterns excreted	% As excreted
4CH	15	2,3	64	9,9
5CH	38	5,9	23	3,8
7 CH	62	9,5	39	6

Similar results confirming the mobilization of metals by VHD of the same metal were reported after with bismuth in guinea-pig (Lapp and coll., 1958) and with arsenic in pigeon (Mouriquand and coll., 1961).

More recent and very rigorous studies were reported by Cazin and coll. (1987) in rats by administration of Na arseniate and radiolabeled arsenic as toxic agent and 7CH arsenic as pretreatment. VHD injection 12 hours after intoxication increases significantly urinary arsenic in rat as compared in control ones.

Similar observations were reported by Fisher in lead intoxicated rats and post-treated a week after with 200 CH lead (Fisher and coll., 1987).

All these experiments, in the same topics, permitted to advance the following conclusions:

- metal high dilutions (even 200 CH) are able to increase the elimination of the same metal used as toxic,
- there are no relationship between dilution height and efficacy,
- the mobilization ability is very potent because released metal patterns are dramatically higher (10^{12} times) than the patterns of metal present in the dilutions.

Moreover, two major studies have reported the effects of VHD in the change of metal-induced lethality and organ toxicity.

Phosphor is a metalloid but is considered here as a "metallic" toxic. These both investigations have shown the efficacy of high dilutions of phosphor and mercury respectively in phosphor-induced hepatotoxicity (Bildet, 1975 - Bildet and coll., 1978) and in mercury-induced nephrotoxicity (Cambar and coll., 1983 - Larue and coll., 1985 -1986).

The efficacy of phosphor VHD has been performed in two murine models of hepatotoxicity induced respectively by phosphor (identical) and CCl_4 (similar).

Serum transaminases (SGOT and SGPT) and LDH are dosed at day 0, 2, 3 and 4 in different lots of 1 g/kg CCl_4 intoxicated rats receiving after either NaCl (control) or 7 and 15 CH phosphor dilutions (treated). A significant decrease in serum enzyme level is noted in 7 CH treated animals (**Figure 2**).

Similar results had been obtained in preventive action by a pretreatment 48 hours before the intoxication.

Demonstration of the effect of P high dilutions was confirmed by double-blind anatomo-pathological observations, in photon and electron microscopy.

The efficacy of mercury VHD were performed in female Swiss mice weighing 18-20g, housed by 15 in a room, in which conditions of lighting remained constant all the year. Two series of experiments were conducted to test the influence of a pretreatment with high dilutions of mercury on $HgCl_2$-induced mortality in mice and also to determine whether this effect is reproducible at different months in the year.

Low concentrations of mercury were prepared from a mother solution to get high dilutions 10^{-18} to 10^{-30} M by serial centesimal dilutions.

Mice received a daily i.p. injection of 10^{-18} and 10^{-30} M $HgCl_2$ dilutions (pretreated groups) or of agitated (successed) distilled water (control groups) throughout a 7-day period. On the 7th day, one hour after the last preventive injection, they were given a single i.p. dose of 5, 6 or 7 mg Hg/kg of $HgCl_2$.

In another experiment, 265 mice were pretreated with 10^{-30} M mercury according to the previous protocol with only 5 mg Hg/kg of $HgCl_2$ at 5 different months in the year (october, december, february, april and june).

In all experiments, mortality was recorded daily during ten days after intoxication to assess the kinetics of the mortality rate.

These different experiments show clearly that high mercury dilutions can provide an important reduction of the letal toxicity of $HgCl_2$, especially when the dose is as close as possible to the letal dose LD50 value.

Indeed, when mice received 5 mg/kg, 73,4% died in the control group whereas only 50% and 26,7% ($p<0,01$) died in groups pretreated respectively by 10^{-18} or 10^{-30} M mercury (**Figure 3**).

Experiments along the year showed a very reproducible protective effect of 10^{-30} M mercury against 5 mg/kg induced-mortality. Moreover, the effectiveness of the mercury high dilution against $HgCl_2$ induced mortality exhibited a circannual rhythm ($p<0,01$) with a maximum located in August.

Effects of metal very high dilutions in in vitro cellular models (table 4)

As already discussed by Poitevin and Aubin (1986), the use of isolated cellular systems in high dilution research present new fair exciting advantages: they permit to use rapidly a great amount of dilutions in standardized and reproducible conditions and to study the numerous possibilities of interaction between two different metals, one used as protective agent and the other as toxic agent.

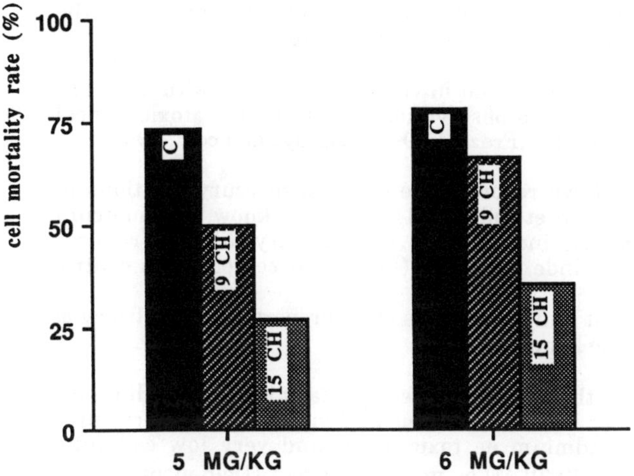

Figure 3 - Lethality rate in $HgCl_2$ intoxicated mice with control (C), 9 and 15 CH mercury dilutions (Cambar and coll., 1983).

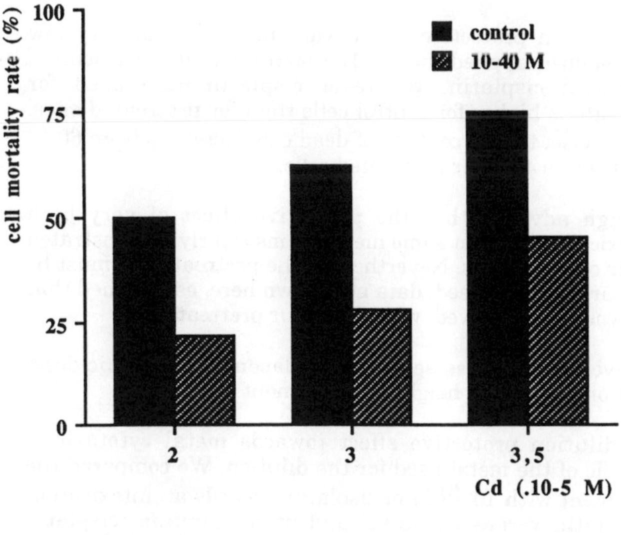

Figure 2 - Cell mortality rate in cadmium intoxicated renal cultured cells in control and 10^{-40} M cadmium very high dilutions (Delbancut, 1994)

All these new investigations will contribute to better understand the effect of the similar or identical metal in the metal protection process.

Moreover, it is noteworthy that in vitro data reported with ultra low doses confirm well the in vivo observations as traditional toxicological studies have already validated. (Frazier, 1990 - Lagroye and coll., 1995).

Previous studies have reported the effect of mercury dilutions in cultured fibroblasts (Boiron et coll., 1981). It is well-know that mercury salts decrease the mitotic index; the 5 CH mercury dilution reduces significantly this mitotic index decrease (-34,8%) as compared in control cultures (-41,4%).

Likewise, Mansvelt and Van Amons (1975) reported such efficacy of mercury dilutions in mice lymphocytes.

In our laboratory, the protective effect of heavy metal high dilutions is assessed with another in vitro model using renal tubular cell cultures (LLCPK1 cell lines), cadmium as toxic agent and very low cadmium concentrations (10^{-10} to 10^{-40} M) as protective agent (Delbancut, 1994 - unpublished data).

Cell mortality towards a ponderal cadmium 24 hour intoxication in pretreated cells for a 120 hour period illustrates fairly the protective effect of the cadmium VHD dilution. Indeed, a very large decrease in pretreated cell mortality, compared to untreated cells one can be shown. For instance, 2.10^{-5} M cadmium induces 51 % of control cell mortality against only 22% for 10^{-40} M cadmium pretreated cells; a mortality reduction more than a half can be seen (**Figure 4**).

In the same way, a protective effect was obtained with very low concentrations of cisplatin applied during 120 hours against intoxication with ponderal dose of cisplatin. Whatever cisplatin dose used for intoxication, death rate is higher for control cells than for pretreated ones. So, at 4.10^{-5} M toxic dose, the percentage of dead cell decreases from 60 % for untreated cells down to 40 % for pretreated cells.

We can though advance that the protective effect of very high dilutions against toxic doses of that same metal seems clearly demonstrated in proximal tubular cells cultures. Nevertheless, the pretreatment must be applied for a 120 hour period. Indeed, data not shown here, established that no protective effect could be observed with a 48 hour pretreatment.

This protective effectiveness seems to be dependant on toxic-dose, pretreatment duration and on the height pretreatment dilution.

Metal high dilution protective effect towards metal cytotoxicity revealed to be specific of the metal used for the dilution. We compared the effect of a pretreatment with 10^{-40} M of cisplatin towards an intoxication with cisplatin (cisplatin versus cisplatin) and with cadmium (cisplatin versus cadmium). Experiment cadmium versus cisplatin shows the total lack of protection.

By indirect immunofluorescence technique, it is possible to compare the aspect and the organization of the network of cytoskeleton proteins in pretreated and control cells towards an intoxication by ponderal high cadmium doses. The pretreatment by high dilution is able to decrease cadmium induced cytoskeleton protein desorganization. Indeed, the actin and vimetin organization is more normal in pretreated cells than in control ones.

We find the same profil of response with our in vitro model as with our previous in vivo one, 10^{-30} and 10^{-40} M dilutions providing a fair protection against cadmium induced cytotoxicity in renal tubular cell cultures. Similar results have recently been described for high dilutions of thymulin (Bastide and al, 1987), bursin (Youbicier-Simo and al., 1993) and silica (Oberbaum and al., 1992).

Our results are confirmed by similar in vitro experiments in other cultured mammalian cells intoxicated by cadmium or arsenic and protected by low dilutions (10^{-6} to 10^{-10} M) of the same metal (van Wijk and Wiegant, 1994).

These authors have studied the stimulation of cellular self-recovery by application of the similia-principle. The self-recovery process occurs after damaging a cell by different factors, as heart shock or metal intoxication.

The aim of this work was to study whether the cellular self-recovery can be stimulated by the application of the similia-principle. They studied the heat shock protein (HSP) induction pattern after exposing hepatoma cultures for one hour to sodium arsenite or cadmium chloride concentrations up to 200 μmoles. Induced synthesis of these HSP was found to be highly specific for each damaging conditions.

Moreover, with respect to the hypothesis on the specificity of the stimulation of self recovery by low doses, cells were exposed for a short time to activate the induction of HSP with arsenic and cadmium for one hour at 100 μmoles. Exposing arsenic pretreated cells to a low dose of arsenic resulted in stimulation of the HSP synthesis. However, exposing arsenic pretreaed cells to a low dose of cadmium had no effect, confirming our personal studies. Indeed, exposing cadmium pretreated cells to low doses of cadmium resulted in a stimulation of the HSP synthesis; but, exposing cadmium pretreated cells to arsenite did not show this stimulation. These results support the hypothesis that self recovery is specifically stimulated by low doses applicated according to the similia-principle.

In conclusion, the protective effect of very high cadmium dilutions against toxic doses of this metal seems clearly demonstrated. A similar protection against cisplatin intoxication is also demonstrated for ultra low doses of cisplatin (personal unpublished data). This protection seems specific since cadmium dilutions do not protect against cisplatin toxicity despite the fact that cisplatin is also a tubulotoxic metal.

Recent experiments carried out in our laboratory have shown that metallothionein (MT) synthesis was increased in Cd intoxicated cells after exposure to very high Cd dilutions for 120 hours.

Similar results have been reported by Pellegrini and coll. (1994) with low doses of cadmium (1 uM to 0,1 nM) in an immature human T cell line. A 24 hour pretreatment with non toxic concentrations of cadmium confered protection against subsequent cytotoxic doses and was able to induce metallothionein - IIa gene expression. An increase of MT synthesis has been already observed with toxic or subtoxic concentrations of heavy metals such as zinc, cisplatin or cadmium. These results are in fair accordance with other ones from our laboratory showing that such a treatment decreases the metal storage in the cell by, for example, changes in membrane permeability. (Delbancut, 1994 - personal unpublished data).

Silicea was studied as modulator of the inflammation during 4 years in Benveniste laboratory (Davenas et coll., 1987).

In many experiments, confirming the previous results, PAF acether synthesis in macrophages is significantly increased in 9 CH Silicea treated mice.

GENERAL CONCLUSION

Heavy metals are largely used in chemical industries and in agriculture. They contribute to risk for health by inducing some environmental pathologies as atmosphere or water contaminants. Indeed, their intake or inhalation, acute or chronic, induces severe lesions at different organs as liver, lung, kidney or nervous system. Heavy metals neurotoxicity or nephrotoxicity are well documented today.

Many strategies for the development of protective agents in metal toxicology offered many substances drug or not as chlorpromazine, sodium overload, furosemide or cold exposure and chronobiology. More mechanistic drugs can be purposed to decrease metal fixation by competition on thiol groups, on SH rich proteins or on over divalent cations. Very high dilutions can play a similar protective effect considering identity or similarity laws as well in animal or vegetal models.

It is noteworthy that many metal high dilutions had been reported as protective in arsenic, lead, cadmium or mercury intoxication in rodents. Similar studies have been reported in plants, in arsenic, copper or mercury intoxication. Such protective effect of very high dilutions can be compared with similar protective effect of metal used in high dilutions as lead, cadmium or mercury.

But, if the results e.g. protective effect are similar whatever the height of the dilution the mechanisms of this protection may be different and the hypothesis for their explanations are unclear.

It is clear that very high dilutions can protect significantly dramatically towards intoxication of similar or identical metals. Nevertheless, it seems very difficult to use the same arguments to explain this protective effect with agents used at 0,1 or 0,005 mg/l or at 10^{-20} to 10^{-40} M.

These numerous studies permit to advance different conclusions:

- similar protective effects of ultra low doses of heavy metals towards heavy metals induce toxicity or cytotoxicity can be observed in animals, in plants or even in isolated cells or in cultured ones; the nervous system cannot be advanced as explanation. It is wellknown that placibo effect is ever advanced but it play in plant or in cell cultures a tiny role!

- significant protective effects can be observed with both protective models "similar" or "identical" metals as well in pretreatment as a curative treatment.

- moreover, this protective effect can be shown with very ultra low dose (10^{-30}, 10^{-40} and even 10^{-200} M).

These results that we considered as true toxicological reports open the way for new exciting investigations to reduce the side effects of metals and also all toxic drugs largely used in clinics by an original approach of detoxification of some toxic agents. Indeed, very high dilutions could decrease metal toxicity by increased of their elimination from living organisms.

Such a protective procedure was already well accepted for pretreatment with milligram or microgram doses in experimental toxicology (Flora and al, 1982a, - 1982b - Stacey and Klaassen, 1981 - Magos and al, 1979 - Yamane and Koizumi, 1982).

REFERENCE

AUQUIERE J.P., MOENS P., MARTIN P.L.
Recherche de l'action de dilutions homéopathiques sur les végétaux. I. Action du sulfate de cuivre 5 CH sur blé intoxiqué au sulfate de cuivre 1 p1000.
J. Pharmaceut. Belg., 36 (1981), 303-320.

AUQUIERE J.P., MOENS P., MARTIN P.L.
Recherche de l'action de dilutions homéopathiques sur les végétaux. II. Action du $CuSO_4$ 0,1 et 0,2%.
J. Pharmaceut. Belg., 37 (1982), 117-134.

BASTIDE M., DAURAT V., DOUCET-JABOEUF M., PELEGRIN A. and DORFMAN P.
Immunomodulatory activity of very low doses of thymulin in mice.
Int. J. Immmunotherapy 3 (1987), 191-200.

BILDET J.
Etude de l'action des différentes dilutions homéopathiques de Phosphore blanc (Phosphorus) sur l'hépatite toxique du rat.
Thèse Pharmacie, Bordeaux II, 1975.

BILDET J., SAUREL M., AUBIN M., CASAHOURSAT L.
Etude de l'action de divers complexes sur l'hépatite expérimentale au tétrachlorure de carbone.
Ann. Homéo. Fr., 4 (1978), 277-286.

BOIRON J., ZERVUDACHI M.T.
Actions de dilutions infinitésimales d'arséniate de sodium sur la respiration de coléoptiles de blé.
Ann. Homéo. Franc., 5 (1963), 738-742.

BOIRON J.
Actions de dilutions homéopathiques de HgCl2 sur la respiration des coléoptiles de blé.
Ann. Homéo. Franc., 5 (1965), 120.

BOIRON J., CIER A.
Elimination provoquée et spécificité d'action des dilutions infinitésimales.
Ann. Homéo. Fr., 10 (1962), 789-796.

BOIRON J., ABECASSIS J., COTTE J., BERNARD A.M.
Etude de l'action de dilutions hahnemanniennes de chlorure mercurique sur l'index mitotique de cultures de cellules animales.
Ann. Homéo. Fr., 3 (1981), 43-49.

BOMPART G., ORFILA C., MANUEL Y.
Cisplatin nephrotoxicity in cadmium pre-treated rats: enzymatic, functional and morphological studies.
Nephron, 58 (1991), 68-74.

CAL J.C., LARUE F., GUILLEMAIN J. and CAMBAR J.
Chronobiological approach of mercury induced toxicity and the protective effect of high dilutions of mercury against mercury-induced nephrotoxicity. In Guillouzo A. (Ed.) Liver Cells and Drugs. John Libbey Eurotext Ltd, Paris, 1988, pp 481-485.

CAMBAR J., DESMOULIERE A., CAL J.C., GUILLEMAIN J.
Mise en évidence de l'effet protecteur de dilutions homéopathiques de Mercurius corrosivus chez la souris.
Ann. Homéo. Fr., 5 (1983), 6-12.

CAMBAR J., DELBANCUT A., DORFMAN P.
Applications de très hautes dilutions en toxicologie.
Cahiers Biother., 125 (1994), 27-34.

CAZIN J.C., CAZIN M., GABORIT J.L., CHAOUI A., BOIRON J., BELON P., CHERRUAULT Y., PAPANAYOTOU C.
A study of the effect of decimal and centesimal dilutions of arsenic on the retention and mobilization of arsenic in the rat.
Human Toxicol. 6 (1987), 315-320.

CIER A., GIRARD A., ROUSSON C., BOIRON J.
Intoxication arsenicale et élimination provoquée d'arsenic.
Ann. Homéo. Franç., 3 (1963), 214-217.

DAVENAS E., POITEVIN B., BENVENISTE J.
Effect on mouse peritoneal macrophage of orally administered very high dilutions of Silicea.
Eur. J. Pharmacol., (1987), 165.

ENDLER P.C., SCHULTE J. (Eds.)
Ultra high dilution. Physiology and Physics. Kluwer Academic Publishers, Dordrecht, 1994.

FISHER P., HOUSE I., BELON P. and TURNER P.
The influence of the homeopathic remedy Pb met on the excretion kinetics of lead in rats.
Human Toxicol. 6 (1987), 321-324.

FLORA S.J.S., BEHARI J.R., ASHQUIN M. and TANDON S.K.
Time dependent protective effect of selenium against cadmium-induced nephrotoxicity and hepatoxicity.
Chem. Biol. Inter., 42 (1982), 345-351.

FLORA S.J.S ., JAIN V.K. , BERAHI J.R. and TANDON S.K.
Protective effect of trace metals in lead intoxication.
Toxicol. Letters 13 (1982), 51-56.

FRAZIER J.M.
Validation of in vitro models.
J. Am. College Toxicol., 9 (1990), 355-359.

GOYER R.A., CHERIAN M.G. (Eds.)
Toxicology of metals. Biochemical aspects, Springer Verlag, Heidelberg, 1995.

GRAVIOU E., BOIRON J.
Action d'une 15 CH de CuSO4 sur la culture de Chlorella vulgaris.
Ann. Homéo. Franç., 7 (1971), 539.

GUILLEMAIN J., CAL J.C., DESMOULIERE D., TETAU M., CAMBAR J.
Effet protecteur de dilutions homéopathiques de métaux néphrotoxiques vis-à-vis d'une intoxication mercurielle.
Cah. Biother., 81 (1984), 27-34.

HALM R.
Actualité et modernité de la recherche expérimentale en homéopathie.
Thèse Université de Montpellier 1 - Doctorat de Pharmacie, 1993.

HERKOVITS J., PEREZ-COLL C.
Zinc protection against delayed development produced by cadmium.
Biol. Trace Elem. Res., 24 (1990), 217-221.

HERKOVITS J. and PEREZ- COLL C.
Potentized microdoses of cadmium reduce the lethal effect of this heavy metal in amphibian embryos.
Berlin J. Res. Homeo. 1 (1991), 93-96.

HUMES H.D. and WEINBERG J. M.
Toxic nephropathy.
In: Brenner B.M. Rector F.C. (Eds). The Kidney, Saunders Ed., Philadelphia, 1986, pp 1491-2008.

ITO T., SAWAUCHI K.
Inhibitory effects on cadmium-induced testicular damage by pretreatment with smaller cadmium dose.
Okajima. Fol. Anat. Jap., 42 (1966), 107-117.

JIN T., NORDBERG G.F.
Cd-toxicity in kidney cells. Resistance induced by short term pretreatment in vitro and in vivo.
Acta Pharmacol. Toxicol., 58 (1986), 137-143.

JONES M.M., CHERIAN M.G.
The search for chelate antagonists for chronic cadmium intoxication.
Toxicololgy, 62 (1990), 1-25.

JONES R.L., JENKINS M.D.
Plant responses to homeopathic remedies.
Br. Hom., 70 (1981), 120-128.

KAJI T., KOYANAGI E., YAMAMOTO C., SAKAMOTO M., HOZUKA H
Possible mechanism for zinc protection against Cd cytotoxicity in cultured vascular endothelial cells.
Toxicology, 76 (1992), 257-270.

LAGROYE I., L'AZOU B., LAKHDAR B., PLANDE J., CAMBAR J. Nephrotoxicity in vitro : modèles glomérulaires.
In Adolphe A. et Guillouzo A. (Eds) - Toxicologie cellulaire in vitro - Editions INSERM, 1995, pp 121-150.

LAPP C., WURMSER L., NEY J.
Mobilisation de l'arsenic fixé chez le cobaye sous l'influence de doses infinitésimales d'arseniate de sodium.
Thérapie, 10 (1955), 625-638.

LAPP C., WURMSER L., KAUTRELE J.
Mobilisation du bismuth fixé chez le cobaye sous l'influence de doses infinitésimales d'un sel de bismuth.
Thérapie, 13 (1958), 438-449.

LAPP C., WURMSER L., NEY J.
Mobilisation de l'arsenic fixé chez le cobaye sous l'influence de doses infinitésimales d'arséniate de sodium.
Thérapie, 13 (1958), 46-55.

LARUE F., CAL J.C., GUILLEMAIN J., CAMBAR J.
Influence du facteur dilution sur l'effet protecteur de Mercurius corrosivus vis-à-vis de la toxicité induite par le chlorure mercurique chez la souris.
Homéo. Fr., 73 (1985), 375-380.

LARUE F., CAL J.C., GUILLEMAIN J., CAMBAR J.
Influence de la durée de prétraitement sur l'effet protecteur de Mercurius corrosivus vis-à-vis de la toxicité induite par le chlorure mercurique chez la souris.
Homéo. Fr., 5 (1986), 275-281.

LINDE K., JONAS W.B., MELCHART D., WORKU F., WAGNER H., EITEL F.
Critical review and meta-analysis of serial agitated dilutions in experimental toxicology.
Hum. Exp. Toxicol., 13 (1994), 481-492.

LITTERST C.L., SCHWEITZER V.G.
Increased tissue deposition and decreased excretion of platinum following administration of cisplatin to cisplatin pretreated animals.
Cancer Chemother. Pharmacol., 12 (1984), 46-49.

LIU X.Y., JIN T., NORDBERG G., RANNAR S., SJOSTROM M., ZHOU Y.
A multivariate study of protective effects of Zn and Cu against nephrotoxicity induced by Cd metallothionein in rats.
Toxicol. Appl. Pharmacol., 114 (1992), 239-245.

MAGOS L., WEBB M., BUTLER G.
The effect of Cd pretreatment on the nephrotoxic action and kidney uptake of mercury in male an female rats.
Brit. J. Exp. Pathol., 55 (1979), 589-594.

MANSVELT J.D., VAN AMONS E.
Inquiry into the limits of biological effects of chemical compounds in tissue culture: low dose effects of mercuric chloride.
Zeitschrift fur Naturforschung, 30 (1975), 643-649.

MERY J.P., FILLASTRE J.P.
Reins et médicaments.
In Hamburger J., Grunfeld J.P. (Eds.), Nephrologie, Flammarion Edition, Paris, 1983, pp 1193-1207.

MESHITSUKA S., OHSHIRO H., NOSE T.
Release of pre-absorbed cadmium from cultured cells by the addition of copper ion.
Experentia, 38 (1982), 1473-1474.

MORITA S.
Defense mechanisms against Cd toxicity. II. Effects of pretreatment with a small oral dose of Cd on absorption, distribution and excretion of Cd after a large oral dose in mice.
Jpn. J. Pharmacol., 35 (1984), 143-151.

MOSS V.A., ROBERTS J.A., SIMPSON H.K.L.
The effect of copper sulfate on the growth of the alga chlorella.
Br. Hom. J., 66 (1977), 160-177.

MOURIQUAND G., CIER A., BOIRON J., EDEL V., CHIGHIZOLA R.
Rétention et mobilisation des toxiques exogènes chez le pigeon. Essais de doses infinitésimales de ces mêmes éléments et variations concomittantes de l'indice chronologique vestibulaire.
Ann. Homéo. Franç., 3 (1961), 766-781.

NETIEN G.
Actions de dilutions homéopathiques sur la respiration du coléoptile de blé.
Ann. Homéo. Fr., 4 (1962), 823-827.

NETIEN G., GIRARDET E.
Expérimentation des hautes dilutions dans le domaine végétal.
Ann. Homéo. Fr., 5 (1963), 729-737.

NETIEN G., BOIRON J., MARIN M.
Action de doses infinitésimales de sulfate de cuivre sur des plantes préalablement intoxiquées par cette substance.
Ann. Homéo. Franç., 8 (1966), 130-136.

OBERBAUM M.,MARKOVITS R., WEISSMAN Z., KALINKEVITS A., BENTWICH Z.
Wound healing by homeopathic silica dilutions in mice.
Harefuah, 123 (1992), 79-82.

PELIKAN W., UNGER G.
The activity of potentized substances. Experiments on plant growth and statistical evaluation.
Brit. Hom. Journ., 60 (1971), 233-266.

PELLEGRINI O., DAVENAS E., EL AZZOUZI B., JURGENS P., BENVENISTE J., MANUEL Y., THOMAS Y.
Stress proteins in human lymphocytes. I. Induction of metallothonein and heat shock protein genes by cadmium in an immature T cell line.
Cell. Pharmacol., 1 (1994), 91-97.

POITEVIN B., AUBIN M.
Méthode d'étude de l'activité des dilutions infinitésimales sur des cellules isolées.
Homeopathie, 2 (1986), 49-52.

POITEVIN B.
Le devenir de l'homéopathie. Eléments de théorie et de recherche. Doin Ed., Paris, 1987.

PROJETTI M.L., GUILLEMAIN J., TETAU M.
Effets curatifs et préventifs de dilutions homéopathiques de sulfate de cuivre appliquées à des racines de lentille pré- ou post-intoxiquées.
Cahiers Biother., 88 (1985), 21-27.

STACEY N.H., KLAASSEN C.D.
Interaction of metal ions with cadmium-induced cellular toxicity.
J. Toxicol. Environ. Helath, 7 (1981), 149.

SUZUKI S.
The protective effects of thiol containing compound on mercuric chloride induced acute inhibition of enzymes from mouse kidney.
Toxicology, 29 (1984), 207-220.

SUZUKI C.A.M., OHTA H., ALBORES A., KOROPATNICK J., CHERIAN M.J.
Induction of metallothionein synthesis by zinc in cadmium pre-treated rats.
Toxicology, 63 (1990), 273-284.

TERHAR C.J., VIS E., ROUDABUSH R.L., FASSET D.W.
Protective effects of low doses of cadmium chloride against subsequent high oral doses in the rat.
Toxicol. Appl. Pharm., 7 (1965), 500.

VISCHNIAC H.
Influence des doses infinitésimales de plomb sur l'évolution de l'intoxication au plomb chez l'animal.
Homéo. Franç., 53 (1965), 21-33.

WEBB M., MAGOS L.
Cd-thionein and the protection against the nephrotoxicity of mercury.
Chem. Biol. interact., 14 (1976), 357-369.

WEIS P., WEIS J.S.
Cd acclimation and hormesis in Fundulus heteroclitus during fin regeneration.
Environ. Res., 39 (1986), 356-363.

WIJK van R., WIEGANT F.A.C. (Eds.)
Cultured mammalian cells in homeopathy research; the similia principle in self-recovery.
Utrecht University Editions (1994).

WURMSER L.
Influence des doses infinitésimales sur la cinétique des éliminations.
Homéo. Fr., 72 (1984), 165-173.

YAMANE Y. and KOIZUMI T.
Protective effect of Mo on acute toxicology of HgCl2.
Toxicol Appl Pharmacol, 15 (1982), 14-221.

YOSHIKAWA H.
Preventive effect of pretreatment with cadmium on acute cadmium poisoning in rats.
Industrial health (Japan), 11 (1973), 13-119.

YOUBICIER-SIMO B.J., BOUDARD F., MEKAOUCHE M. , BASTIDE M., BAYLE J.D.
Effects of embryonic bursectomy and in ovo administration of highly diluted bursin on adrenocorticotropic and immune responses of chickens.
Int. J. Imunotherapy, 9 (1993), 169-180.

FUNDAMENTAL RESEARCH ON HIGH DILUTION EFFECTS. A CLASSIFICATION OF NON-CLINICAL RESEARCH TOPICS

P. CHRISTIAN ENDLER
Institute for Holistic Medicine, University of Urbino, Italy and Ludwing Boltzmann Institute for Homoeopathy, Graz

PIERRE DORFAMAN
Paris, France

1. Introduction

A survey of the present 'state of the art' of research on homoeopathy is presented by M. Bastide in this volume. The complementing paper presented here aims at describing the 'building' of homoeopathy as a complex system with a certain number of doors.

Basic tenets of fundamental research on homoeopathy

Observation of physiological effects; 'similarity' and 'potency' rule

A large variety of studies with regard to fundamental research on homoeopathy has been presented up to now. The studies on the interaction of an organism and a test substance can in principle be divided into two complementary parts.

Studies were performed *on the reactional cybernetics of living systems* with application of either conventionally prepared (diluted) test substances or test substances prepared according to a 'homoeopathic' protocol of step by step dilution and agitation ('potencies').

This type of studies is related to the *similarity principle* which all clinical homoeopathy is based on. The similarity principle states that a substance which may cause symptoms in a healthy organism, may be applied to a diseased or ill organism in a curative sense if this organism shows identical symptoms (symptoms similar to the former). The stimulation of an irritated autoregulative recovery by application of the *similarity principle* is considered to be the essence of homoeopathy. In a basic (reduced) form, this rule concerns the curative effect of an agent at high dilution after intoxication of a living system with the same agent at low dilution - 'isopathic principle' in a modern sense, van Wijk and Wiegant, 1994 -).

Research on homoeopathy, however, may focus more *on the test substance applied* (*potency* research). Biological studies were either performed in order to demonstrate the difference between 'conventionally' diluted substances and homoeopathic potencies of the same substances, mostly at the level of low dilutions, or in order to describe biological effects caused by (often extremely) high potencies tested versus neutral control.

The traditional *potency principle* assumes that, provided the similarity rule is met, a homoeopathic drug is more active ('sharp'), the more the initial mother substance is diluted in the standardised process of stepwise dilution (e.g. 1 : 10 or 1 : 100, respectively) and agitation (intermediate succussion). Furthermore, typical effects (e.g. stimulation versus inhibition of a physiological process) seem to be related to typical ranges of dilution.

More studies on a few reliable models linking the gap between the areas of low and high dilution are needed in order to have more knowledge about the interconnection of the *similarity principle* and the *potency principle*.

Biophysical explanation

Further studies concern *the biophysical interaction* of an organism and a test substance.

Physics approach

The *physical properties* of the potencies themselves are subject of further studies.

Common aims of fundamental research

It is obvious that a synopsis of all research results should ideally point towards a consistent description of the reaction cybernetics of the living system *and* of the changes occurring during the dilution process of the test substance.

2. Main Part

In the following, existing studies are classified with regard to the stimulus or stimuli applied: this may be either only one single stimulus or two consecutive stimuli. In the latter case, the respective mother tinctures may be identical or different, if a diluted substance is used. Furthermore, the dilutions may be prepared in different ways (see Tab. 1).

This differentiated way of presentation is chosen in order to show that homoeopathy is a genuine type of science in itself, with its own research strategies. Furthermore, it was chosen in order to stimulate more research along these lines, perhaps leading to a wider understanding of the laws of nature in general.

In this text, the symbol '##' indicates that a study met the criteria for publication in conventional (indexed) scientific journals, i.e. that the respective paper underwent a peer-review process. This classification may exclude important work that, for different reasons, was not submitted to conventional media. The symbol '#' indicates that a study met the criteria for discussion in conventional journals, i.e. that the content of the study was explicitely discussed in such media or that a relevant abstract was printed in a peer-reviewed journal.

It has to be pointed out that this list of studies is not exhaustive. Furthermore, the paper remains incomplete with regard to research on the preparation process of homoeopathic potencies, quality control and storage conditions.

Tab. 1

	classification of study	*source of 1st and 2nd stimulus:*	*1st*	*2nd*	*type of observation*
A	conv. pharmaceutics		**L**	-	typical effect
B	hormesis		**H**	-	inversion of effect
C	tolerance	**i**	**L**	**L**	protection
I	homoeopathy: general		**P**	-	exploration of effects
II	"	**i**	**L**	**H**	cure
III	"	**i**	**L**	**P**	detoxification
IV	"	**i**	**P**	**L**	protection
V	"	**i**	**P**	**P, P**	provocation of symptom (hom. drug proving)
VI	homoeopathy: special	**d**	**L**	**L**	protection
VII	"	**d**	**L**	**H**	cure
VIII	"	**d**	**S**	**L**	α) symptom
				L	β) cure
IX	"	**d**	**L**	**L**	α) symptom
				H	β) cure
X	"	**d**	**L**	**L**	α) symptom
				P	β) cure
XI	"	**d**	**S**	**L**	α) symptom
				H	β) cure
XII	"	**d**	**S**	**P, P**	α) symptom
				P	β) cure

For legend, see next page

Legend to Tab. 1: Effects described after application of one stimulus or two consecutively applied stimuli on a living system.

A,B,C; I-XII, different types of studies. For classification, see below.

Source of 1st and 2nd stimulus: in case a dilution is used, the respective mother tinctures may be i, identical, or d, different.

Type of stimulus: the 1st and /or the 2nd stimulus may be at low dilution (crude substance or mother tincture, L); or at high dilution (any degree of dilution of the mother tincture, H); or at high dilution in potentised form (P); or the initial state may be a stressed or pathological state (illness) (S). Theoretically, 24 combinations (types of studies) are possible.

The different types of studies found in literature are:

A **conventional pharmaceutical research**

occurrance of a typical effect of an agent as known from conventional pharmacology

B **research on hormesis**

inversion of this effect at high dilution of this agent, known as 'hormesis'

C **research on self-tolerance**

protective effect of a 1st stimulus with regard to a consecutive 2nd application of the same noxious agent

general fundamental research on homoeopathy including the empirical homoeopathic intoxication design in drug proving

I different effects of potentised agents that remain to be further explored

II stimulation of recovery of the living system by the 2nd stimulus

III a detoxification by the 2nd stimulus

IV a protective effect of the 1st stimulus

V the provocation of substance specific (illness) symptoms by frequent repetition of one and the same stimulus (empirical homoeopathic intoxication design for drug proving)

special fundamental research on homoeopathy and therapeutic application of homoeopathy

VI a protective effect of the 1st stimulus, even if this stimulus is of a different nature from the 2nd stimulus

VII a curative effect of the 2nd stimulus, even if this stimulus is of a different nature from the 1st stimulus

VIII the initial finding of S. Hahnemann on 'homoeo'-pathy:

(α) in two different individuals 1 and 2, similar typical symptoms may occurr due to a pathological process (stimulus 1) or due to an intoxication (stimulus 2), respectively

(β) stimulus 2, e.g. a crude substance, exerts a curative effect on the individual 1

IX recent cell biological finding. As in (VIII), with stimulus 2 applied at a high dilution

X immunological finding. As in (VIII), with stimulus 2 applied at a high dilution in potentised form

XI examples from modern conventional medicine on the use of diluted agents

XII the frame of homoeopathic medicine, based on the similarity between stimulus 1 and 2 (simile principle) and on the use of the agent (stimulus 2) at high dilution in potentised form (potency principle). P, P means the previous provocation of symptoms in a drug proving (see V).

For further details, see text.

A Conventional pharmaceutical research

Conventional pharmaceutical research describes typical physiological effects of agents at relatively low dilutions. Normally, these effects are either stimulatory or depressive.

B Research on hormesis

The Arndt-Schulz principle states that weak stimuli stimulate vital activity, medium stimuli may support this activity, while the strongest stimuli may annul vitality (Martius 1923). Put simply, hormesis means that in many cases a stimulatory effect occurs in living systems after exposure to a high dilution of an otherwise toxic agent, whereby stimulation may even mean an improvement of health or an enhancement of longevity (Oberbaum and Cambar 1994). In various fields of biology, such dose-dependent reverse effects (biphasic dose /response relations) have been described and a large number of papers has been presented in conventional media (for survey see Linde 1991; Faiderbe et al. 1993; Popp 1994; Oberbaum and Cambar 1994).

Such reversed biological effects in various ranges of concentration of the same agent represent the basic principles of the *similarity* and the *potency principle* (Popp 1994). There is evidence that these effects become more marked on special preparation (homoeopathic potentisation) of the test substance (Larue et al. 1985 ##; Cal et al. 1988 ##; Lauppert and Endler 1995.

C Research on tolerance

These studies concern the protective effect of an agent at low dilution with regard to a consecutive second application of the same agent at the same concentration.

In the so-called 'self-tolerance' studies, the same stressor conditions are applied twice in the same strength with a time interval in between. In various fields of biology, protective effects of the 1st stimulus have been described and a large number of papers has been presented in conventional media (see van Wijk and Wiegant 1994).

Aspects specially relevant for homoeopathy are, firstly, that one of the stimuli may be applied at high dilution (see II, III) and, secondly, the 1st and the 2nd stimulus may be of a different nature (see VII, 'cross-tolerance').

General fundamental research on homoeopathy

I Different effects of agents at high dilution in potentised form

A large number of studies has been performed in order to test the assumption that agents can exert biological activity even if applied at high and extremely high dilution, often beyond Avogadro's limit of theoretical O-molarity. The structure of these experiments is simple in comparison to the other models, since only one stimulus is applied. In all cases, the agent underwent a potentisation process.

In vitro and ex vivo studies

Boyd 1954 studied the influence of a mercury potency on the starch splitting activity of malt-diastasis. In this meticulous study performed with a very large number of samples over a period of several years, a stimulation of this reaction was found (#).

Harisch and Kretschmer 1988a studied the influence of potentised zinc on peritoneal cells of rats. An affection of histamin release was found (##).

Bonavida 1992, Bonavida et al. 1993 studied the influence of synergistic compounds on tumor cell growth. Typical effects were found (##).

Doutremepuich et al. 1987, 1990, 1994 studied the influence of potentised acetylsalicylic acid on blood platelet aggregation. In this type of study that was repeated over several years, a reduction of bleeding time was found (##).

Belon 1987; Boiron and Belon 1982; Poitevin et al. 1986, 1988; Sainte Laudy 1989; Sainte Laudy et al. 1986; Sainte Laudy and Belon 1993 studied the influence of potentised histamin or bee venom on the degranulation of basophils. Donors were either allergic persons or donor blood was sensitised with anti IgE antibodies (see also VIII). In these mutually confirmative studies, an inhibition of basophil degranulation was found (##).

Davenas et al. 1988 (positive); Benveniste et al. 1991 (positive); Ovelgonne et al. 1992 (negative); Litime et al. 1993 (positive); Hirst et al. 1993 (negative) studied the influence of potentised anti IgE antibodies on the degranulation of basophils. In some, but not in all of these studies, a stimulation of basophil degranulation was found (##).

Weismann et al. 1992 a,b studied the influence of potentised antigen on IgM and IgG antibodies. A modulation of the immune response to KLH was found (see also VIII).

Petit et al. 1989 studies the influence of potentised substances on subcellular enzymatic activity. No effects were found (##).

Sukul and Zaghlool 1990 studied the influence of traditional homoeopathic remedies on the isolated ileum of rats. Hadji et al. 1991; Benveniste et al. 1992 studied the influence of different hormones on the activity of isolated perfused hearts. In these studies, an enhancing effect on the organs was found (#).

In vivo studies or zoological studies

Bastide et al. 1985, 1987; Bastide and Doucet-Jaboeuf 1985; Doucet-Jaboeuf et al. 1982, 1984, 1985 a,b studied the influence of potentised thymulin on humoral and cellular immune response in (in some studies thymectomized) mice. Daurat et al. 1988 a,b studied the influence of potentised interferon on the immune response in mice. Youbicier-Simo et al. 1993 studied the influence of potentised bursin on (bursectomized) chickens. In all three types of the studies, a stimulation of the immune response was found (##).

Davenas et al. 1987 studied the influence of potentised quartz, a traditional homoeopathic remedy, on mice macrophages. A stimulation of their activity was found (##).

Koenig 1926 studied the influence of potentised metal salts on amphibian metamorphosis. Endler et al. 1991 a,b, 1994 a-d, 1995 a,b studied the influence of potentised thyroxine on amphibian metamorphosis and activity. In these mutually confirmative studies, stimulation and inhibition peaks were found at different degrees of dilution. The studies on the influence of thyroxine (see also III) were multicentrically repeated (##).

Sukul et al. 1986 studied the influence of traditional homoeopathic remedies on experimentally induced catalepsy in mice. A prolongation of the cataleptogenic effect was found (##).

Botanical studies

Kolisko 1926; Pelikan and Unger 1965; Jones and Jenkins 1981, 1983; Netien and Graviou 1978; Dutta 1989; Steffen 1984; Pongratz and Endler 1994; Betti 1994, Lauppert 1995 studied the influence of potentised metal salts and other compounds on the growth of wheat and yeast, respectively. In these mutually confirmative studies, stimulation and inhibition peaks of growth were found at different degrees of dilution. Popp found effects at moderate, but not at high degrees of dilution. Two studies (Speciani 1988, negativ; Lauppert 1995, positive) were published as thesis. A survey is given by Dorfman 1993.

II Curative effect of an agent at high dilution after intoxication with the same agent at low dilution

In vitro studies

In this type of studies, a stressor condition (heavy metal, heat shock, oxidative stress) is applied to cell cultures in order to induce cellular recovery processes. Consecutively, it is determined whether after exposure to a damaging condition, cells can be stimulated by a high dilution of the same stressor, which would be ineffective in control conditions.

The aim of these experiments on cellular recovery processes is to analyze the stimulatory effect of low doses (= relatively high dilutions) of compounds which are applied according to the similia-principle in its most elementary form (isopathic principle, van Wijk and Wiegant 1994).

Van Wijk and Wiegant 1994; Ovelgönne et al. 1995 studied the influence of various stressors on cell cultures. In these mutually confirmative studies, a stimulation of autoregulative recovery was found (##).

III Curative effect of an agent at high dilution in potentised form after intoxication with the same agent at low dilution

In these studies as in (II), the similia-principle in its basic form is tested.

Different from (II), the agent used for stimulation of recovery underwent a potentisation process.

The reason for this is that it is assumed that the curative effect only occurs if or is more marked if the agent is potentised (potency principle).

In vivo or zoological studies

In this type of study, animals are detoxicated after experimental intoxication. Usually enhanced excretion of the toxin in the urine and stools is described. A metaanalysis is given by Roth 1991.

Cazin 1986; Boiron 1988; Boiron and Cier 1962 a,b, 1971; Boiron et al. 1968, 1982; Cazin and Gaborit 1982; Cazin et al. 1987; Chaoui 1988; Gaborit 1987 a,b studied the influence of potentised arsenite after arsenite intoxication of rats. Mouriquand et al. 1961, 1962, 1975 used pigeons and Lapp et al. 1955, Lapp and Wurmser 1958; Wurmser 1984 used guinea pigs. In these mutually comparable studies, enhanced detoxification was described (##).

Fisher and Capel 1982 (positive); Fisher 1987 (negative, - ## -) studied the influence of potentised lead on lead-intoxicated rats.

Herkovits et al. 1988, 1990 studied the influence of potentised cadmium on cadmium-intoxicated frog spawn. Decrease of letality was found (##).

Endler et al. (1994b, 1995b, see I) discussed their multicenter results on the inhibition of metamorphosis by potentised thyroxine with regard to the naturally enhanced thyroxine level of the amphibia. Lauppert and Endler 1995 studied the influence of potentised thyroxine on thyroxine hyperstimulized amphibia. In this study, a curative effect was found.

Botanical studies

Projetti et al. 1985 studied the influence of potentised copper salt lentils previously intoxicated with copper salt . A curative effect was found.

Human studies

Patterson 1944 studied the influence of potentised mustard gas on mustard gas burns of the skin. Therapeutic (as well as prophylactic) effects were found.

Reilly et al. 1986 studied the influence of potentised grass pollen on hypersensitive persons exposed to those allergens. A therapeutic effect was found (##).

Gardes 1989 studied the influence of the potentised chemotherapeutic substance nalidixine acid on persons treated with nalidixine acid. Enhanced excretion was found. The study was published as a thesis.

IV Protective effect of an agent at high dilution in potentised form with regard to a consecutive second application of the same agent at low dilution

In these studies as in (III), the similarity principle in its basic form as well as the potency principle are investigated. In contrast to (III), the potentised agent is applied before the noxious agent.

In vitro study

Delbancut et al. 1993 studied the effect of potentised cadmium on cadmium-intoxicated renal tubular cell cultures. A protective effect was found.

Zoological studies

Cambar 1983 a,b; Guillemain et al. 1984; Larue and Cal 1985, 1986; Cal et al. 1986, 1988 a,b; Larue et al. 1985 studied the influence of potentised mercury on mercury-intoxicated mice. In these mutually comparable studies, a protective effect was found.

Souza Magro et al. 1986 studied the influence of the potentised antibiotic gentamycin on gentamycin-treated rats. A protective effect on nephrotoxicity was found.

Labonia et al. 1986 studied the influence of a potentised snake poison on mice intoxicated by this poison. A protective effect was found.

Botanical studies

Projetti et al. 1985 studied the influence of potentised copper salt on copper salt intoxicated lentils. A protective effect was found.

V Provocation of substance specific (illness) symptoms by frequent repetition of the application of one and the same agent at high dilution in potentised form

In these studies, again as above, one and the same agent is applied at different times. However, all stimuli (usually much more than two in this type of study, and usually diluted beyond Avogadro's value) underwent a potentisation process. Interestingly, while the effect of one single application of the potency was curative or protective, i.e. opposite to that of the agent at low dilution in (III) and (IV), in this type of study, frequent application led to disease or illness symptoms or an aggravation of symptoms.

Zoological study

Endler et al. 1994a studied the influence of potentised thyroxine on amphibia. When the extremely diluted agent was applied at short intervals, a stimulation of thyroxine-controlled metamorphosis and hyperactivity was found (see also I).

Human studies

are described in detail in contributions to drug provings (empirical homoeopathic intoxication studies). The intoxication symptoms are the basis of homoeopathic manuals.

Special fundamental research on homoeopathy

In these studies, again as above, two stimuli are applied. In contrast to the protocols concerning the similarity principle in its basic form (i.e. the isopathic principle, see II-IV), here, the 1st and 2nd stimulus are of a different nature. The studies thus concern the similarity principle in its complex form.

VI Protective effect of an agent at low dilution with regard to a consecutive application of a different agent at low dilution

Different to the so-called 'self-tolerance' studies (C), in these 'cross-tolerance' studies, the protective effect is exerted by a 1st stimulus that is of a different nature than the 2nd stimulus. A very large number of conventional articles is present on this topic (see van Wijk and Wiegant 1994).

In vitro study

These studies are especially relevant to homoeopathy when an observed stressor-specificity is shown to result in a stressor-specific development of tolerance, such as was outlined by van Wijk and Wiegant 1994.

VII Curative effect of an agent at high dilution in potentised form after intoxication with a different agent at low dilution

These studies are closely related to the detoxification protocols (III) on the similarity and potency principle. However, the curative agent is not a potency of the noxious agent, but one of an entirely different agent.

Zoological study

Aubin et al. 1979, 1980; Bildet et al. 1975, 1977, 1978, 1984 a,b,c; Andresen 1985; Harisch and Kretschmer 1988 a,b, 1990; Harisch et al. 1986, 1987, 1990; Harisch et al. 1984, 1986; Sur et al. 1990 studied the influence of traditional homoeopathic remedies (e.g. phosphorus) on CCl_4 induced liver cell damages in rats. A curative effect is reported (Harisch et al., some ##). One study (1985) was published as a thesis.

Santini et al. 1991 studied the influence of potentised copper on neostigmine induced intestinal hypertransit. A reduction of the neostigmine action was found (copper administered alone showed no effect) (##).

VIII Curative effect of an agent at low dilution that uses to cause symptoms in a healthy organism ,

on a diseased or ill organism if the latter shows identical symptoms

This type of human study was used by Hahnemann in the early phase of detection, exploration and first clinical evaluation of 'homoeo'-pathy. See the contributions to clinical research.

IX Curative effect of an agent at high dilution (that, at low dilution, uses to cause symptoms in a healthy organism),

on an organism intoxicated by a different agent at low dilution if the latter shows identical symptoms

In vitro study

The aim of these experiments is to study the specificity of the similia-principle. The most effective low-concentration (= relatively high dilution) stimulation of recovery processes in damaged cells was shown to occur when a high similarity was found between the damaged condition and the effect of the low dilution of the 2nd stimulus (van Wijk and Wiegant 1994). These observations confirmed the aspect of specificity of the similia-principle.

X Curative effect of an agent at high dilution in potentised form (that, at low dilution, uses to cause symptoms in a healthy organism),

on an organism sensitised by a different agent at low dilution if the latter shows identical symptoms

These studies are related to (IX). In contrast to (IX), they concern not only the similarity principle in its complex form, but also the potency principle.

In vitro study

This type of study is closely linked to some immunological studies, i.e. where basophils are pretreated with antigens: Poitevin et al. 1988 studied the influence of traditional homoeopathic remedies on anti IgE antibody - treated basophils. An inhibition of basophil degranulation was found (##).

Zoological studies

Dorcsi 1974 studied the influence of a traditional homoeopathic remedy on experimenal edema in rats. A curative effect was found. Bildet et al. 1989 studied the influence of potentised bee venom on UV-induced cutaneous erythema in guinea pigs (both a bee sting as well as ulraviolet rays can induce an inflammatory reaction). In this successively repeated study a curative effect was found (##).

Practical homoeopathy claims that homoeopathic dilutions of bee venom are capable of inhibiting cutaneous erythema induced both by bee venom as well as by ultraviolet rays.

Botanical studies

Speciani 1988 (thesis) performed several related botanical studies, with positive results.

XI

Examples from modern conventional medicine

Mercury salts, as part of their toxicologic picture, cause oliguria and anuria; many of the classical diuretic drugs are mercury derivatives (e.g. mersalyl and chlormerodrin); the same holds true for the diuretics of the sulfonamide family (chlorthalidone, mefrusid, and chlorthiazide); digitalis, which in toxic quantities causes a picture of arrhythmic tachycardia, is indicated for the treatment of atrial fibrillation; at a high dosage, emetin causes a picture of diarrhea and ulcers of the colon, resembling amebiasis; in conventional medicine it is used as a drug against amebiasis (Oberbaum 1994)

XII Curative effect of an agent at high dilution in potentised form (that, at frequent application at high dilution in potentised form, uses to cause symptoms in a healthy organism),

on a diseased organism, if the latter shows identical symptoms

This type of treatment marks the specific frame of late Hahnemannian and modern homoeopathic medicine that is based α) on the similarity between the symptom picture (see V) of the agent and the symptom picture of the disease (*similarity principle*) and β) on the use of the agent at an adequate high dilution in potentised form (*potency principle*).

In vitro study

Poitevin et al. 1986 studied the influence of traditional homoeopathic remedies on basophils of hypersensitive (allergic) persons. An inhibition of basophil degranulation was found (##).

Zoological studies

Bildet et al. 1989 studied the influence of a traditional homoeopathic remedy on UV-irritated guinea pigs. This was done because of the similarity of the respective symptom pictures. A curative effect was found (see also XI) (##).

Khuda-Bukhsh and Banik 1990 studied the influence of a traditional homoeopathic remedy on damages in X-irritated mice. This was done due to the similarity of the respective symptom pictures. A curative effect was found (##).

Oberbaum et al. 1992 studied the influence of a traditional homoeopathic remedy on wound healing in mice. This was done because of the similarity of the respective symptom pictures. A curative effect was found.

Veterinary studies

Wolter 1985 studied the influence of a traditional homoeopathic remedy on labour pain in pigs. This was done because of the similarity of the respective symptom pictures. A curative effect was found.

Further veterinary studies are discussed in the contribution to clinical research.

Human studies

are discussed in the contributions to clinical research.

Biophysical explanation

Ludwig 1987 and Smith 1987, 1994 described typical resonance frequencies related to potencies prepared according to the instructions of homoeopathy. At each step of the stepwise dilution and agitation process, more typical frequencies use to occurr. This leads to the idea of an energetic interrelationship of a potency and an organism, possibly including positive and negative resonance phenomena.

Van Wijk and Wiegant 1989, 1994 and Endler et al. 1994a,b, 1995a,b studied the influence of potencies versus controls that were sealed in hardglass vials in order to make a direct mechanical contact between the dilution and the organism impossible. In these mutually comparable studies, significant effects were found on humans and on other vertebrates. The study of 1995b was performed multicentrically (##).

Senekowitsch et al. 1995 studied the influence of frequency patterns that were electronically scanned from a potency versus a control and stored on data carrier (CD). A significant effect, comparable to that of the original potency, was found (#).

Citro et al. 1991, 1994, 1995; Aissa et al. 1993, 1995; Endler et al. 1995a,b; Thomas et al. 1995 studied the influence of test liquids that were prepared by means of electronic transmission of information from mother substances. In these mutually comparable studies, significant effects, compared to analogously prepared control, were found (#; 1995b ##).

Physics

Studies on physical properties of potencies (aqueous or on lactose) are discussed in Resch and Gutmann 1986; Luu-d-Vingh 1974 (##), Larue 1986 (##); Righetti 1988; King 1988, Majerus 1990, Weingärtner 1992; Antonchenko et al. 1992 Demangeat et al. 1992 (##), 1996; Bischof and Rohner 1992; Anagnostatos 1994; Endler and Schulte 1994; Endler et al. 1996. Recent physics research revealed that water dipoles may develop phase coherent oscillations through radiation coupling (del Giudice et al. 1988, del Giudice 1994). It is proposed that these could be modulated as a time-ordered pattern of signals and induce a coherent wave propagation (electron propagation). Further theory suggests that the phase coherent oscillations may originate information pattern formation through isotopicity effects in the dilutions (Berezin 1994, Schulte 1994).

3. Acknowledgements

The two main lines of research on homoeopathy, namely on the test substance applied (potency research) and on the reactional cybernetics of living systems (research on the similarity principle) are approached in [1-4]. The authors thank all those who gave their advice for establishing the classification frame, especially M. Haidvogl, P. Andersch, R. van Wijk, M. Bastide and F.A. Popp; those who helped in monitoring the discussed studies, especially M. Righetti, H. Walach, P. Fisher, G. King, J. Hornung, M. van Wassenhoven, M. Gooch and the group „Forschung im Dienste der Gesundheit"; those who helped in classifying the studies, especially F.A.C. Wiegant and B. Poitevin; H. Walach, C. Heckmann and T. van Asseldonk for their redactional help and E. Lauppert for his secretary work. The material presented in this paper is part of an internal working paper of the European Committee for Homoeopathy, Brussels.

4. References

[1] Bastide M., this volume

Bastide M. (ed.). Signals and Images. Kluwer, Dordrecht, 1997.

[2] Endler PC and Schulte J. Ultra High Dilution. Physiology and Physics. Kluwer, Dordrecht 1994.

[3] Endler PC and Schulte J. Ultra High Sensitivity. Kluwer, Dordrecht, in preparation.

[4] Van Wijk R and Wiegant FAC. Cultured Mammalian Cells in Homoeopathy Research: the Similia Principle in Self-Recovery. Utrecht University, Utrecht 1994.

The extensive list of references to this article can be ordered from: European Committee for Homoeopathy. Brussels, Belgium, chée de Bruxelles 132. An updated version of this paper including this reference list is presented in [3].

STIMULATION OF SELF-RECOVERY BY THE APPLICATION OF LOW DOSES. STUDIES WITH CELLS

R. VAN WIJK and F.A.C. WIEGANT
Department of Molecular Cell Biology, Utrecht University
P.O Box 80.056, 3508 TB Utrecht, The Netherlands

ABSTRACT

The aim of our studies is to understand the stimulation of self-recovery processes at the cellular level by compounds which are applied according to the similia-principle. In order to induce self-recovery, cells are exposed to mildly damaging conditions. In our research a non-lethal concentration of sodium arsenite or a mild heat shock was used which induces the so-called heat shock- or stress response. This response includes the synthesis of a specific class of proteins (the heat shock proteins or stress proteins), the stimulation of proliferation, and the development of tolerance. Based on several lines of experimental evidence, we have proposed that self-recovery can be evaluated in terms of the development of tolerance and the enhanced production of heat shock proteins. With respect to the aim of our studies, it is of interest to determine whether, after an exposure to a damaging condition, cells can be stimulated in their development of tolerance and the production of heat shock proteins, by stressor doses that are not effective in undamaged (control) cells. In the present paper, we summarize the experimental evidence, in which it was shown that moderate stressor conditions (such as sublethal temperature elevation or arsenite concentration) transiently stimulate cell proliferation. Furthermore, it will be shown that these conditons were not only able to stimulate growth, but also induced the development of tolerance. Then, the evidence is summarized which indicates that the hsps can be used as a molecular parameter for cellular recovery. Finally, the experiments are described which were performed to study the effect of low doses of arsenite as well as of mild heat shock temperatures on the synthesis of hsps and on the development of tolerance in arsenite- and in heat shock-sensitized cell cultures. Along this line we offer an explanation for the capacity of low doses to stimulate molecular processes that are essential for self-recovery.

1. Introduction

The study of cell growth and self-recovery of a damaged cell population has been one of the most challenging problems, which has been approached from many different points of view, such as biochemistry, biophysics, developmental biology and oncology. A few other fields, mostly clinical, show interest in the related phenomenon of regeneration. Research in these fields is particularly aimed at understanding the ways to stimulate the process of self-recovery.

Cells of many organs are in a state of quiescence. However, they retain the capacity to proliferate and re-enter the cell cycle. This occurs for instance in the process of wounding and also after application of mild stressor conditions. These conditions not only may lead to perceptible damage to organs when present during a prolonged period of time, but they also stimulate the intrinsic regenerative capacity, aimed at abolishing the inflicted damage. To be effective, the process of regeneration involves two main sequences of events: (a) stimulation of cell proliferation, and (b) induction of tolerance towards the damaging condition. Due to the relatively greater resistance of the cells, the spread of injury is limited on the one hand, and on the other hand the process of overall recovery is speeded up through stimulation of cell proliferation. For understanding of the regenerative growth response there is a definite advantage in describing cell reactions in terms of a coordinate expression of the sequence of events which involve the **stimulation of proliferation** and the **induction of resistance or tolerance** after stress.

A central issue of our studies is the investigation of the regulatory mechanism of self-recovery and its stimulation in diseased or disturbed biological systems by the application of low doses of compounds according to the similia-principle. According to this principle, a substance, capable of provoking symptoms in a healthy organism, acts as a curative agent in diseased organisms in which the same symptoms are manifested. The aim of our studies is to understand the stimulation of self-recovery processes at the cellular level by compounds which are applied according to the similia-principle to disturbed cell cultures. However, before being able to study the validity of the similia-principle, first the phenomenon of self-recovery will be described.

2. Outline of the experimental approach.

In order to induce self-recovery, cells are exposed to mildly damaging conditions. In our research a non-lethal concentration of sodium arsenite or non-lethal heat shock was used which induces a response that includes the synthesis of specific classes of proteins, the stimulation of growth and the development of tolerance. In detail we will analyze the following series of questions:
1. Are quiescent mammalian cells stimulated to proliferate by a threatening condition?
2. Are quiescent mammalian cells triggered to undergo a transition to a physiological state in which they are able to tolerate increased threats?
3. Is there a coordinate manner in which proliferation is stimulated and tolerance is induced?

With respect to the aim of our studies, it is of interest to determine whether, after an exposure to a damaging condition, cells can be stimulated by low stressor doses that are not effective in undamaged (control) cells. To analyze this we considered it as an essential and first step to define a molecular marker for self-recovery. Thus, the question was asked whether there is a common denominator which induces proliferation and tolerance, or whether it is possible to define a simple molecular marker for the induction of the recovery process?

In the present paper, we summarize the experimental evidence, in which it was shown that moderate stressor conditions (such as sublethal temperature elevation or arsenite concentration) transiently stimulate cell proliferation.

Furthermore, it will be shown that these conditons were not only able to stimulate growth, but also induced the development of tolerance. Then, the evidence is summarized which indicates that the hsps can be used as a molecular parameter for cellular recovery, Finally, the experiments are described which were performed to study the effect of low doses of arsenite as well as of mild heat shock temperatures on the synthesis of hsps and on the development of tolerance in arsenite- and in heat shock-sensitized cell cultures.

3. The growth stimulatory (hormetic) effects of heat shock: Stimulation of DNA synthesis by subliminal amounts of EGF and by PGF2α in quiescent Swiss mouse 3T3 fibroblastic cells.

Our first studies on growth hormesis were performed with cells subjected to a moderate enhancement of the ambient temperature, known as hyperthermia or heat shock. We presented evidence that a mild heat shock of confluent quiescent 3T3 cultures can enhance the induction of G1-S progression by EGF and PGF2α (Van Wijk et al.,1984 a,b; 1985). We considered heat shock as a very attractive stressor or toxic principle for several reasons. Unlike chemicals there are no problems associated with the uptake or compartmentalization of the compounds. The temperature reached in the cell is equal to that in the media surrounding the cells and after lowering the temperature, an immediate normalization of the ambient temperature occurs. To study a possible growth-stimulatory (hormetic) effect of heat shock on confluent resting Swiss 3T3 cell cultures, the effect of incubation at different temperatures on DNA synthesis stimulated by EGF was measured (Van Wijk et al., 1985). Preincubation at 42°C did not have any effect on cells stimulated by EGF. Preincubation at 44°C for different times increased the values of the labelling index stimulated by EGF from 7% to a plateau of 25% after 30 min. A further increase of temperature to 46°C increased the stimulating effect of EGF to a value of 20% when the treatment lasted for only 20 min, while preincubation for a longer period inhibited the G1-S progression. The mitogenic effect of another growth factor, prostaglandin (PGF2α), was also enhanced by a heat shock. Adding PGF2α at a dose suboptimal to obtain its full mitogenic effect, results in a labelling index of about 6% within 24 hours. A 15 min incubation at 46°C, which hardly affected the survival of cells, resulted in a labelling index of 18%. Moreover, a 25 min incubation at 46°C of cells which had been treated with PGF2α enhanced the labelling index to about 25%. A heat treatment of 35 min abolished the mitogenic response (Van Wijk et al., 1984b). More recent results show that sodium arsenite addition to confluent quiescent C3H10T1/2 cultures can enhance the mitogenic effect of serum (Van Wijk et al., 1993). This potentiation was observed following an arsenite treatment that was subthreshold for loss of clonal capacity of these cells.

For this phenomenon the term 'hormesis' was proposed (Southam and Ehrlich, 1943) to describe "a stimulatory effect of subinhibitory concentrations of any toxic substances on any organism". In a few earlier studies, authors used cell cultures to investigate the toxicity of nonmetal compounds and they also found stimulation of growth. Creech (1939) reported that 1,2,5,6 dibenzanthracene choleic acid significantly increased the growth of fibroblast cells derived from mouse embry connective tissue, and Gabliks et al. (1967) demonstrated that O,O-dimethyl S-methylcarbamoylemethylphosphorodithioate (Cygon), disulfoton O,O-diethyl S-(2-ethylthio)ethylphosphorodithioate (Di-Syston), and O,O-dimethyldithiophosphate of diethylmercaptosuccinate (Malathion) stimulated the replication of mouse liver cells. A more recent example is 2,3,7,8-tetrachlorodibenzo-p-dioxin (TCDD), a prototype for a group of halogenated aromatic hydrocarbons increased DNA synthesis in neonatal-derived epidermal cells under conditions of confluency (Milstone and LaVigne 1984).

4. Stimulation of growth and induction of tolerance.

A question of major importance for understanding the recovery response can now be stated. Are sublethal stressor-conditions capable to stimulate proliferation and induce tolerance? This question has been discussed for quiescent Swiss 3T3 fibroblasts, by comparing the effects of a heat shock on the induction of thermotolerance and potentiation of EGF-induced DNA synthesis using two hyperthermia regimens (Figure 1). The first was a 20 minutes incubation at different temperatures up to 46°C, the second was a 44°C treatment for different time-periods up to 60 minutes. It was concluded that temperature dependent potentiation of EGF induced G1-S progression and the induction of thermotolerance show a close resemblance (Van Wijk et al. 1984a,b; 1985)..

5. Search for a molecular marker

We considered it as an essential and first step to define a molecular marker for self-recovery. Thus, the question was asked whether there is a common denominator which induces proliferation and tolerance, or whether it is possible to define a simple molecular marker for the induction of the recovery process? For this reason the molecular basis of the stimulation of growth as well of tolerance will be shortly described.

The effect of heat and arsenite on cell proliferation can be explained by their stimulating effect on c-fos gene expression.

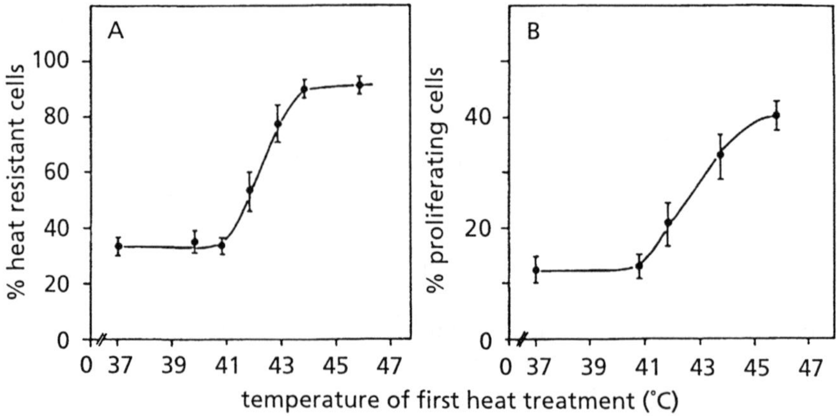

Figure 1 - The effect of temperature in a 20 minutes preincubation period on the percentage of quiescent Swiss 3T3 cells which are heat-resistant after a 30 minutes 46°C treatment as determined by the percentage of morphologically heat-resistant cells (A) and the percentage of proliferating cells (B). The 46°C treatment follows 3 hours after the preincubation. The percentage of morphologically resistant cells was determined by fixation duplicate cultures at 2 hours after the 46°C treatment. The percentage of the cells with proliferative capacity was determined by clonal assay trypsinization and plating out triplicate cultures at 4 hours following the 46°C treatment. Plating efficiency was 78 per cent.

In recent years the kinetics of cell cycle regulated oncogenes has been well investigated (Adamson, 1987; Kovary and Bravo, 1991a). The expression of different members of the fos- and jun-families of transcription factors is rapidly induced following serum stimulation of quiescent fibroblasts. Several arguments suggest that c-fos and c-jun serve a role in coupling mitogens to long-term responses, leading to cell proliferation. Thus, microinjection of anti-fos (Riabowol et al., 1988a; Kovary and Bravo, 1991b) and anti-jun (Kovary and Bravo, 1991b) family antibodies efficiently blocked the entrance to the S-phase of serum-stimulated cells. Furthermore, the production of c-fos antisense RNA inhibits 3T3 cell proliferation (Holt et al., 1986) and blocks renewed growth of quiescent 3T3 cells (Nishikura and Murray, 1987).

The effect of heat and arsenite on tolerance development can be explained by their stimulating effect on hsp synthesis. There is much evidence of the involvement of hsps in tolerance (Li and Laszlo, 1985; Landry et al., 1987; Hahn and Li, 1990; Parcell and Lindquist, 1993,1994). By microinjecting into rat embryo fibroblasts a mixture of affinity-purified monoclonal antibodies to hsp70, Riabowol et al. (1988b) observed a reduction of viability of cells exposed to hyperthermia. Using the technique of gene transfection Li et al (1991) demonstrated the involvement of hsp70 in thermotolerance.

It is shown that the concentrations of arsenite that enhance the mitogenic effect of serum increase the mRNA levels of c-fos, hsp68 and hsp84 and induce the specific synthesis of heat shock proteins (Van Wijk et al., 1993).
Similar data have been obtained studying the effect of different heat shocks on the mRNA levels of c-fos, hsp68 and hsp84 (Tuijl et al.,1991; 1993).

6. Induction of hsp synthesis as marker of cellular self-recovery

Based on several lines of experimental evidence, we have proposed that self-recovery can be evaluated in terms of the development of tolerance and the enhanced production of hsps (Van Wijk and Wiegant 1994). Hsps facilitate repair and refolding of stress-damaged proteins and also prevent irreversible denaturation and aggregation of cellular proteins (for review see Welch 1992; Parsell and Lindquist 1994). In this latter function they may act as 'protector proteins', which may be one of the underlying molecular mechanisms by which cells acquire tolerance.

7. Do low doses of arsenite stimulate hsp synthesis (self-recovery) in arsenite-intoxicated cells?

This study has been focused on the regulation of cellular sensitivity to arsenite-intoxication. There is evidence for the existence of a state of tolerance to arsenite (Lee and Hahn 1988; Lee *et al.* 1989) and recently the phenomenon of an arsenite-induced transition from enhanced (sensitization) to reduced (tolerance) sensitivity to arsenite has been demonstrated (Wiegant *et al.* 1993). Previously, arsenite has been shown to be an effective activator of hsp gene expression in the cell line used in our studies (Wiegant *et al.* 1987). We have investigated whether a treatment of H35 rat hepatoma cells with arsenite sensitizes cells to arsenite. It can be argued that when a threat is applied which is just strong enough to stimulate cells to start the replenishment of hsps, that this replenishment can be stimulated by a lower doses of the stressor, provided that it is given directly following the first threat. For this reason we incubated arsenite-treated cells with low doses of arsenite, according to the step-down protocol, and studied the occurrence of enhanced induction of the major hsps. In addition, the effect of low concentrations of arsenite on the induced development of arsenite-tolerance has been studied in arsenite-sensitized cells. According to previous definitions, 'tolerance' is described as a stress-induced cellular resistance to a lethal stress treatment, whereas 'sensitization' is defined as an increase in cellular susceptibility for a stress treatment (Wiegant *et al.*

1993).

7.1. Heat shock proteins

An 1-hour treatment with 0 or 100 μM arsenite was either or not followed by a continuous exposure to low concentrations of arsenite (1, 3 or 10 μM). At 3-hour intervals samples were incubated with [^{35}S]-methionine for 1 hour, and subsequently prepared for one-dimensional gel-electrophoresis. In figure 2 an example is given of a 1 hour treatment with 100 μM followed by either 0, 1, 3 or 10 μM arsenite. The relative rate of synthesis of each hsp was quantified and the time course of the different hsps, expressed relative to actin synthesis, was determined and presented in figure 3. A treatment with 100 μM, followed by an incubation in an arsenite-free medium shows a transient increase in the synthesis of hsp60, 68, 70, 84 and 100. This condition is considered to be a 'mild' treatment. The increase in relative synthesis of most hsps reaches a maximum at 3 hours after the pretreatment. When cells pretreated with 100 μM are incubated in media containing a low concentration of arsenite (1, 3 or 10 μM), an enhanced and prolonged induction of several hsps can be observed. Incubation of cultures in low concentrations of arsenite only, did not induce the synthesis of hsps. As the treatment with 3 or 10 μM arsenite alone (without pretreatment) did not cause any induction of hsps, it is concluded that a step-down arsenite treatment enhances the relative synthesis of hsps. Apparently, the sensitizing effect of an arsenite pretreatment can be determined in terms of an enhanced synthesis of specific hsps.

7.2. Survival characteristics and development of tolerance in arsenite-pretreated cells

We investigated whether the induction of arsenite-tolerance in arsenite-pretreated H35 cell cultures could be modulated by a post-incubation with low concentrations of arsenite. First, the H35 cell population was incubated for 1 hour with 100 μM arsenite. Subsequently the culture was incubated with a lower concentration of arsenite (up to 10 μM) for 4 h. Then the sensitivity or the degree of tolerance was tested by establishing the survival after an exposure for 2 hour to 300 μM arsenite. In Figure 4A it is shown that when cells have been pre-treated with 100 μM arsenite and post-incubated at 1, 3 or 10 μM, a decreased sensitivity is observed in comparison with cells that have been post-incubated in the absence of a low doses of arsenite. Apparently, pre-treated cells are able to develop a higher level of tolerance in the presence of low concentrations of arsenite. In cultures without pretreatment the presence of low concentrations of arsenite for up to 4 hours did not induce tolerance (data not shown).

In order to investigate whether the rate of cell killing and/or the capacity to withstand a stress treatment was affected (represented by D_0 and D_q respectively); a pretreatment with 100 μM arsenite was followed by an incubation with either 0 or

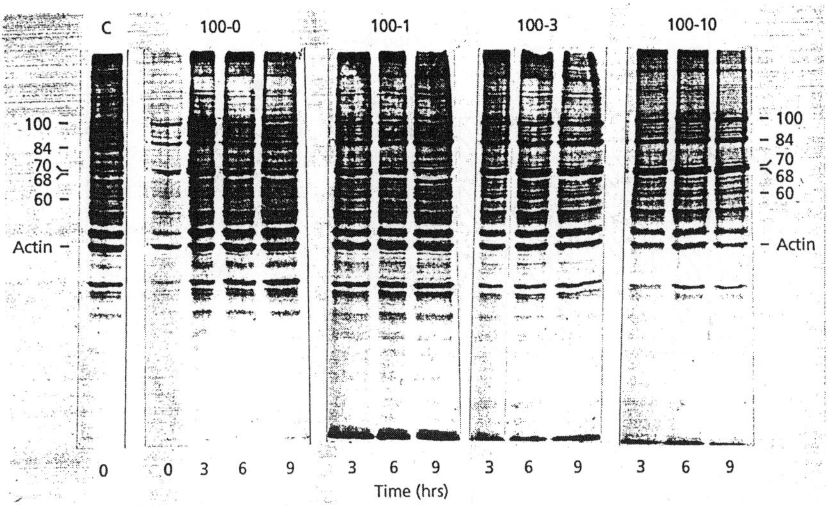

Figure 2 - Induction of hap-synthesis in H35 cells, pre-treated with 100, μM arsenite during 1 hour was followed by a continuous treatment with 0,1, 3 or 10 μM arsenite during 9 hours. At various moments after pretreatment (0, 3, 6 or 9 hours), cultures were pulse-labelled with [^{35}S]-methionine for 1 hour and prepared for 1-D PAGE. The same amount of protein was applied to each lane. (C= control cells).

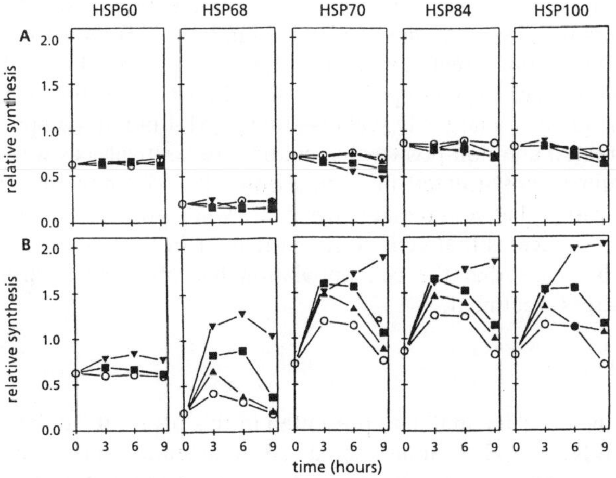

Figure 3 - The relative synthesis of the major hsps (hsp60, 68, 70, 84 and 100) after a treatment with 0 μM (A) or 100 μM (B) arsenite for 1 hour, subsequently followed by an incubation in 0 μM (circle), 1 μM (triangle-up), 3 μM (square) or 10 μM (triangle-down) arsenite during the indicated time (in hours). The data are expressed as the ratio of the optical density between the indicated hsp and actin.

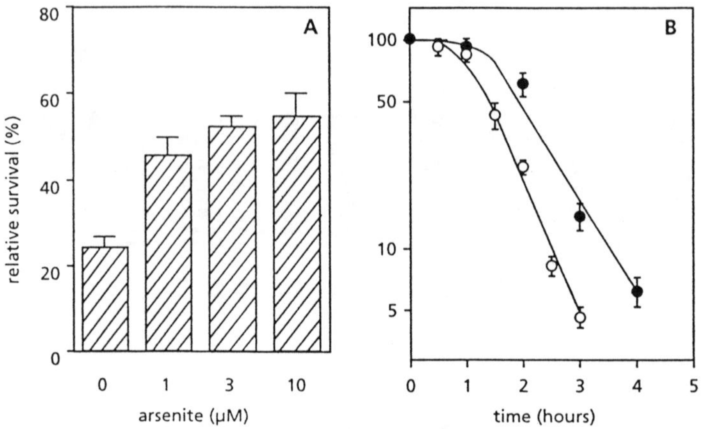

Figure 4 -A The percentage survival after a step-down arsenite treatment, indicating the level of tolerance obtained after a pretreatment with 100 μM arsenite, followed by a 4 hour post-treatment with either 0, 1, 3 or 10 μM arsenite and finally a test treatment with 300 μM arsenite for 2 hours. **B** Survival curves of H35 cells subjected to a step-down arsenite treatment. Cell cultures were incubated during 1 hour in 100 μM arsenite, followed by a post-treatment in either 0 μM (open circle) or 10 μM (closed circle) arsenite for 4 hours, subsequently followed by a test-treatment with 300 μM during a period of up to 4 hours. Data were corrected for the effect of pretreatment and presented as the surviving fraction of pre-treated cells.

10 μM for 4 hours. After this incubation a test-treatment with 300 μM over a period of up to 4 hours was given. As can be observed in figure 4B the survival curve of the cell culture which is post-treated with 10 μM, shows a slope which is less steep than found in cultures which were not post-treated and starts to decline at a later moment in time. Consequently, a larger D_0-value is found (51.4 min in comparison with 34.7 min in post-treated and non-posttreated cultures respectively) as well as a larger D_q value (89.0 min in post-treated in comparison with 65.2 min in non-post-treated cultures), indicating that cells die at a lower rate when post-treated with 10 μM. The larger D_q value indicates that cells start to die later in time when post-treated with 10 μM, which means that the cell population has obtained a larger capacity to withstand a stress treatment.

8. Discussion

From the experimental data presented in this paper, it can be concluded that an exposure of Reuber H35 rat hepatoma cells to arsenite concentrations of 100 μM for 1 hour, rapidly changes the sensitivity to a second arsenite treatment with respect to the induction of the heat shock proteins and the development of tolerance. The present experiments demonstrate that after a step-down arsenite treatment and under appropriate conditions, an increase in the synthesis of arsenite-induced heat-shock proteins could be observed, as compared to the rate of hsp synthesis in cultures which received arsenite pretreatment only. Also, an increase in the level of hsp68-mRNA by low doses of arsenite has been shown in arsenite-sensitized cell cultures (Ovelgönne et al 1995). An incubation of cell cultures in low concentrations of arsenite only (without arsenite-pretreatment), did not result in any detectable induction of hsp-synthesis. It was also observed that under arsenite step-down conditions, the development of tolerance was enhanced in comparison with the level of tolerance obtained by a single treatment with arsenite.

These results are in agreement with our previous experiments using heat as the stressor condition. It has been shown that during step-down heating under appropriate conditions, the relative rate of hsp-synthesis as well as of hsp68-mRNA was increased to a greater extent in comparison with the values observed in cultures to which only the heat pretreatment was applied (Schamhart *et al.* 1992; Van Wijk *et al.* 1994). In addition, an enhanced development of tolerance during step-down heating conditions was described (Delpino *et al.* 1992; Van Wijk *et al.* 1994).

In conclusion, a stimulation of the self-recovery has been demonstrated upon the application of the isopathic principle (the similia principle in its most elementary form). For the next step in our research program, the model presented here can be used for the study of the specificity of the similia-principle. The results of the experiments aimed to identify the specificity of the similia-principle are described elsewhere in this volume.

Acknowledgement
The financial support of the HomInt organisation is gratefully acknowledged.

9. Literature

Adamson E.D. (1987). Oncogenes in development. *Development* **99**: 449-471.

Creech E.M.H. (1939) Carcinogenic and related non-carcinogenic hydrocarbons in tissue culture. *I.Am.J.Cancer* **35**: 191-202.

Delpino A., Gentile F.P., Di Modugno F., Benassi M., Mileo A.M. and Mattei E. (1992). Thermosensitization, heat shock protein synthesis and development of thermotolerance in M-14 human tumor cells subjected to step-down heating. *Radiat.Environm.Biophys.* **31**: 323-332.

Gabliks J., Bantug-Jurilla M. and Friedman L. (1967). Responses of cell cultures to insecticides. IV. Relative toxicity of several organophosphates in mouse cell cultures. *Proc.Soc.Exp.Biol.Med.* **125**: 1002-1005.

Hahn G.M. and Li G.C. (1990). Thermotolerance, thermoresistance and thermosensitization. In: *Stress proteins in biology and medicine*, edited by Morimoto R.I., Tissieres A. and Georgopoulos C. Cold Spring Harbor Laboratory Press, pp.79-100.

Holt J.T., Venkat Gopal T., Moulton A.D. and Nienhuis A.W. (1986). Inducible production of c-fos antisense RNA inhibits 3T3 cell proliferation. *Proc.Natl.Acad.Sci.U.S.A.* **83**: 4794-4798.

Kovary K. and Bravo R. (1991a). The jun and fos protein families are both required for cell cycle progression in fibroblasts. *Mol.Cell.Biol.* **11** (**9**): 4466-4472.

Kovary K. and Bravo R. (1991b). Expression of different Jun and Fos proteins during the G0 to G1 transition in mouse fibroblasts: in vitro and in vivo associations. *Mol.Cell.Biol.* **11 (5)**: 2451-2459.

Landry J., Lamarche S. and Chrétien P. (1987). Heat shock proteins; a lead to the understanding of cell thermotolerance. In: *Thermotolerance; volume I*, edited by Henle K.J. CRC Press, Boca Raton, Florida. pp.146-173.

Lee K.J. and Hahn G.M. (1988). Abnormal proteins as the trigger for the induction of stress responses: heat, diamide and sodium arsenite. *J.Cell.Physiol.* **136**: 411-420.

Lee T.C., Wei M.L., Chang W.J., Ho I.C., Lo J.F., Jan K.Y. and Huang H. (1989). Elevation of gluthatione levels and gluthatione s-transferase activity in arsenic-resistant chinese hamster ovary cells. *In Vitro Cell.Dev.Biol.* **25**: 442-4480.

Li G.C. and Laszlo A. (1985). Thermotolerance in mammalian cells: a possible role for heat shock proteins. In: *Changes in eukaryotic gene expression in response to environmental stress*, edited by Atkinson B.G. and Walden D.B. Academic Press, New York. pp.227-254.

Li G.C., Li L., Liu Y.K., Mak J.Y., Chen L. and Lee W.M.F. (1991). Thermal response of rat fibroblasts stably transfected with the human 70-kDa heat shock protein-encoding gene. *Proc.Natl.Acad.Sci.U.S.A.* **88(5)**: 1681-1685.

Milstone L.M. and Lavigne J.F. (1984). 2,3,7,8,-Tetrachlorobenzo-p-dioxin induces hyperplasia in confluent cultures of human keratinocytes. *J.Invest.Dermatol.* **82**: 532-534.

Nishikura K. and Murray J.M. (1987). Antisense RNA of proto-oncogene c-fos blocks renewed growth of quiescent 3T3 cells. *Mol.Cell.Biol.* **7**: 639-649.

Ovelgönne J.H., Souren J.E.M., Van Rijn J., Van Wijk R. and Wiegant F.A.C. (1995). Enhancement of the stress response by low concentrations of arsenite in arsenite-pretreated H35 hepatoma cells. *Toxicol.Appl.Pharmacol.* **132**: 146-155.

Parsell D.A. and Lindquist S. (1993). The function of heat-shock proteins in stress tolerance: degradation and reactivation of damaged proteins. *Annu.Rev.Genet.* **27**: 437-496.

Parsell D.A. and Lindquist S. (1994). Heat shock proteins and stress tolerance. In: *The biology of heat shock proteins and molecular chaperones*, edited by Morimoto R.I., Tissières A. and Georgopoulos C. CSHL Press, New York. pp.457-494.

Riabowol K.T., Mizzen L.A. and Welch W.J. (1988a). Heat Shock is lethal to fibroblasts microinjected with antibodies against hsp70. *Science* **242**: 433-436.

Riabowol K.T., Vosatka R.J., Ziff E.B., Lamb N.J. and Feramisco J.R. (1988b). Microinjection of fos-specific antibodies blocks DNA synthesis in fibroblast cells. *Mol.Cell.Biol.* **8**: 1670-1676.

Schamhart D.H.J., Zoutewelle G., van Aken H. and Van Wijk R. (1992). Effects on the expression of heat shock proteins by step-down heating and hypothermia in rat hepatoma cells with a different degree of heat sensitivity. *Int.J.Hyperthermia* **8(5)**: 701-716.

Southam C.M. and Ehrlich J. (1943). Effects of extract of western red-cedar heartwood on certain wood decaying fungi in culture. *Phytopathology* **33**: 517.

Van Wijk R., Otto A.M. and Jimenez de Asua L. (1984a). Effect of serum and growth factors on heat sensitivity in Swiss mouse 3T3 cells. *J.Cell.Physiol.* **119**: 155-162.

Van Wijk R., Otto A.M. and Jimenez de Asua L. (1984b). Hyperthermia can enhance the initiation of DNA synthesis stimulated by growth factors in Swiss mouse 3T3 fibroblasts. *Exp.Cell Res.* **153**: 522-527.

Van Wijk R., Otto A.M. and Jimenez de Asua L. (1985). Increase of epidermal growth factor-stimulated cell-cycle progression and induction of thermotolerance by heat shock: temperature and time relationship. *Int.J.Hyperthermia* **1**: 147-156.

Van Wijk R., Welters M., Souren J.E.M., Ovelgönne J.H. and Wiegant F.A.C. (1993).

Serum-stimulated cell cycle progression and stress protein synthesis in C3H10T½ fibroblasts treated with sodium arsenite. *J.Cell.Physiol.* **155**: 265-272.

Van Wijk R., Ovelgönne J.H., de Koning E., Jaarsveld K., Van Rijn J. and Wiegant F.A.C. (1994). Mild step-down heating causes increased transcription levels of hsp68 and hsp84 mRNA and enhances thermotolerance development in Reuber H35 hepatoma cells. *Int.J.Hyperthermia* **10**: 115-125.

Van Wijk R. and Wiegant F.A.C. (1994). Cultured mammalian cells in homeopathy research; the similia-principle in self-recovery. Utrecht University.

Van Wijk R. and Wiegant F.A.C. (1994). Is disturbed self-recovery stimulated by applying the similia principle? 1. Mode of testing in fundamental research. *Br.Hom.J.* **(in press)**

Welch W.J. (1992). Mammalian stress response: cell physiology, structure/function of stress proteins, and implications for medicine and disease. *Physiol.Rev.* **72**: 1063-1081.

Wiegant F.A.C., Souren J.E.M., Van Rijn J. and Van Wijk R. (1993). Arsenite-induced sensitization and self-tolerance of Reuber H35 hepatoma cells. *Cell Biol.Toxicol.* **9**: 49-59.

Wiegant F.A.C., van Bergen en Henegouwen P.M.P., Van Dongen G. and Linnemans W.A.M. (1987). Stress-induced thermotolerance of the cytoskeleton of mouse neuroblastoma N2A cells and rat Reuber H35 cells. *Cancer Res.* **47**: 1674-1680.

COHERENCE IN BIOLOGICAL SYSTEMS AND WATER

CYRIL W. SMITH
Department of Electronic and Electrical Engineering, University of Salford
Salford, M5 4WT, England

ABSTRACT

The concept of coherence in biological systems is introduced and related to electromagnetic theory, energetics and the dimensions of systems, and to bio-information. The phenomenon of coherence in water is considered with its thresholds and limits. This is related to the properties of water "memory".

1. Introduction to Coherence in Biological Systems

It is now clear that coherent frequencies of non-ionizing radiation from the sub-Hertz of circadian rhythms to at least the ultraviolet of bio-photons are involved in biological systems. There are various techniques available for the detection and measurement of bio-electromagnetic fields and the coherent frequencies and the patterns of frequencies generated by biological systems. The frequency "window" is the most important "window" for biological systems. With most biological systems and materials, the lowest and highest of the coherence frequencies are stimulatory, between these, depressive and stimulatory frequencies alternate. There are a few exceptions to this alternation of stimulatory and depressive response. There is also an intensity "window", for this the lower threshold may correspond to that magnetic field at which a single quantum of magnetic flux links the cross-sectional area of a biological cell, and the upper intensity threshold may correspond to that field sufficient to break the coherence.

Fröhlich showed that a dielectric material polarized electrically acquires an energy dependent upon its shape and that deformation of the material will cause an electrical energy change, positive or negative, together with a change in the elastic energy which is always positive for a material previously in equilibrium. Interaction between electrical and elastic displacements leads to the establishment of metastable highly polarized states. He also remarked that magnetic flux quantization is not restricted to superconductors but, is a general property of all materials although, in the usual energy distributions single flux quanta phenomena do not appear due to the small energy differences in neighbouring flux quantum numbers. However, if a system is able to respond to the magnetic flux quantum then it will also have the Josephson effect available.

2. Electromagnetic Theory

Electric charge at rest exhibits the properties of static electricity: attraction and repulsion, the formation of chemical bonds. Electric charge moving at a constant velocity additionally gives rise to magnetism (a very second-order effect based on relativity theory). Electric charge under acceleration radiates energy into space as a transverse electromagnetic wave travelling at the velocity of light.

A wire carrying an electric current generates a magnetic field (B webers/m^2) as closed loops of magnetic flux (F webers) in the space around the wire at right angles to the direction of the current (i amps). Associated with this current is a magnetic vector potential (A webers/m) in the direction of the current which can influence the "wave function" of electrons. A changing magnetic vector potential generates an electric field. A good article on understanding electromagnetic field theory has been written by Carpenter [1], who instead of using Maxwell's equations and the concept of flux, starts from electric and magnetic potentials.

The geometry of a magnetic field is determined by the way that the wire carrying the current is wound into a coil. If it is wound as a long solenoid this will generate a uniform magnetic field (B) inside the solenoid. There will be both B and A components in the space surrounding it. A toroid contains the B component completely within the coil but generates an A component in the surrounding space. A Möbius loop cancels out both the B and A components and does not potentise water. The Helmholtz coil arrangement (2 equal coils separated by their radius) gives a very uniform field (B) at the centre, close to the axis.

3. Energetics

There has been much progress in understanding the scientific fundamentals of the ways that living systems use and are influenced by electromagnetic fields. In physics, the size of regions within which collective action can be effective can be estimated from considerations of the physical energy involved. Energy calculations only depend on the initial and final energy states and are independent of the physical processes by which this is achieved. For a field to be effective in establishing any order, it must be able to overcome the randomising effect of thermal fluctuations. The energy contained in a field is proportional to the volume it occupies. Equating the energy in the field to the thermal energy gives the threshold volume within which the field can impose its order. The cube-root of this volume gives an idea of its linear dimensions [2]. An electromagnetic field comprises both electric and magnetic field components which are mutually at right angles as well as both being at right angles to the direction in which they propagate, at the velocity of light. From these components the incident power density (joules/sec/m^2 or watts/m^2) can be calculated and also related to the critical dimensions which can be ordered by incoming radiation.

For atomic dimensions (0.1 nm, 1 Å), the theoretical threshold electric field is above 10^9 V/m, the order of field used in field emission microscopy to break chemical bonds; a whale (10 m long) can sense electric fields of about 1μV/m, which is approaching its theoretical threshold. The geomagnetic field ought to be significant to single biological cells, and certain (magnetotactic) bacteria do make use of it. The corresponding theoretical threshold for effects of radiation on a single biological cell comes out equal to the 100 W/m^2, which is a guideline level for incident non-ionising radiation based on thermal effects, because of course, both calculations are based on the same fundamental criterion. Any bio-effects observed at weaker fields must be due to cooperative sensing and action within multi-cellular organisms. Oscillations both in chemical reactions and electrical charges are a natural phenomenon in living systems [3], and both can be affected by external electromagnetic fields.

Thus, if a system is physically large and well organised like a radio-telescope, very small energy fields can theoretically be detected by it, particularly if there is an energy store and one is prepared to wait long enough. Then, very small densities of incident power (*energy per second*) can accumulate until enough energy is available to produce an effect. A signal integrates in proportion to exposure time, the noise is proportional to the square-root of time, so the signal-to-noise ratio improves as the square-root of the integration time.

4. Bio-Information and Coherence

Every physical system contains both pure energy and pure information. Even the high energy, hot gases in the cylinders of a car engine carry information as to their chemical constituents and temperature. Biological systems operate in water at temperatures little different from ambient, so they cannot be powered thermally (like a steam engine). They use the *Free Energy* changes resulting from chemical reactions. The structure and function of biological systems is determined by very specific and closely regulated biochemical reactions which are controlled by enzymes (catalysts). Living cells can do work because they do not attain a state of equilibrium; the thermal equilibrium state is one of no work, a passive state. Only systems that are not in equilibrium can be subjected to control or regulation of their work activity.

For the meaningful control or regulation of any system, and this includes its repair or healing, some degree of order or coherence must pre-exist within it in space and time. At its simplest, the term *coherence* means that two or more separate oscillators maintain a constant phase relationship (i.e. they march in step) for a prolonged period of time (the coherence time). If their oscillations propagate as waves, there will be a distance (equal to the velocity multiplied by the coherence time) over which this constant-phase relationship is found. This is the coherence length, and within it wave interference effects are found.

Any physical process which coherently controls a chemical reaction must be operating within "quantum" physics because there is no chemistry in "classical" physics. This statement must exclude such general parameters as temperature, pH, or pressure, which may affect the general activity of an ensemble of chemically reacting molecules through incoherent (time-averaged) information without providing control at the level of coherent molecules. "Biological coherence and response to external stimuli" has been the subject of a book edited by Fröhlich [4]. He considered that long-range phase correlations can give rise to the sort of order present in superconductors, super-fluid helium, and lasers. Living systems can achieve the necessary degree of coherence in the ground state so, they can behave as room-temperature superconductors, sensitive to single quanta of magnetic flux. Thus, they would acquire the frequency-voltage interconversion (500 MHz/μV) provided by the Josephson effect even though they would not exhibit the zero resistance and superfluidity phenomena of low-temperature superconductivity [5].

In the absence of any convincing theoretical explanation for the effects of weak electromagnetic fields on biological systems, it is difficult even to try to determine the thresholds for observed effects. Since any electric field must be generated across a capacitance, the displacement current will produce magnetic effects. Magnetic fields are generated by currents in coils and in the direction of the current there is a magnetic vector potenetial. A high alternating electric field produced at a sharp point would not potentise water even on succussion so, it is unlikely that the induced electric field $E = - dA/dt$ is involved. The only other way a magnetic vector potential can affect water or a biological system is through its involvement in the wave equation or order parameter of what must be regarded as a quantum system.

5. Coherence in Water

Water is essential for life as we know it and, water from certain sources has had healing properties attributed to it from ancient times. The water in a living cell is highly structured [6] and mostly does not exchange by diffusion. Ice can form in crystals orientated to a magnetic field which which is evidence of magnetic effects on spatial long-range order in supercooled water [7]. Water is a liquid and its ordering must be in the time domain. Del Giudice and Preparata and co-workers [8] have continued to investigate the Quantum Electro Dynamics (QED) of various systems in 'Condensed Matter Physics' and, in particular, they have examined the dynamics of the rotational levels of water molecules. This requires the solution of the Schrödinger Wave Equation coupled with the Maxwell Electromagnetic Equations and leads to a model for liquid water which is a condensed vapour, rather than a molten solid.

E. Del Giudice [9,10] has presented in simple language an outline of the conceptual breakthrough by G. Preparata. In the course of a general programme of research on the coherent dynamics of matter, Preparata was able to show that the universally accepted quantum theory of gauge fields applied to quarks shows that they have a collective existence and the property of being non-isolatable. In condensed matter, which includes liquids and living systems, the relevant gauge field is the electromagnetic field. The quanta of these fields can interfere positively among themselves so that the particles move within a self-produced and self-trapped electromagnetic field of minimum energy in the ground state. This always occurs for quarks and may extend to higher levels. This is not the result of excitations in a far-from equilibrium dissipative system, the coherence is fundamental.

It seems that the bio-information is carried on the alternating magnetic vector potential while the magnetic field performs a formatting function [2,3]. This means that bio-information exists at the quantum level. It is clear that the "window" of frequency is the most important, the intensity "window" has a lower threshold which has been found to correspond to the magnetic field at which a single quantum of magnetic flux links the cross-sectional area of the biological cell. The upper threshold is probably that field which is sufficient to break the coherence of the system.

According to QED, water consists of two distinct intermixed phases; incoherent water molecules and clusters of coherent water molecules or domains about 75 nm in size, which act coherently towards any electromagnetic field, internal or external. These domains are stable against thermal agitation because of an energy gap of 12.06 eV against which the room temperature thermal energy of 1/40 eV has little effect. This model can account for the anomalies in the physical properties of water and in general can provide a theoretical foundation to the "flickering" hydrogen-bonded network of water molecules envisaged by Linus Pauling.

Water coherence can be destroyed by strong enough fields or by vaporization. As a simple example, Del Giudice originally considered the 50 cm^{-1}, 1.5 THz resonance (1THz = 1 TeraHertz = $1x10^{12}$ Hz) in water, which is to be seen in Raman scattering and in neutron experiments [11]. This gives a coherent domain of 100 μm, which would comprise 10^{17} water molecules in coherent oscillation. The number of water molecules that could be involved in a quantum fluctuation that would not break the coherence is the square root of this figure, ie. 3×10^7 water molecules. For radiation to interact with coherent water, it has to give its energy not to a single water molecule, but to the whole mass of 3×10^8 molecules. As a result, the 1.5 THz oscillation slows down to 5 kHz and the field develops a longitudinal component. Subsequent work has suggested that the 12.06 eV energy gap in water is more fundamental. This gives the smallest size for a coherence domain as 75 nm, although the 100 μm coherence domain may be superimposed.

6. Water "Memory"

Water can be 'imprinted' with a frequency using an alternating magnetic field, or with an alternating magnetic vector potential plus either succussion (a mechanical shock wave or a vortex) or a strong magnetic field. Water will slowly acquire an imprint by contact with a sample of imprinted water or metal; the higher frequencies imprint more quickly than the lower frequencies.

The instrumental detection of coherence frequencies in water has been achieved (but, with problems of repeatablity) using a technique whereby a pair of gold wire electrodes is dipped into a cuvette containing the sample of 'frequency imprinted' water, These electrodes are connected to a low-noise, high-gain, narrow-band amplifier, or to a Fourier Transform Spectrometer with signal averaging over a large number of measurements. The frequencies measured in this way were the same as those originally 'imprinted' into the water and the same as those measured by a subjective method.

7. Thresholds and Limits to Coherence

As a result of this work, much is now known about how to erase 'water memory', albeit as the result of trying to measure erased samples. Heating water above 70°C will erase its memory, as will an inter-electrode current greater than about a microamp, or even the tube being handled by the 'wrong' person. But, most importantly, placing the water in a hypomagnetic container (e.g. a closed mu-metal box) erases not only 'water memory' but will also erase potentised medicaments (both liquids and lactose tablets). The threshold steady magnetic field necessary for the maintenance of 'water memory' is about 400 nT. At this field, the mean separation between magnetic flux quanta (2.07×10^{-15} Wb) is 71 μm. A strong magnetic field (>250 μT) must only suppress 'water memory' effects, because, in this case the original 'imprint' is recoverable by succussion. The 'imprinted' resonances in water will propagate as waves of coherence (Wüst Waves). They will not propagate through water in tubes of 2mm diameter or less. Insulated wires can be arranged to conduct a coherence wave from 'imprinted' water and by this means, one can provide the basic arrangement for an Aharanov and Bohm experiment, which does work. These above figures indicate some of the critical conditions for the coherence.

8. Conclusions

It is difficult but, essential to try to determine the thresholds for macroscopic coherence effects. Since any electric field must be generated across a capacitance, the displacement current will produce magnetic effects. Magnetic fields are generated by currents and in the direction of the current there is a magnetic vector potential. There is the possibility of coherent frequencies of alternating magnetic vector potential fields affecting biological systems both directly, and through the imprinting of the biological systems environmental water 'memory'. The magnetic vector potential can affect water or a biological system through the wave equation or order parameter. This carries the implication that a biological system must be regarded as a macroscopic quantum system, and that bio-information is modulated on to a coherent frequency.

9. References

[1] J. Carpenter J, "Understanding electromagnetism", *IEE Engineering Science and Education Journal*, 2: 273-280, 1993.

[2] C.W. Smith, "Coherent electromagnetic fields and bio-communication", in Popp F-A, Warnke U, König H L and Peschka W (eds.) *Electromagnetic Bio-Information*. Munich: Urban & Schwarzenberg, pp.1-17, 1989.

[3] P.E. Rapp, "An atlas of cellular oscillators". *J. exp. Biol.* 81: 281-306, 1979.

[4] H. Fröhlich (ed.), *"Biological Coherence and response to External Stimuli"*, Berlin: Springer-Verlag, 1988.

[5] E. Del Giudice, S. Doglia, M. Milani, C.W. Smith and G.Vitiello, "Magnetic flux quantization and Josephson behaviour in living systems". *Physica Scripta* 40: 786-791, 1989.

[6] J.S. Clegg, "Intracellular water, metabolism and cell architecture: part 2". In: Fröhlich H and Kremer F (eds) *Coherent excitations in biological systems.* Berlin, Heidelberg: Springer-Verlag. pp.162-177, 1983.

[7] C.W. Smith, "Electromagnetic bio-information and water", in Endler PC and Schulte J (eds.) *Ultra High Dilutions - Physiology and Physics.* Dordrecht: Kluwer Academic. pp.187-201, 1994.

[8] E. Del Giudice, "Is the "memory of water" a physical impossibility?", in Endler PC and Schulte J (eds.) *Ultra High Dilutions - Physiology and Physics.* Dordrecht: Kluwer Academic. pp. 117-119, 1994.

[9] E. Del Giudice, 1993. "Coherence in Condensed and Living Matter". *Frontier Perspectives*, 3(2), 16.

[10] E. Del Giudice, 1995. "On Coherence in Matter". *Frontier Perspectives*, 4(2), 4-5.

[11] J.B. Hasted, 1973. *Aqueous Dielectrics*. London: Chapman & Hall.

[12] R. Arani, I. Bono, E. Del Giudice and G. Preparata, 1995. "QED and the Thermodynamics of water", Milan Reprint MITH 93/3 (to be published in *Intl. J. Mod. Phys. B.*).

COMPARISON OF DIFFERENT DILUTIONS OF THE SAME SUBSTANCE

KARL W. KRATKY
Institute for Experimental Physics, University of Vienna
Boltzmanng. 5, Vienna, Austria

ABSTRACT

The effect of a remedy cannot be traced back to the active substance alone. There is also an intricate connexion with the degree of dilution, e.g., in water, and the way the dilution has been produced. To get more insight, a global model of water has been developped. It consists of several levels which are associated with different degrees of dilution. In the present paper, this relation is made more precise, and some new conclusions are drawn.

1. Hormesis and Homeopathy: The Generalized Simile Principle

In biophysics and medicine, one compares the effects of different substances (drugs, remedies) on organisms. The details of dispensing one of these drugs are not so important at first sight. However, this is not the case. For instance, it is known from hormesis that there exist dose-dependent reverse effects [1,2]. A hypothetical drug which decreases the blood pressure at high dose may increase it at low dose, the effect presumably vanishing at all when the dose becomes too small. Thus, the relation between cause and effect is not as simple as it seems to be. A possible explanation is the following: At high dose, the drug lowers the blood pressure in a more or less direct chemical way. At low dose, it acts indirectly by stimulating the autoregulatory system of the body, which in turn raises the blood pressure. Thus the same substance may have opposite effects. Conventional medicine uses drugs which act "against" certain diseases (allopathy), while homeopathic remedies are thought to be "similar" to the diseases (simile or similia principle). When using homeopathic drugs, there is often a short-term allopathic effect called initial aggravation. It takes some time until the "paradoxical" positive effect of homeopathic drugs prevails. Thus, there is also some kind of time-dependent reverse effect [3]. For an introduction to homeopathy, see Ref. [4].

Hormesis concentrates on the dose, homeopathy on the condition of the treated person. In healthy people, the homeopathic remedy usually creates the symptoms of those diseases which can be cured by the same remedy. This may be called a condition-dependent reverse effect. The effect of a drug may also vary with daytime and the season (chronobiology [5]) or depend on the manner it is presented and commented by the physician (placebo /nocebo effect [6]). Furthermore, it may depend on the details of the preparation of the drug, e.g., using different potencies in homeopathy, see below. In 1992, I proposed the generalized simile principle (or principle of resonance) to cover all kinds of reverse effects: If there is a specific effect of a drug, this effect varies in strength or may even be reversed depending on the way of preparation (dilution, exact proportion of substances, etc.), the daytime, the way of dispensing and the actual condition of the person who is given the drug (healthy/sick, nervous/calm, etc.).

Usually, people associate two strange things with homeopathy: that it works "the other way round" (simile principle) and that it uses high dilutions where it should have "no effect at all". The basic point of homeopathy, which has found scientific support recently [7], is the simile principle. That low doses (or high dilutions) are dispensed in many cases, orginates from the fact that homeopathy often uses toxic substances which would be dangerous in high doses. The problem that the desired effect is vanishing for high dilutions is overcome by the special technique of potentizing. Instead of just diluting a substance, it is also succussed at every step. Hahnemann, the discoverer of homeopathy, found this by experience, a theoretical explanation is still missing. We stick to the D-potencies (D: decimal). Starting from the mother tincture (usually D0), each potentization step means diluting by a factor 10 in a water /alcohol mixture and shaking the liquid in a specific manner. Six such steps have to be accomplished in order to obtain D6, the dilution factor being 1 million. For further details, see Ref. [8].

In potencies beyond about D23, no molecule of the original substance is present any more. The apparent efficacy of such high potencies is of course a challenge to science. This will be treated in Section 2. Furthermore, it is interesting to compare the effects of the different potencies. In the literature, there are several hints that the effect of homeopathic drugs oscillates when the dilution is varied continuously [9,10]. Thus, the drugs may also have an allopathic effect if the wrong potency is used. The initial aggravation may also vary for different potencies. In fact, there are only a few typical potencies which are usually prescribed. They have been found out by experience, a systematic scientific investigation would be important.

2. Global Model of the Structure and Dynamics of Water

The efficacy of very high potencies may reflect the properties of water itself (being synonymous to the water /alcohol mixture in the following). The diluted substance just helps the water to get into a structure which has the desired effects. Thus, I introduced a global model of the structure of water [11,12] which was also used for mathematical calculations in the spirit of chaos research. In this model, an energy landscape is defined which represents the occurrence of large-scale structures in water. The potential energy Epot (in units of room temperature) depends on two angle-like variables x and y (defined between -180 and +180°, the two limits being identified). In nature, many forms turn out to be fractal. Thus, the energy landscape of water is also assumed to be self-similar, see Fig. 1, where one looks at the interesting part of the landscape from the side. Starting

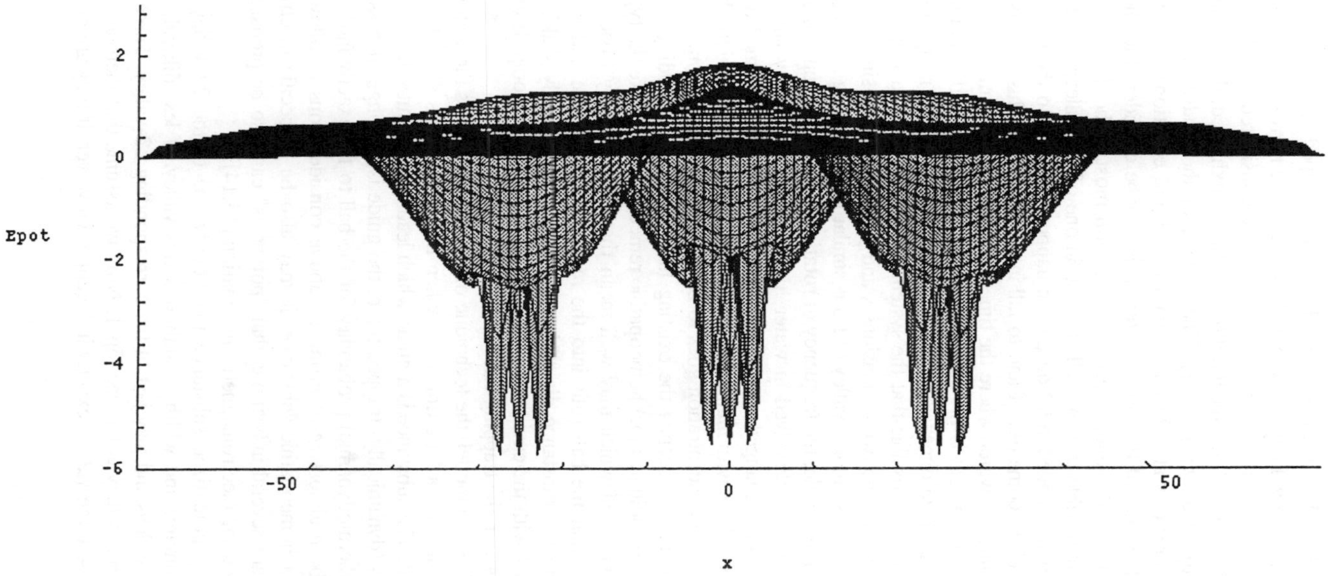

Figure 1. The global energy landscape of water. A look at the basin-shaped valleys from the side.

from the "highland" (level L0), there are three large basin-shaped valleys (level L1) in the vicinity of 0^0 for both x and y. Each of these hollows has three narrower valleys at its base (L2) and so on. In the actual model, three such levels are considered (L1, L2 and L3). In Fig. 1, only the 3 hollows of L1 and the 9 hollows of L2 are exhibited, the 27 valleys of L3 would be too thin to be drawn properly. The minima of the landscape correspond to specific global structures of water. From the choice of Epot, equations of motion follow which result in the dynamics of the model water. This may be considered as the (frictionless) movement of a ball in the landscape. By construction, most of the time the ball spends in L0 and L1. It is improbable that the ball is found in one of the valleys of L2 or L3. On the other hand, it may be trapped in L2 od L3 if it happens to fall into the respective level [12]. Thus it is difficult to come out of and to fall into the valleys. This reflects the possibility of metastable states of water, where the time scale may range from seconds to years.

The proposed model is the simplest version which meets the requirements of reasonable calculations. Two dimensions x, y are necessary to exhibit nonperiodic (chaotic) motion which makes it possible that the system can be trapped in a valley. Three valleys are necessary to utilize the two dimensions x and y. From the self-similarity it follows that the dynamics of the system in a valley of L3 is similar to that in, e.g., L2, but it moves much faster in L3. For example, the frequency of rotation in a valley is much higher.

If substances are dissolved in water, they change the energy landscape. The properties of the liquid are changed accordingly. In our model, it is assumed that specific substances make new valleys with minima at specific places (x^*, y^*), the width, but not the depth vanishing when the concentration goes to zero. Special substances may fit the landscape in the sense that they deepen the existing hollows without building new ones. I suppose that they are (candidates for) homeopathic remedies. They make the typical water landscape more pronounced which may well mean that they are very toxic. Now, it is much more probable that the ball rolls into the resulting deeper hole and gets trapped there. Simply diluting the substance diminishes the effect. Additionally shaking means a better thermal contact with the environment. This is included in the model. It changes the equations of motions and the trapping effect.

In the proposed model, the technique of potentization fixes the water structure in the appropriate valley for a long time even when the substance is diluted or (for very high dilutions) absent. The substance is a guide which leads the ball into the desired valley. Then the structure is (dynamically) trapped there, the guide is no longer necessary. Without the guide, the (thermodynamic) probability for the ball to fall exactly into the desired valley would have been almost zero. From the above considerations it follows that the guide need not be a homeopathic substance, it may also be a specific electromagnetic field. In fact, there are several indications that "potencies" can also be prepared without any substances, only using electromagnetic information [13,14].

If a highly potentized substance fits exactly one of the 27 valleys of L3 in location (of the minimum) and width, it stabilizes this valley. If less diluted, it is much broader than the L3-valleys and may just fit the corresponding L2-valley. Three valleys of L2 belong to a common valley in L1, see Fig. 1. Now we assume that there are three substances which fit these three valleys exactly in location. However, if they are concentrated so much that their width is as large as the common L1-valley, they stabilize it in the same way, the information about the small difference in location being lost. In the other direction, a shift from L1 via L2 to L3 in the water structure corresponds to increasing potencies. The substance is necessary during potentization, but then it just disturbs the water landscape (which can be circumvented using the "cleaner" electromagnetic way).

It would of course be possible to extend the model to further levels (L4, ...). There are several reasons why this is not done. First, many natural fractals are self-similar over three orders of magnitude. Second, including L4 would make the calculations very time-consuming. Other modifications like improvement of the shape of the valleys should come earlier. Last, L4 and so forth only come in for high dilutions above D23, where there is no material effect at all. Missing a guide, it is not clear at all in which finer valleys the ball will be trapped (a electromagnetic guide, however, may exist for those levels) - or maybe the potency remembers the exact location (x^*, y^*) of a substance which is no longer there? The presented model is restricted to a material guide, thus it makes no sense to consider more levels, may they exist or not. The model can, however, predict that homeopathic remedies can be collected in groups depending on their locaton in the x-y-plane. Remedies which stabilize the same L1-valley are indistinguishable in low potencies, but split up in higher potencies. Then, the difference of similar remedies unfolds.

The counterpart of the water in the homeopathic remedy is the water in the human body. If (part of) it is fixed in one of the valleys of the landscape for some time, this probably corresponds to a certain disease. According to the simile principle, the appropriate remedy ("simile") is a homeopathic drug which has the same valley fixed. How the healing works, cannot be deduced from the model. However, another comment is advisable. The choice of the number three in connexion with the valleys may well have a real counterpart since Hahnemann thought that the health problems can be traced back to the so-called miasms, three basic (chronic) diseases.

Another task would be to connect the above global model with more concrete local models of cluster formation [15]. The idea behind the global model is that according to the hydrogen bonds supramolecular structures are formed. The angles between hydrogen bonds being very restricted, this induces restrictions on the global structure of water clusters and superclusters, symbolized by the (sharp) valleys in the landscape of angle-like variables x and y. Hydrogen bonds play an important role, e.g., in water, alcohol and sugar. The proposed global model does not distinguish between the three solvents. In fact, it is not only possible to potentize substances in liquid form by diluting in water or water/alcohol mixtures, but also in solid form using sugar.

By the way, the individual clusters are not fixed structures. It is more a collective dynamics of the fluctuating clusters which characterizes the water globally. Accordingly, it is very difficult to find these clusters by measurements using conventional experimental devices. This may be compared with the collective effect of wind and the individual velocity of the molecules. At a global level, we can only feel the wind. At a molecular level, the high velocities (typically 2000 km/h) veil the collective velocity (20 km/h). Another example is gravity, which is dominant at astronomic scale, but negligible at microscopic scale.

3. Different Properties of Different Potencies

3.1 Identification of the Levels

The detailed relation between the levels and the potencies has been left open so far. Now, we try to bring them together quantitatively. We start with the mother tincture (D0). It is so concentrated that the water landscape is disturbed considerably. This is also the case after one or two steps of dilution. Now we dilute the substance until (in our model) the width just corresponds to the proper valley in L1. Then, the disturbance of the water structure has become relatively small. The substance may even "vanish" for our eyes and tongue in the water structure, and the liquid seems to be pure water. For many substances, this is typically the case for D5 [16]. This is confirmed in the following way: If a substance cannot be dissolved in water/alcohol, it has to be potentized (by diluting and triturating) in sugar up to D4. From D5 on, however, it is possible to potentize it in liquid form [4]. Thus, potency D5 fits the valleys of L1. N steps later, the effect of the substance shall correspond to L2, N further steps to L3. What is N? Due to several hints in the literature, N seems to be about 9, see below. This corresponds to an equivalence of potencies D5, D14, and D23, where the drug would fit the valleys of levels L1, L2 and L3, respectively. By the way, D23 just corresponds to a dilution where about one molecule of the substance is present.

There is much evidence that the effect of a homeopathic drug oscillates when the potency is increased [10, pp. 77-81]. Comparing the electromagnetic spectrum of a series of potencies yields the result that the spectrum is sharp for some potencies, broad for others [17]. If one looks at the curves of Ref. [18], one can guess the period N to be about 9 to 10. Schramm [19] is most explicit: He speaks of different vibrations of D1-D9 which are repeated for D10-D19 and so forth (the same resonances occur, e.g., for D7 and D16). Thus, N=9 according to his investigations using some kind of electro-acupuncture. There is also a connection to the so-called LM-potencies. The dilution step is 50000 instead of 10 in the case of D-potencies. The LM-potencies are sometimes used in order to minimize the initial aggravation. Maybe the aggravation is also an oscillatory effect, being zero twice during one period. The square of 50000 is 2.5x10**9=2.5E9, the resulting estimate of N being again close to 9.

We combine the above considerations and make the following hypothesis: The potencies D5, D14 and D23 fit "perfectly" L1, L2 and L3, respectively. Now, the other potencies are associated with the nearest level, i.e., L1: D1-9, L2: D10-18, L3: D19-27. Thus there is a jump between the relevant levels first between D9 and D10 and second between D18 and D19. A further jump to a possible L4 would take place from D27 to D28. In homeopathy, one loosely speaks of low, medium and high potencies, but the expressions are used in various ways. Now it makes sense to associate these classes to the groups of potencies specified above. From D28 on, one could speak of very high or ultra high potencies. By the way, the proposed relation between the levels and potencies slightly differs from the preliminary relation suggested in Ref. [11], where N was assumed to be 8.

As already mentioned, up to D4 the substance disturbs the water landscape on a large scale. Thus it may well be that the proposed taxonomy only makes sense from D5.

Table 1. Various Interpretations of Different Regions of Potencies.

Potencies	Level	Time Scale	Frequency	Functions [20]	Process [11]
Low	L1	short (acute)	low	metabolic (body)	stationary
Medium	L2	medium	medium	rhythmic (soul)	periodic
High	L3	long (chronic)	high	nervous (mind)	chaotic

In another interpretation, D0 to D2 (or D3) could be associated with L0 and not with L1. Some homeopaths even think that the very low potencies are not "homeopathic" at all. There may be some connexion with the results of Knapp [21]. He used a variant of Kirlian photography to get pictures of many homeopathic drugs. He found that the type of picture usually tipped over at very low potencies which, however, varied from substance to substance.

In homeopathy, not only the classification of potencies remains open, but also their interpretation. One can hear the following claims: Higher potencies have a higher "frequency" (in the physical or metaphorical sense?); they refer more to the chronic than acute diseases; they have to be dispensed more seldom than the low potencies; they act more on the psyche than on the body (since the material basis is vanishing for high dilution?). In fact, there is a connexion between homeopathy and psychiatry [22]. The anthroposophic interpretation [20] of homeopathy is even more explicit, see Table 1, last but one column. The lower potencies are associated with the body and its metabolic functions. Medium potencies correspond to the soul represented by rhythmic functions like breathing. High potencies correspond to the mind (spirit?) which is represented by the senses and the (central) nervous system. Furthermore, the action of the potencies in the three regions are often symbolized by a zigzag line. I do not know how many of these statements are based on scientific work and how many on intuition. Anyway, I was not convinced. The terms are strange, as is usual the case in complementary medicine. There, it is often not clear what is really meant by words like frequency, energy or information.

Looking over the terminology used by the anthroposophs, the term "rhythmic" sounded more familiar to me than the others. This corresponds to the periodic oscillations which are often studied in chaos and systems research, a new branch of science. There, three types of processes are investigated (in brackets, other terms which fit are displayed):
1) Stationary state (fixed point, steady state, equilibrium, constant)
2) Periodic motion (oscillation, vibration, rhythmic process, cycle)
3) Chaos (aperiodic or nonperiodic motion).
It really makes sense to associate not only rhythmic with periodic, but also metabolic with stationary and nervous with chaotic, see the last column of Table 1. The metabolism refers to the more or less stationary state of the body (homeostasis!), and the chaotic behaviour of the central nervous system is an important topic of current research (which, however, deals with the brain and usually circumvents possible connexions with the mind).

3.2 New Interpretation of the Work of Smith and Endler

At this stage, I looked for a further support of the three regions and the quantitative hypothethis of mine and remembered a paper of Smith and Endler [14]. There, all potencies from D5 to D30 of thyroxine were investigated. The authors looked for resonance frequencies between a potency and a coil fed by a laboratory oscillator (the range of frequencies being 0.01 Hz =1.E-2 Hz to 10 MHz =1.E7 Hz). To do this, it was necessary to measure the corresponding microtremor of a test person ("biological sensor"). The result was quite interesting. For D5, a basic resonance frequency of 0.07 Hz occurred which remained for all other potencies. From D6 on, at each potency two further frequencies appeared which also remained for the higher potencies (the lower and higher frequencies being called frequency 1 and 2, respectively). Thus the believe that higher potencies are related with higher frequencies was supported, which is also consistent with the proposed energy landscape of water structure. Much more could not be said from the Figure and Table presented in Ref. [14]. I tried to get more information out of it using a double-logarithmic plot for frequencies 1 and 2 from D6 to D30. And really, an interesting pattern emerged, see Fig. 2.

My first impression was that the zigzag line seemed to appear which had been described by anthroposophy. A more detailed inspection shows that the breaks of the curve close to D10 and D19 match the transitions from one level to the next in my hypothesis. Averaging over frequencies 1 and 2, the three peaces of the zigzag line can be approximated by straight lines. Extrapolating the low-potency line to the left yields about 0.01 Hz for D4 and even less for D3 and so on. Since only frequencies from 0.01 Hz on were investigated in [14], a possible continuation of the straight line to very low potencies could not be found.

Now we turn to the higher potencies. At D18 and D19, something happened with the thyroxine so that the two additional frequencies did not appear properly. Is it just by accident that this occurred at the (predicted) transition from level L2 to L3? Even more can be deduced from the above data. The next break (to a hypothetical level L4) should take place at about D28. Looking at Fig. 2, this can only be suspected. In the original Figure of [14], however, a break at D28 can clearly be seen, afterwards the slope is larger again. Anthroposophy sticks to the three levels body, soul and mind and does not consider potencies beyond D30. In classical homeopathy, there seems to be no practical limit, even potencies D1000 are in use. If one assumes that the zigzag line of Fig. 2 goes on and on, from the medium slope one can guess at which potencies very high frequencies are obtained which may correspond to the inter- and intramolecular vibrations of water. Taking 1.E13 Hz as a typical value, this would correspond to D100.

Coming to the molecular level means that some physical limit is reached beyond which the storage capacity of water is probably exhausted. This has to be seen in connexion with the water clusters. In nature, larger structures are often related with lower velocities and frequencies. One may think again of the movement of molecules embedded in a macroscopic wind. Accordingly, the occurrence of higher frequencies in higher potencies presumably means that smaller and smaller clusters are involved, which has an end

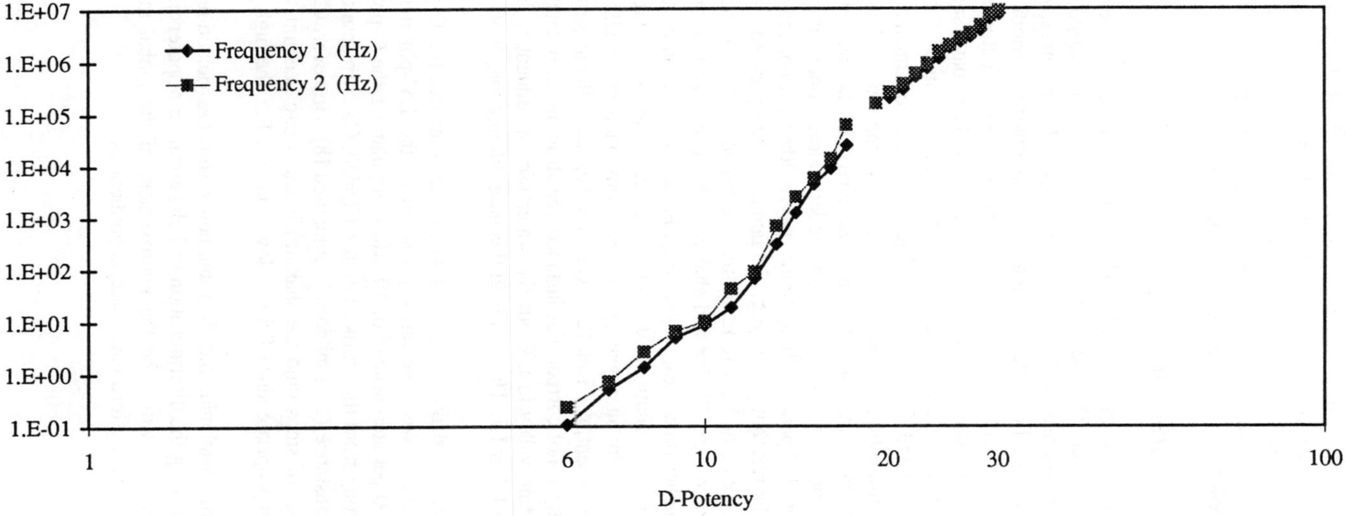

Figure 2. Frequencies 1 and 2 as functions of the potency of thyroxine, double-logarithmic plot. Range of potencies: D6 to D30.

somewhere (at D100?). On the other hand, one may think about the problem of finding physical measurement devices to "prove" the difference between high potencies and pure water, see Ref. [23]. Knowing (and extrapolating) from Fig. 2 the frequencies associated with certain potencies, the latter will be easier brought together with measurement methods which work best at the corresponding frequencies.

3.3 Ultra High Dilutions and the Effect of Succussion

If it makes sense to go beyond L3, what is the continuation in terms of Table 1? In chaos and systems research, there are some indications that one can go beyond chaos. In the so-called cellular automata [24], some kind of special computers, four classes of results are distinguished. The first three are similar to stationarity, periodicity and chaos. The fourth class is not well-understood and has not yet been investigated enough. There, processes occur which are even more complex than chaos. They have something to do with evolutionary processes. In an analogous way, the continuation of Table 1 would go beyond the individual level and enter the social groups or even the human race as a whole. The terms in the other interesting columns would be "very long time scales" (thousands to hundred thousands of years) and "social (collective mind)" in the last but one column. This extension to sociology is consistent with modern ideas concerning an integrated biopsychosocial treatment of patients. In systemic (psycho)therapy, not only the individual patient is treated, but the relevant group (e.g., the family) is incorporated.

At the end, we come back to the difference between dilution and potency. The ingredients of potentization are diluting and shaking. The former and latter are more important at lower and higher potencies, respectively. Probably, D2 cannot be distinguished from the corresponding mere dilution. At D30, however, it may be doubted if the act of "dilution" (of pure water in pure water) has any consequences at all. Probably, the succussion would be sufficient, see Ref. [25]. On the other hand, the argument of diluting pure water in pure water is not correct. For instance, the structure of potency D26 may be characterized by a certain valley in L3, but the water used as diluent is "normal water" which is somewhere in L0 or L1. Thus, even in this case diluting may have some effect in addition to successing.

The relevance of shaking is crucial when one compares the D-potencies with other kinds of preparation. We have already spoken of the LM-potencies, where the dilution factor is 50 000 per step instead of 10. More popular are the C-potencies, whĕre this factor is 100. Starting from the mother tincture C0=D0, C3 corresponds to D6 if the concentration of the substance (1 : 1 million) is compared [8]. However, C3 is equivalent to D3 if the the number of steps (and the shaking) is taken into account. Here, the first point of view seems appropriate since D6 is a low potency. For the high potencies, the question remains open.

In conclusion, combining old ideas and new results has yielded new insight into several questions concerning the discrimination and classification of potencies. Taking this into account, a further step would be the improvement of the global model of water structure, which could in turn afford more precise predictions.

4. Acknowledgement

The author is grateful to Mag. Thomas J. Milavec for fruitful discussions and his assistance when preparing this paper.

5. References

[1] M. Oberbaum and J. Cambar, "Hormesis: Dose Dependent Reverse Effects of Low and Very Low Doses", in: P.C. Endler and J. Schulte (Eds), *Ultra High Dilution. Physiology and Physics*, Dordrecht: Kluwer, 1994, pp. 5-18.

[2] K. Linde, *Dosisabhängige Umkehreffekte. Eine differenzierte Literaturbetrachtung,* München: Doctoral Thesis, 1991.

[3] K.W. Kratky, "The Generalized Simile Principle as a Joint Framework of Allopathy and Homeopathy", in: C. Bornoroni (Ed), *OMEOMED 92. First International Congress,* Bologna: editrice compositori, 1993, pp. 49-54.

[4] F. Dellmour, *Homöopathie. Eine Einführung in die Grundlagen der homöopathischen Medizin,* Wien: Facultas, 1992.

[5] G. Geyer and A. Stacher (Eds), *Chronobiologie und ihre Bedeutung für die Therapie. Behandlung nach Tageszeit*, Wien: Facultas, 1992.

[6] A. Stacher (ed), *Placebo und Placebophänomen,* Wien: Facultas, 1995.

[7] R. van Wijk and F.A.C. Wiegant, *Cultured Mammalian Cells in Homeopathy Research. The Similia Principle in Self-Recovery,* Utrecht: University of Utrecht, 1994.

[8] W.F. Daems, "Die Entwicklung des Potenzierverfahrens", in: W.F. Daems (Ed), *Was sind potenzierte Heilmittel? Zum Verständnis der homöopathischen und anthroposophischen Medizin,* Stuttgart: Freies Geistesleben, 1993, pp. 11-36.

[9] J. Himmelsbach, "Siebzig Jahre Potenzforschung. Ein Überblick", in: W.F. Daems (Ed), *Was sind potenzierte Heilmittel? Zum Verständnis der homöopathischen und anthroposophischen Medizin,* Stuttgart: Freies Geistesleben, 1993, pp. 92-142.

[10] M. Righetti, *Forschung in der Homöopathie. Grundlagen, Problematik und Ergebnisse*, Göttingen: Burgdorf, 1988.

[11] K.W. Kratky, "Moderne Naturwissenschaft und Homöopathie - Offenere Einstellung durch neue Entwicklungen?", in: R.G. Appell (Ed), *Homöopathie 150 Jahre nach Hahnemann. Standpunkte und Perspektiven,* Heidelberg: Haug, 1994, pp. 90-125.

[12] K.W. Kratky and T.J. Milavec, "Besteht ein qualitativer Unterschied zwischen Hochpotenz und reinem Lösungsmittel?", in: O. Bergsmann (Ed), *Struktur und Funktion des Wassers im Organismus. Versuch einer Standortbestimmung*, Wien: Facultas, 1994, pp. 40-56.

[13] M. Citro, C.W. Smith, A. Scott-Morley, W. Pongratz and P.C. Endler, "Transfer of Information from Molecules by Means of Electronic Amplification", in: P.C. Endler and J. Schulte (Eds), *Ultra High Dilution. Physiology and Physics,* Dordrecht: Kluwer, 1994, pp. 209-214.

[14] C.W. Smith and P.C. Endler, "Resonance Phenomena of an Ultra High Dilution of Thyroxine - Preliminary Results", in: P.C. Endler and J. Schulte (Eds), *Ultra High Dilution. Physiology and Physics,* Dordrecht: Kluwer, 1994, pp. 203-207.

[15] S. Anagnostatos, P. Pisses, K. Viras and M. Provata, "Physico-Chemical Study of Homeopathic Dilutions", in: C. Bornoroni (Ed), *OMEOMED 92. First International Congress,* Bologna: editrice compositori, 1993, pp. 77-86.

[16] M. Moser, private communication.

[17] P. Kokoschinegg, private communication.

[18] M. Moser and L. Schwarz, "Phänomene des Potenzierens und der Kreislauf des Wassers", in: O. Bergsmann (Ed), *Struktur und Funktion des Wassers im Organismus. Versuch einer Standortbestimmung,* Wien: Facultas, 1994, pp. 57-69.

[19] E. Schramm, "Biologische Testmedizin, ein Beitrag zur Elektrodiagnostik", in: O. Bergsmann (Ed), *Elektrodiagnostik,* Wien: Facultas, 1992, pp. 63-68.

[20] E. Marti, "Vom Wesen des Potenzierens", in: W.F. Daems (Ed), *Was sind potenzierte Heilmittel? Zum Verständnis der homöopathischen und anthroposophischen Medizin,* Stuttgart: Freies Geistesleben, 1993, pp. 37-66.

[21] D. Knapp, "Die bioenergetische Strahlung und homöopathische Medikamente", *Allgemeine Homöopathische Zeitung,* vol. 230, pp. 4-15, 1985.

[22] R.G. Appell (Ed), *Homöopathie, Psychotherapie und Psychiatrie. Hahnemanns weiterwirkender Impuls,* Heidelberg: Haug, 1993.

[23] O. Weingärtner, *Homöopathische Potenzen. Wunsch und Wirklichkeit bei der Suche nach der therapeutisch wirksamen Komponente,* Berlin: Springer, 1992.

[24] S. Wolfram, "Universality and Complexity in Cellular Automata", *Physica D,* vol. 10, pp. 1-35, 1984.

[25] W. Pongratz, A. Nograsek and P.C. Endler, "Highly Diluted Agitated Silver Nitrate and Wheat Seedling Development - Effect Kinetics of a Process of Successive Agitation Phases", this volume.

EFFECTS OF CADMIUM VERY HIGH DILUTIONS IN RENAL TUBULAR CELL CULTURES

CAMBAR J., DELBANCUT A. and BARROUILLET M.P.
Groupe d'Etude de Physologie et Physiopathologie Rénales
Faculté de Pharmacie - 3, place de la Victoire - 33000, Bordeaux (France)

INTRODUCTION

Chronic exposure to heavy metals, major environmental toxicants as cadmium or lead, can be implicated in a variety of pathologies such as renal, hepatic or nerve damages (Mery and Fillastre, 1983 - Humes and Weinberg, 1986)- For decades, many authors have tried to reduce *in vivo* metal toxicity using chelating agents or other metals as competing agents.

In this field, success has been obtained with pre- or post-treatment using low concentrations of the same toxic inducer agent (Terhar and al, 1965 - Ito and al, 1966- Yoshikawa, 1973) or high dilutions (Fisher and al, 1987 - Cazin and al, 1987- Herkovitz and Perez- Coll, 1991). Critical rewiews on effectiveness of serial diluted agitated agents in experimental toxicology has been recently published (Linde and al, 1994- Cambar and al, 1994).

For twelve years, our laboratory begun studies about the evidence and probation of heavy metal very high dilutions in experimental toxicology. Our previous experiments *in vivo* in rodents had clearly demonstrated the fair protective effect of mercury high dilution in mercury-induced intoxication (Cal, 1986-1988 - Larue and al, 1985).

The recent development of cell cultures permitted to largely reduce the animal experimentation and propose a convincing alternative method for a better approach of molecular mechanisms at cellular level. Renal cell cultures can be considered today as a fruitful tool to study the nephrotoxic potenncy of numerous substances (Bruggeman and al, 1989 - Lagroye and al, 1995 - Toutain and al, 1995).

The last years, our laboratory using renal cell cultures, aimed at detecting *in vitro* an eventual protective effect of pretreatment by very highy diluted cadmium or cisplatin against high cytotoxic concentrations of these same metals and at correlating these data with our previous *in vivo* results. A recent book presented similar *in vitro* results in other cultured mammalian cells (Van Wijk and Wiegant, 1994).

EXPERIMENTAL RESULTS

I/ *In vitro* determination of heavy metal high dilution protective effect by cadmium cytotoxicity assessment in renal cell cultures

Protective effect of heavy metal high dilutions is determined with an *in vitro* model using renal cell cultures. Cadmium or vehicle is hundredfold serially diluted in buffer (culture medium) and agitated for 30 s. We use very low cadmium concentrations by solution deconcentration (10^{-10} to 10^{-40} M).

Cultured cells issued from proximal tubular cell line (LLCPK 1) are incubated during 120 hours with either diluted cadmium (pretreated cell group) or with diluted medium (control cell group) and intoxicated by large dose of cadmium. Cell viability is assessed by neutral red test ; mortality rate is determined by comparison to non intoxicated cells. During all the procedure (pretreatment and intoxication) calf fetal serum (CFS) concentration is only 2 % to reduce any possible chelation of the metal with various proteins of the CFS. A protective index of very high dilution is determined comparing the death ratio of the pretreated cell group with the control one. First experiment is summarized in the **figure 1**.

The **figure 1** shows LLCPK 1 cell mortality towards ponderal cadmium, 24 hours after intoxication in case of pretreated cells for a120 hour period either with highly diluted solutions of cadmium (pretreated cell group) or with diluted culture medium solutions (control cell group).

This curve shows very significant differences between these both cell groups; illustrating fairly the protective effect of the 10^{-40} M cadmium dilution. Indeed, we could see a very large decrease of pretreated cell mortality, compared to untreated one. For instance, 2.10^{-5} M cadmium induces 51 % of control cell mortality against only 22% for pretreated cells: a reduction of mortality more than a half can be seen.

This protective effect remains whatever the toxic dose cadmium used, with a little less efficiency for high toxic intoxication concentration. For 5.10^{-5} M cadmium, the protective effect is quite low and not statistically significant.

Result significativity is determined by ANOVA global analysis and t Student test. Similar results have been obtained at five different times in the year, showing a good reproducibility of the described protective effect.

In conclusion, we can advance that protective effect of very high dilutions against toxic doses of that same metal seems clearly demonstrated in proximal tubular cells cultures. Nevertheless, the pretreatment must be applied for a 120 hour period. Indeed, data not shown here, established that none protective effect could be observed with a 48 hour pretreatment.

In the same way, a protective effect has been obtained with very low concentration of cisplatin applied during 120 hours against intoxication with ponderal dose of cisplatin. Whatever cisplatin dose used for intoxication, death rate is higher for control cells than for pretreated ones. So, at 4.10^{-5} M toxic dose, the percentage of dead cell decreases from 60 % for untreated cells down to 40 % for pretreated cells. Nethertheless, the index of protection is quite more low for cisplatin than for cadmium.

So, the protective effectiveness seems to be dependant on toxic-dose, pretreatment duration and, as we can see in the **figure 2**, also on the height pretreatment dilution. For a 120 hour period, we had incubated cells with serial centesimal dilution of cadmium from 10^{-10} to 10^{-40} M. Then, cells were intoxicated with our usual ponderal range of cadmium.

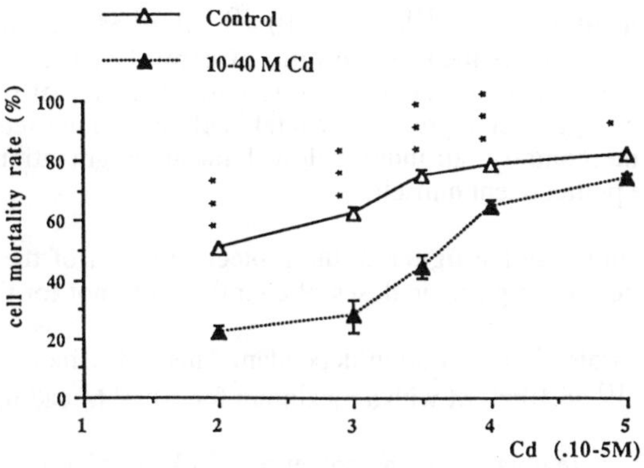

Figure 1 - Protective effect of Cd 10-40 M dilution against Cd cytotoxicity in renal tubular cultured cells.

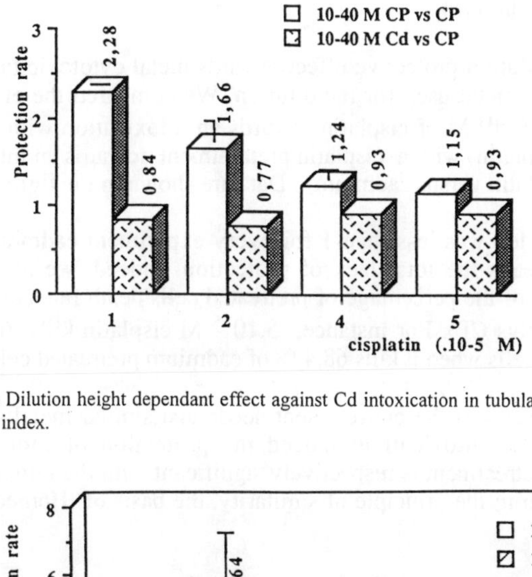

Figure 2 - Dilution height dependant effect against Cd intoxication in tubular cell cultures assessed by protection index.

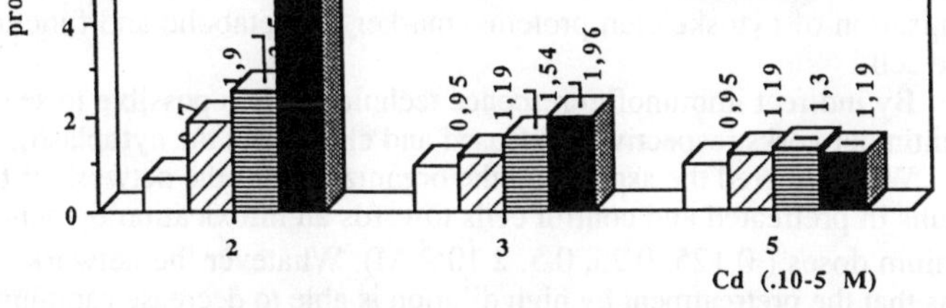

Figure 3 - Protection index of cisplatin and cadmium very high dilution in intoxicated in renal tubular cultured cells by the same metals.

For those four dilutions (10^{-10}, 10^{-20}, 10^{-30}, 10^{-40} M) a protective index was determined. It is defined by the ratio of control cell death rate to pretreated cell one. If this index is equal to 1, it represents a lack of protective effect; if it is upper to 1, then it expresses favourable effect and consequently a protective one. At the contrary, an index below 1 means aggravation of cell death induced by the pretreatment himself.

First, as we can see in the **figure 2**, the protective effect of the highest dilutions is not efficient for high toxic doses of cadmium, but not for the lower ones.

The index of protection is dilution dependant This index increases with the dilution from 10^{-10} to 10^{-40} M with a maximum for 10^{-40} M cadmium.

The same data were obtained with another metal like cisplatin. Protection index was also dilution height dependant. However, the effectiveness of this protection is less important for the cisplatin than for cadmium. So, the most important index for cisplatin was 1,68 (2.10^{-5} M cisplatin) when for cadmium is 4,70 (2.10^{-5} M cadmium).

Metal high dilution protective effect towards metal cytotoxicity revealed to be specific of the metal used for the dilution. We compared the effect of a pretreatment with 10^{-40} M of cisplatin towards an intoxication with cisplatin (cisplatin versus cisplatin) with a cisplatin pretreatment towards an intoxication with cadmium (cisplatin versus cadmium) . Data are shown in the **figure 3**.

Protection index was less than 1 for every experiment cadmium versus cisplatin, which means the total lack of protection. Indeed, we even note a significant decrease of the percentage of pretreated cells death rate compared to the control one (CP vs CP). For instance, 5.10^{-5} M cisplatin kills 63,1 % of cisplatin pretreated cells when it kills 68,4 % of cadmium pretreated cells.

The effectiveness of the pretreatment needs that similar metal should be used for dilution and intoxication; indeed the protection of cadmium and cisplatin dilution pretreatment is respectively significant with the intoxication of cadmium and cisplatin; the principle of similarity, the basis of Homeopathy, is confirmed.

After evidence of such differences in mortality between pretreated cells or not, we have decided to study intracellular structures, especially the organisation of cytoskeleton proteins, marker of metabolic and functional state of the cell.

By indirect immunofluorescence technique , it is possible to see actin and vimentin networks respectively diffused and clearcut in the cytoplasm.

We compared the aspect and the organization of the network of these both proteins in pretreated and control cells towards an intoxication by ponderal high cadmium doses (0,125, 0,25, 0,5, 2.10^{-5} M). Whatever the network studied, it seems that the pretreatment by high dilution is able to decrease cadmium induced cytoskeleton protein desorganization.

Indeed, for the same intoxication concentration, network aspect of actin and vimetin is more normal in pretreated cells than in control ones. Vimentin network seems yet clearcut, well-organized and developed even in 24 hour intoxicated cells by 0,25 .10^{-5} M cadmium.

II/ Heavy metal high dilution protection: mechanistic approaches

1/ Effects of cadmium very high dilutions on intracellular cadmium penetration

Flame spectrophotometry permits us to determine precisely cadmium concentration. It can be deducted the penetration rate of cadmium inthe cell by estimation of cadmium present in the cells compared to the sum of cadmium present in the mixture and in the supernatant.

Cells have been intoxicated at the end of the 120 hour pretreatment with ponderal concentrations of cadmium, done directly in the cell medium from cadmium standard solutions.

As we can see in the **figure 4**, cadmium penetration in the pretreated cell group is 35 % more low than in the control one; for the lowest intoxication concentration, at 3,75 mg Cd, penetration rate decreases from 0,36 to 0,22. It seems that for this concentration, the very high dilution can reduce the cadmium penetration and so lower the intracellular cadmium concentration which induces less dramatic cell damages.

Nonetheless, except at this concentration, no significant difference between pretreated and control group could not be observed. Moreover, these results are confirmed by atomic absoption, a more sensitive method for cadmium dosage.

2/ Effects of cadmium very high dilutions on metallothionein synthesis by tubular cultured cells

As reported by many authors, cadmium is well-known to be a metallothionein (MT) synthesis stimulating agent; it seemed us obvious to determine if MT could interfer to explain the very high dilution protection. MT had been detected and quantified with radioimmunoassay (RIA) techniques by Drs Lamboeuf and Carrera from the laboratory of " Signalisation et Differenciation des Macrophages" (Unité I..N.S.E.R.M. du C.H.U de Rangueil à Toulouse).

The **figure 5** shows clearly that cadmium induces well metallothionein synthesis in renal tubular cell culture, and that the pretreatment with cadmium very high dilutions enhances metallothionein biosynthesis after a 24 hour cadmium intoxication. The 120 hour pretreatment with 10^{-30} M dilution increases significantly the concentration of metallothionein compared to the control group.

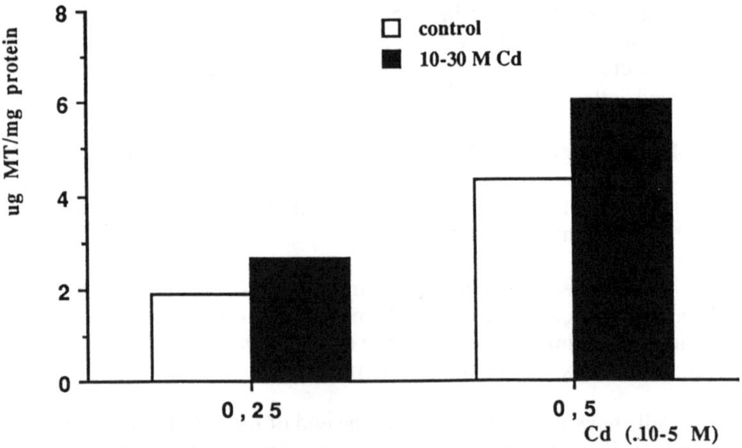

Figure 4 - MT Syntesis in cadmium intoxicated and cadmium pretreated cadmium intoxicated renal tubular cultured cells.

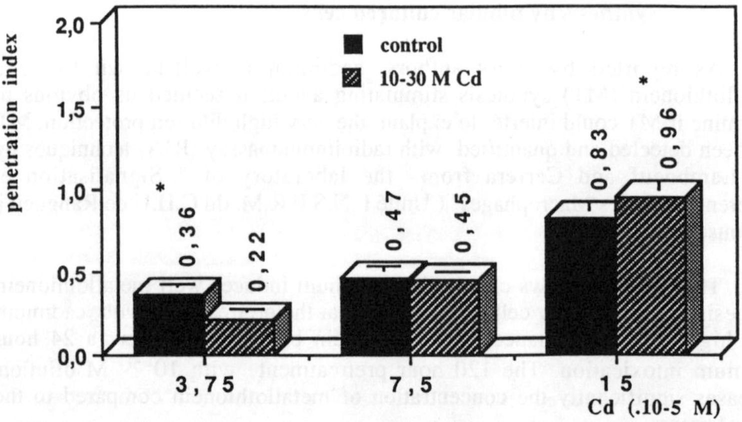

Figure 5 - Cd penetration ratio in Cd intoxicated and Cd pretreated and Cd intoxicated renal tubular cultured cells.

After 24 hour incubation with $0,25.10^{-5}$ M cadmium, concentration of MT decreases from 1,93 mg MT / mg protein for control cells to 2,67 mg MT / mg protein for pretreated cells. In the same way, $0,5\ .10^{-5}$ M cadmium induces the synthesis of 4,0 mg of MT / mg protein in the control cell group against 6,03 mg of MT / mg protein in the pretreated cell group.

So, it seems that cadmium very high dilutions would enhance metallothionein biosynthesis consecutively to a ponderal intoxication with cadmium; this MT biosynthesis seems to remain dilution-dependant. Indeed, data not shown here, revealed that a solution of 10^{-10} M cadmium still induce metallothionein sythesis but less than cadmium 10^{-20} and 10^{-30} M solutions. The pretreatment with 10^{-20} M favours MT synthesis but not enough for protecting against an intoxication with $0,5.10^{-5}$ M cadmium.

In fact, it seems that only the pretreatment with cadmium 10^{-30} M dilution has a protective effect which could explain its ability to enhance MT biosynthesis consecutively to ponderal intoxication of cadmium.

GENERAL DISCUSSION

We have already shown, in vivo in rodents, that high dilutions of mercury can significantly reduce the death rate induced by toxic doses of the same metal. Thus, death rate of mice intoxicated with high doses of mercury was markedly reduced by a 7 day pretreatment with 10^{-30} M concentrations of that metal. For example, death rate following a single injection of 5 mg/kg HgCl2 was 73,4% in control, 50% for those pretreated with 10^{-18} M and only 26,7% for those pretreated with 10^{-30} M.

We find the same profil of response with our in vitro model, 10^{-30} and 10^{-40} M dilutions providing a fair protection against cadmium induced cytotoxicity in renal tubular cell cultures. Similar results have recently been described for high dilutions of thymulin (Bastide and al, 1987), bursin (Youbicier-Simo and al., 1993) and silica (Oberbaum and al., 1992).

The classical neutral red test permits a standardization of cytotoxicity evaluation procedures using renal cell cultures intoxicated by toxic cadmium solutions. Our results are reproducible (5/5 for 10^{-30} M and 4/4 for 10^{-40} M). This in vitro model using cell cultures eliminates numerous sources of variability, especially chronobiological ones, present in *in vivo* tests. Our results are confirmed by similar *in vitro* experiments in other cultured mammalian cells intoxicated by cadmium and protected by low dilutions of the same metal (van Wijk and Wiegant , 1994).

The protective effect of very high cadmium dilutions against toxic doses of this metal seems clearly demonstrated. Similar protection against cisplatin intoxication is also demonstrated for ultra low doses of cisplatin.

This protection seems specific since cadmium dilutions do not protect against cisplatin toxicity despite the fact that cisplatin is also a tubulotoxic metal.

Recent experiments carried out in our laboratory have shown that MT synthesis was increased in Cd intoxicated cells after exposure to very high Cd dilutions (10^{-20}, 10^{-30} and 10^{-40} M) for 120 hours. Similar results have been reported by Pellegrini and al (1994) with low doses of cadmium (1 uM to 0,1 nM) in an immature human T cell line. A 24 hour pretretment with non toxic concentrations of cadmium confered protection against subsequent cytotoxic doses and was able to induce metallothionein - IIa gene expression. An increase of MT synthesis has been already observed with toxic or subtoxic concentrations of heavy metals such as zinc, cisplatin or cadmium. These results are in fair accordance with other ones from our laboratory showing that such a treatment decreases the metal storage in the cell by, for example, changes in membrane permeability.

These *in vitro* data support our i*n vivo* results suggesting that very high dilutions of metals may be useful in decreasing or even eliminating the toxic effects of nephrotic metals in patients chronically exposed to such products. Such a protective procedure was already well accepted for pretreatments with milligram or microgram doses in experimental toxicology (Flora and al, 1982a, - 1982b - Stacey and Klaassen, 1981 - Magos and al, 1979 - Yamane and Koizumi, 1982).

These data demonstrated not only the relevance of i*n vitro* models in this area of toxicology but also confirm the specific protective effects of high dilutions of cadmium or cisplatin against subsequent cytotoxic doses of the same metals.

ACKNOWLEDGMENTS

We thank Dr P. Dorfman for assistance with statistical analysis and helpful and kind discussion. These studies were supported by a grant of Laboratoires Dolisos with a CIFRE contract with le Ministére de la Recherche et de laTechnologie.

REFERENCES

BASTIDE M., DAURAT V., DOUCET-JABOEUF M., PELEGRIN A. and DORFMAN P. (1987). Immunomodulatory activity of very low doses of thymulin in mice. Int. J. Immmunotherapy 3, 191-200.

BRUGGEMAN I.M., MERTENS J.J., TEMMINK J.H., LANS M.C., VOS R.M. and VAN BLADEREN P.J. (1989). Use of monolayers of primary rat kidney cortex cells for nephrotoxic studies. Toxicol. Vitro 3, 261-269.

CAL J.C., LARUE F., GUILLEMAIN J. and CAMBAR J. (1986). Chronobiological approach of protective effect of Mercurius corrosivus against mercury induced nephrotoxicity. Ann. Rev. Chronopharmacol. 3, 99-103.

CAL J.C., LARUE F., GUILLEMAIN J. and CAMBAR J.(1988). Chronobiological approach of mercury induced toxicity and the protective effect of high dilutions of mercury against mercury-induced nephrotoxicity. In Guillouzo A. (Ed.) Liver Cells and Drugs. John Libbey Eurotext Ltd, Paris, pp 481-485.

CAMBAR J., DELBANCUT A. and DORFMAN P (1994). Applications des trés hautes dilutions en toxicologie. Cahiers de Biothérapie, 125, 27-34.

CAZIN J.C., CAZIN M., GABORIT J.L., CHAOUI A., BOIRON J., BELON P., CHERRUAULT Y., PAPANAYOTOU C.(1987). A study of the effect of decimal and centesimal dilutions of arsenic on the retention and mobilization of arsenic in the rat. Human Toxicol. 6, 315-320.

FISCHER P., HOUSE I., BELON P. and TURNER P. (1987) .The influence of the homeopathic remedy plumbum met on the excretion kinetics of lead in rats. Human Toxicol. 6, 321-324.

FLORA S.J.S., BEHARI J.R., ASHQUIN M. and TANDON S.K. (1982a). Time dependent protective effect of selenium against cadmium-induced nephrotoxicity and hepatoxicity.
Chem. Biol. Inter., 42, 345-351.

FLORA S.J.S ., JAIN V.K. , BERAHI J.R. and TANDON S.K. (1982b). Protective effect of trace metals in lead intoxication. Toxicol. Letters 13, 51-56.

HERKOVITS J. and PEREZ- COLL C. (1991). Potentized microdoses of cadmium reduce the lethal effect of this heavy metal in amphibian embryos. Berlin J. Res. Homeo. 1, 93-96.

HUMES H.D. and WEINBERG J. M.(1986) Toxic nephropathy. In Brenner B.M. Rector F.C. (Eds). The Kidney, Saunders Ed., Philadelphia, pp 1491-2008.

ITO (T.) and SAWAUCHI (K.). (1966) Inhibitory effects on cadmium-induced testicular damage by pretreatment with smaller cadmium dose. Okajima. Fol. Anat. Jap., 42,, 107-117.

LAGROYE I., L'AZOU B., LAKHDAR B., PLANDE J.. and CAMBAR J. (1995) Nephrotoxicity in vitro : modèles glomérulaires. In Adolphe A. et Guillouzo A. (Eds) - Toxicologie cellulaire in vitro - Editions INSERM, pp 121-150.

LARUE F., CAL J.C., DORIAN C., GUILLLEMAIN J. and CAMBAR J. Influence du prétraitement de dilutions infinitésimales deMercurius corrosivus sur la mortalité induite par le chlorure mercurique. Néphrologie, 6, 86.

LINDE K., JONAS W.B., MELCHART D., WORKU F., WAGNER H. and EITEL F. (1994). Critical review and meta-analysis of serial agitated dilutions in experimental toxicology. 5.°, pp 481-492.

MAGOS L., WEBB M. and BUTTLER M. (1979). The effect of Cd pretreatment on the nephrotoxic action and kidney uptake of mercury in male and female rats. Brit J Pathol, 55, 589-594.

MERY J.P and FILLASTRE J.P.(1983) - Reins et médicaments. In: Hamburger J, Grunfeld J.P (Eds), Nephrologie, Flammarion Editions, Paris, 1193-1207.

OBERBAAUM M., MARKOVITS R., WEISSMAN Z., KLINKEVITS A. and BENTWICH Z.(1992) - Wound healing by homeopathic silica dilutions in mice. Harefuah, 123;79-82.

PELLEGRINI O., DAVENAS E., EL AZZOUZI B., JURGENS P., BENVENISTE J., MANUEL Y. and THOMAS Y. (1994) Stress proteins in human lymphocytes. Induction of metallothionein and heat shock protein genes by cadmium in an immature T cell line. Cellular Pharmacology, 1,91-97.

STACEY M.A. and KLAASEEN C.D.(1981). Interaction of metal ions with cadmium-induced cellular toxicity. J. Toxicol. Environ. Health, 7, 149.

TERHAR C.J., VIS E., ROUDABUSH R.L. and FASSET D.W. (1965) - Protective effects of low doses of cadmium chlorideagainst subsequent high oral doses in the rat. Toxicol Appl Pharm, 7, 500.

TOUTAIN H.J.- Néphrotoxicité in vitro: modèles de cultures de cellules tubulaires rénales (1995). In Adolphe M. et Guillouzo A. (Eds) - Toxicologie cellulaire in vitro - Editions INSERM, Paris, pp 151-189.

WIJK van R. and WIEGANT F.A.C.
Cultured mammalian cells in homeopathy research; The similia priniple in self-recovery. Utrecht University Editor (1994).

YAMANE Y. and KOIZUMI T. (1982). Protective effect of Mo on acute toxicology of HgCl2. Toxicol Appl Pharmacol, 15, 14-221.

YOSHIKAWA H. - Preventive effect of pretreatment with cadmium on acute cadmium poisoning in rats. (1973) Industrial health (Japan), 11, 13-119.

YOUBICIER-SIMO B.J., BOUDARD F., MEKAOUCHE M. , BASTIDE M. and BAYLE J.D. (1993) Effects of embryonic bursectomy and in ovo administration of highly diluted bursin on adrenocorticotropic and imune responses of chickens. Int. J. imunotherapy, 9, 169-180.

BIOPHOTONS. WEAK LIGHT EMISSION OF CELLS

F.A. POPP
Institute of Biophysics, The Technology Center, Sauerwiesen 6
Kaiserslautern, Germany

1. Introduction

Biophoton emission is a general phenomenon of living systems [1]. It concerns a weak photon emission from a few up to some hundred photons per second and per square centimeter surface area, at least within the spectral region from 200 to 800 nm. In contrast to bioluminescence, biophoton emission is a constant radiation reflecting sensitive changes in the physiological and biological state of the system under study, i.e. cell growth, differentiation, health and disease as well as all possible external influences such as temperature and chemicals. Despite quasi-stationary biophoton emission, so-called "delayed luminescence" is at present more and more of scientific interest. So, instead of stationary biophoton emission, re-emission of the system after light-illumination (with white or monochromatic light) is the subject of this kind of investigation. It turns out that all living systems display a rather long-lasting reemission kinetics which cannot be described by an exponential function but follows rather accurately a hyperbolic relaxation behavior. Since delayed luminescence enters after relaxation continuous biophoton emission, it is likely that both phenomena can be explained by the same mechanism. The large increase in the literature about biophoton emission reflects a growing interest and a slowly increasing understanding, and manifold new and powerful applications of this non-invasive bio-indicator-technique.

2. Results

Of basic importance for the understanding of weak biophoton emission are the following recent results:

1. The total intensity of a relaxed living system lies within the range of a few up to some hundred photons/(s cm^2) which is the intensity range of "non-classical light." This means that biophoton emission is a quantum phenomenon. It concerns light with the highest possible signal/noise-ratio [2].

2. The spectral distribution never displays narrow peaks around definite frequencies. Rather, it is quite flat and can be truthfully assigned to a f=constant-behavior which means that the probability of occupying the phase space does not depend on the wavelength. This shows evidence that the biological system is "far away" from thermal equilibrium [3] (Fig. 1).

Fig. 1

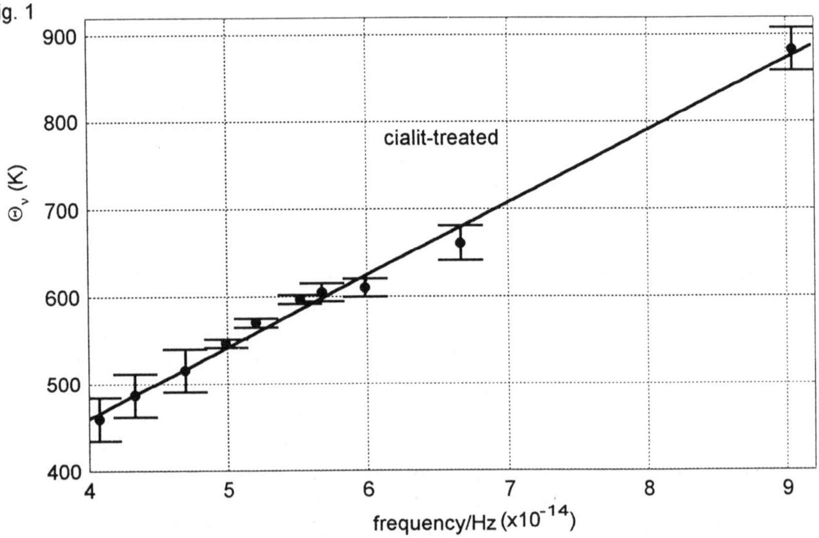

Fig. 1: Measurement on outgrown cucumber seedlings treated with cialit, with the aid of interference filters show that $\Theta(\nu)$ increases proportionally with the frequency ν.

3. The photocount statistics (PCS) accounts for the frequency of measuring n photons (where n=0,1,2,...) in a preset time interval Δt. It turns out that for a quasi-stationary state the PCS follows with surprisingly high agreement a Poissonian distribution. This shows that the biophoton field is either (almost) fully coherent or, just the opposite, rather chaotic with a coherence time much smaller than $1/\langle\dot{n}\rangle$ [1,3] (Fig. 2).

4. After excitation by white or monochromatic light, every biological system (including the human skin, see Fig.3) relaxes to its original biophoton emission quite slowly and with high accuracy according to a $1/t^\alpha$ - function (where t is the time and $1 \leq \alpha \leq 2$), rather than according to an exponential (exp $-\lambda t$) function. Since an exponential relaxation function turns into a 1/t-function as soon as we move from a chaotic to a fully coherent field, this result confirms the statement that biophotons originate from a fully coherent field [2,4,5] (Fig. 3).

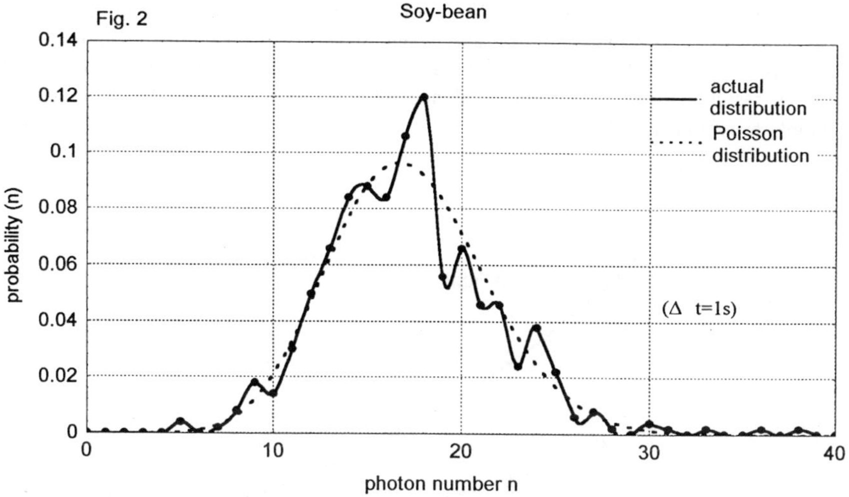

Fig. 2: The photocount statistics (=probability p(n,Δt) of registering n counts in a preset time interval Δt, where n=01,2,..) is very similar for all biological systems. If Δt is sufficiently small such that the mean number of photons in the field becomes lower than about 100, p(n, Δt) displays a Poissonian (and sometimes even a sub-Poissonian) photocount distribution.

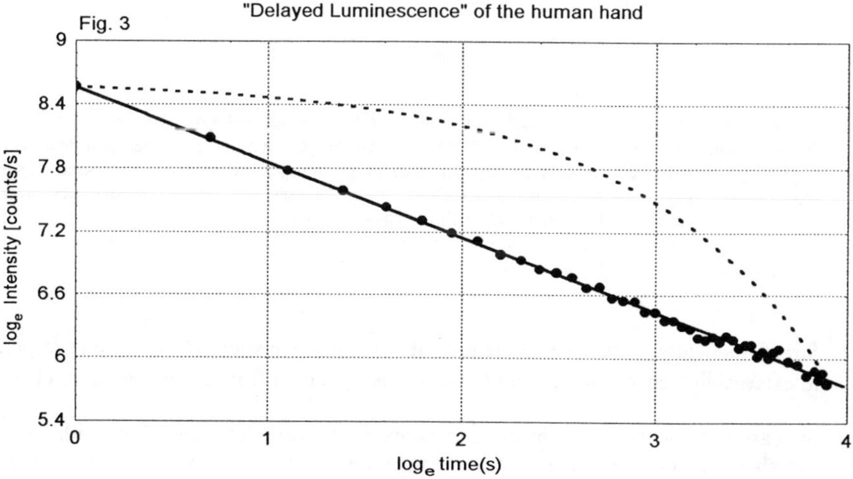

Fig. 3: In a double-log-plot a hyperbolic (1/t)-function gets a straight line. The measurement of delayed luminescence of a healthy human skin (•••) displays fairly high agreement to a hyperbolic function rather than an exponential, which is plotted as the broken line.

5. The optical extinction coefficient of biophotons passing through thin layers of sand and soya cells of various thickness can take values of at least one order of magnitude lower than that of artificial light with comparable intensity and spectral distribution. Besides these results, there is further evidence of nonlinear optical effects concerning biophotons [6] (Fig. 4).

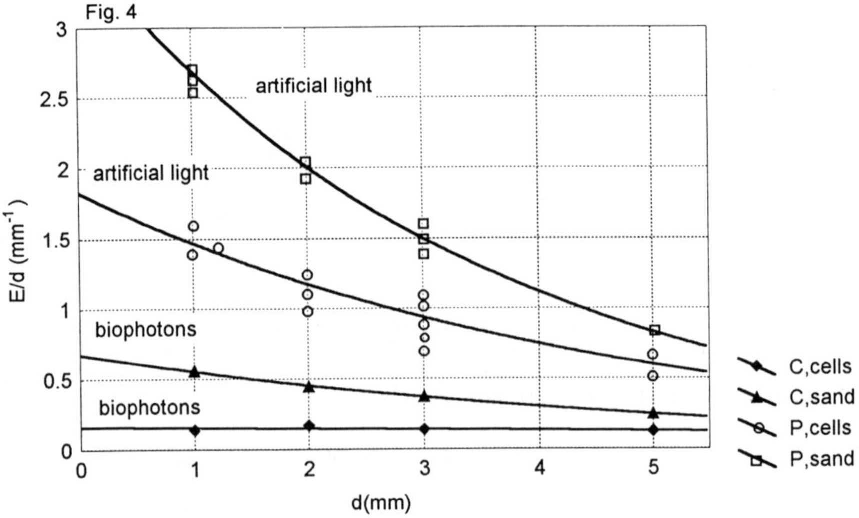

Fig. 4: The extinction coefficient E/d (mm^{-1}) of sea sand (P,sand) and soya cells (P,cell) of various thickness d(mm) was first measured at λ=550 nm in a Gilford 250 spectrophotometer, and then the E/d of photons emitted by cucumber seedlings, and passing through just the same layers of sand (C,sand) and cells (C,cells), respectively, was determined. It can be seen that (i) the transparency of both artificial and "biological" light increases with increasing thickness of the dispersive media, since E/d decreases with increasing d, and (ii) the transparency of "biophotons" is at least two orders of magnitude higher than that of artificial light (of comparable wavelengths) .

6. Stress frequently increases the intensity of biophoton emission. However, the response to external influence is sometimes quite confusing and far from being understood [7].

7. Increasing temperature induces an increase in biophoton emission with typical overshoot reactions, while decreasing temperature may lead to an undershoot response. Actually, this temperature dependence is rather characteristic of physiological functions such as membrane permeability, glycolysis, and many others. It indicates that biophoton

emission either originates from physiological processes or is its essential regulator [8] (Fig. 5).

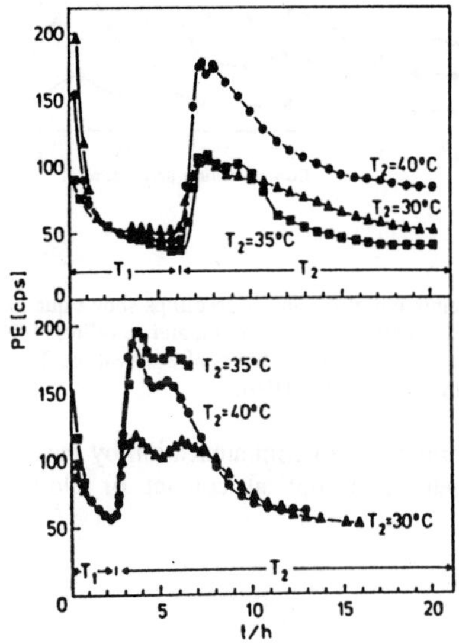

Fig. 5: Temperature dependence of photon emission from cucumber seedlings. The biophoton emission at T_1 = 19°C has been normalized to the value 50 cps for all samples after 2 hours of observation. the decay during the first hours corresponds to "delayed luminescence" and has nothing to do with the temperature effect under study. Then, after different time intervals (upper curves 6hrs., lower curves 3hrs.), the temperature was increased to T_2 within one hour. the measured biophoton emission is a function of T_2, the time interval before heating and therefore of the pretreatment of the seedlings. The picture is a typical example of temperature dependence of a physiological process. One can see clearly the overshoot-reaction, some oscillatory behavior and the late dropping down.

8. There is evidence that the conformational states of DNA influence biophoton emission. This has been demonstrated, for instance, by a clear correlation between biophoton emission and the intercalation of ethidium bromide into DNA [9]. The results indicate that DNA is strongly involved in biophoton emission. Red blood cells which do not contain DNA do not emit biophotons (Fig. 6).

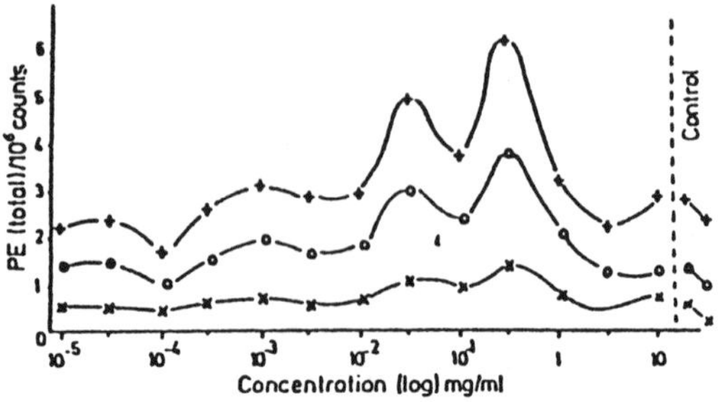

Fig. 6: The total photon emission of cucumber seedlings shows qualitatively the same dependence on the concentrations of ethidium bromide as uncoiling and recoiling of DNA. The three curves show the increase of photon emission after 1,2 and 5 hours of observation. This change is compatible with the dynamics of the conformational change of DNA.

9. There are some indications of bio-communication by means of biophotons, i.e. mutual influence of dinoflagellates in optical contact or blood, performing phagocytosis [10,11].

3. Discussion

The results do not favor the "radical-hypothesis" of biophoton emission. Although one could interpret low-level luminescence and the "white-noise-like" spectral distribution as indicating a stochastic variety of electronic excitations, the physiological temperature dependence, as well as the high sensitivity of biophoton response to external influences including nonlinear optical properties, contradict the assumption of radical emission or, more generally, "imperfection" theories. For many years now we have enunciated the coherence theory of biophoton emission which is in line with all the experimental results obtained up to now. The basic statements of this theory [1] may be summarized as follows:

1. Biophotons originate from an (almost) fully coherent field within the living system. Experimental evidence exists for hyperbolic relaxation under ergodic conditions. Further indications come from the Poissonian photocount statistics as well as from the spectral distribution which can be assigned to an "ideal open system."

2. One, if not the only, necessary condition for the production of biophotons is the permanent capacity of the living system to keep its high order: it has to produce and to remove permanently entropy. For this reason, the living organism constantly emits radiation. A striking example concerns the germination capacity of seeds which emit biophotons over thousands of years. Their photon emission reflects, on the other hand, the capacity of germination. Consequently, the low entropy (high-order) states of the system are then the main sources of emission. It is reasonable to assume that DNA

works as the source of lowest entropy (highest order). These assumptions are supported by experimental results [12].

3. The exciplex formation provides a possible mechanism for biophoton emission. This kind of physico-chemical coupling is well known in biological systems, mainly due to the properties of the DNA. This mechanism can explain not only the intensity and frequency distribution of biophotons, but also the high degree of coherence including relaxation kinetics [13].

4. In a very general sense the high degree of coherence can be traced back to the Dicke theory of constructive and destructive interference [1,14] which provides that the molecular absorbers of photons have to have much smaller distances than the wavelength of light. This condition is fulfilled, for instance, in the exciplex system of DNA. The Dicke theory also explains some biological functions, including the results of biocommunication experiments [1].

4. References

[1] F.A. Popp, K.H. Li and Q. Gu, *Recent Advances in Biophoton Research and its Applications*, World Scientific, Singapore, 1992.

[2] F.A. Popp, Q. Gu and K.H. Li, "Biophoton Emission: Experimental Background and Theoretical Approaches", *Modern Physics Letters B,* Vol. 8, pp.1269-1296, 1994.

[3] F.A. Popp, B. Ruth, W. Bahr, J. Böhm, P. Grass, G. Grolig, M. Rattemeyer, H.G. Schmidt and P. Wulle, "Emission of visible and ultraviolet radiation by active biological systems", *Coll. Phenomena* 3, pp. 187-214, 1981.

[4] K.H. Li and F.A. Popp, "Non-exponential decay law of radiation systems with coherent rescattering", *Phys. Lett. A* 93, pp. 262-266, 1983.

[5] F.A. Popp and K.H. Li, "Hyperbolic relaxation as a sufficient condition of a fully coherent ergodic field", *Int. J. Theor. Phys.* 32, pp. 1573-1583, 1993.

[6] F.A. Popp, W. Nagl, K.H. Li, W. Scholz, O. Weingärtner and R. Wolf, " Biophoton emission - New evidence for coherence and DNA as source", *Cell Biophysics* 6, pp. 33-51, 1984.

[7] R. van Wijk, et.al., "Biophoton emission, stress and disease", *Experientia* 48, pp. 1029-1102, 1992.

[8] J. Slawinski and F.A. Popp, "Temperature hysteresis of low level luminescence from plants and its thermodynamical analysis", *J. Pl. Physiol.* 130, pp. 111-123, 1987.

[9] M. Rattemeyer, F.A. Popp and W. Nagl, "Evidence of photon emission from DNA in living systems", *Naturwissenschaften* 68, pp. 572-573, 1981.

[10] F.A. Popp, Q. Gu and M.W. Ho, in *Bioelectrodynamics and Biocommunication*, eds. M.W. Ho et.al.,World Scientific, Singapore, 1994.

[11] X. Shen, "Experimental study on photocount statistics of the ultraweak photon emission from some living organisms. *Experientia* 49, pp. 291-295, 1993.

[12] B. Zeiger and F.A. Popp, in preparation.

[13] Q. Gu, "Quantum theory of biophoton emission", in *Recent Advances in Biophoton Research and its Applications*, eds. F.A. Popp, World Scientific, Singapore, 1992, pp. 59-112.

[14] R.H. Dicke, "Coherence in spontaneous radiation processes", *Phys. Rev.* 93, pp. 99-110, 1954.

EXPERIMENTAL PHARMACOLOGICAL RESEARCHES CONCERNING VEGETAL EXTRACTS IN HIGH DILUTIONS: I. CHAMOMILLA, *IN VIVO*, ON CENTRAL NERVOUS SYSTEM, AT THE "SENSITIVE" MICE

AURELIA CRISTEA

Department of Pharmacology, Faculty of Pharmacy, Str. Traian Vuia 6, Bucharest, Romania

ABSTRACT

Low degree centesimal dilution, 5 CH, and high degree centesimal dilution, 30 CH, have been prepared, starting from tincture of Chamomilla (Cham), by succesive water dilution at a 1 : 100 rate and dynamisation (D), according to Hahnemann technique. The treatment was made *per os*. The effects of C dilutions, on central nervous systems (CNS), were studied. The blind screening methodology, used in experimental pharmacology, for CNS action investigation, on mice, was utilised. The effects on both "normal" and "sensitive" healthy individuals were investigated. In order to select the "sensitive" to Cham drug-type, an original procedure was used. The actometric test, which has lead to divide the animal collectivity, according to Gauss bell-shaped distribution curve, in individuals with medium spontaneous motor activity ("normals") and in individuals with hypo- or hypermotility ("sensitive"), was chosen. The effects noticed in these researches have had, without exception, a highly statistical significance ($p < 0.01$) in "sensitive" extreme individuals and have been statistically insignificant ($p > 0.05$) in "resistant" extreme individuals. In the author's opion, the extreme individuals, within a population with normal Gauss-like biological variability, represent psycho-neuro-endocrine typologies with opposite dominants, which should make the object of future biological and pharmacological studies.

1. Introduction

Taking into consideration the fact that the effect of a drug is the final result of the significance that an organism confers to the information it decodifies from the drug signal, according to the program and its time-space particularities, within our concept of informational pharmacology [1], we have recognized an outstanding role of particularities of the organism, regarded as receiver of the message conveyed by the drug signal.

This informational concept of approaching the drug effect, casts a new light on the importance of the biological variability and of the individual reactivity for this drug effect and pharmacotherapy.

Therefore in the last years, we extended pharmacological researches even over the extremes, represented by healthy individuals, "sensitive" and "resistant" to a drug, selected from a Gauss-distributed normal population. The trials were made based on a simple original method [2]. Results of those studies were extremely interesting [3,4,5].

In our opinion, healthy individuals "sensitive" to certain drugs represent psychoneuroendocrine typologies which should become the object of further biological and pharmacological studies. Anyway, the question as to ignore or to analyze biological variability is today's concern in pharmacology [6].

It may be worth stressing the fact that for centuries, assesment of homeopathic drugs' effect was made on healthy humans "sensitive" to that remedy. That is why we suggest pursuing studies on individual reactivity using also a homeopathic remedy. Both low degree and high degree dilutions were tested.

The vegetable species Chamomilla was especially selected by us, in order to assess its effect on CNS; these effects of Chamomilla are not acknowledged by allopathic medicine, while being known and used by homeopathic therapeutic method.

In allopathic therapy, the extracts of *Matricaria chamomilla* flowers are used for the following pharmacodynamical actions: analgesic, antispastic, antiinflamatory, antihystaminic and antiseptical. In homeopathic therapy, the dilutions of Chamomilla tincture are used not only in unbearable colics and algia in different locations, accompanied by an excitation of the central nervous system (CNS) which results in agitation, irritability, fury, violence and aggresion (including verbal), but also in convulsions consecutive to a state of furious agitation.

In this lecture we present our experimental pharmacological research concerning the effects of Chamomilla at low and high degree dilutions on CNS, *in vivo*, in "sensitive" mice [7].

2. Methods

2.1. Testing solutions

Low degree dilution, 5 CH, and high degree dilution, 30 CH, of Chamomilla (Cham), have been prepared starting from Cham tincture by succesive water dilution at a 1 : 100 rate and dynamisation (D), according to Hahnemann homeopathic technique. It was processed comparatively to control groups (not-treated and treated with the solvent represented by the corresponding dynamised (D) solution of 70^0 alcohol in distilled water).

The treatment with Cham remedy was made *per os*, four times at a one day-period with low degree dilutions and two times at three days-period with high degree dilutions. The tests were made three hours after the latest administration.

The administration of reference drugs (amphetamine, diazepam, clonazepam, noraminophenasone) was made *per os*, in a single dose and the tests were made at a one hour-latence. Distilled water was used as a solvent for the reference drugs.

The volumes administered *per os* were 0.1 ml / 10 grams weight of mouse and 1 ml / 100 grams weight of rat.

2.2. Animals

The chosen animal species were male albino mouse and male wistar rat with an average weight respectively of 20 and 200 grams.

2.3. Experiments

The blind screening methodology, known in allopathic pharmacology, was utilised for CNS action investigation.

Both biological actions in healthy animal and therapeutical effects in animals submitted to certain experimental morbide models (omologated in allopathic pharmacology) were studied for Cham.

Statistical analysis

N = 100	Minimum = 2	S.E.M. = 0.522
Mean = 16.684	Maximum = 32	Sum = 1585
Median = 17	Standard deviation = 5.089	Variance = 25.899

Gauss's normal distribution	Theoretical	Experimental
Individuals between -1S.D. and +1S.D.	67 %	65 %
Individuals between -2S.D. and +2S.D.	95 %	96 %

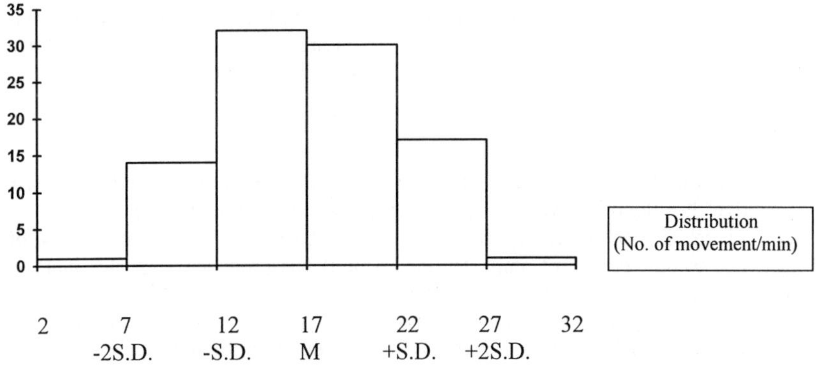

Figure 1. Gauss bell shaped curve in actometric test

2.4. Procedures of Selection of the Sensitive Type

We found and developed a simple and safe methodology for selecting the healthy type "sensitive" to the action of an allopathic or homeopatic drug. This methodolgy consists of the following:

- establishing the physiological function appropriate for the pharmacological investigation of the drug being studied ;
- determining the level of that physiological function in every individual from the group suitable for the experimental or clinical study (at least 20 individuals and optimally 100 individuals);
- tracing the Gauss bell-shaped curve, using the distribution frequency of the values of the investigated physiological function in the group subject to selection;
- dividing the individuals into normal ones, the function being at level f = average f ± 1 (about 67 % of individuals) and extreme ones with f > average f + 1 and f < average f - 1 (33 % of the individuals);
- selecting the healthy individuals "sensitive" to the drug under study, either those

with the function f > average f + 1 or those with f < average f - 1 (16.5 % of the individuals);

- clasifying the "sensitive" healthy individuals by groups for the purpose of the pharmaceutical research.

We tested the described methodology for the selection of the "sensitive" healthy individuals, in both allopathic and homeopathic pharmacology.

We have chosen the actometric test, in order to select the "sensitive" type to Cham remedy, which test has lead us to divide the animal collectivity, according to Gauss bell-shaped distribution curve, in animals with medium spontaneous motor activity ("normals") and in animals with hypo- or hypermotility ("sensitives") (Fig. 1).

2.5. Statistics

The results were statistically processed by t Student, Wilcoxon and Chi square test. Only statistical-significant results ($p < 0.05$) are mentioned. The effects (%) were calculated compared to a control group treated with the solvent.

3. Results

The low degree dilution, 5 CH, of Cham has an action in the *stimulatory sense* on "sensitive" healthy mice, with spontaneous hypomotility, that is (Fig. 2):

- increasing the motor activity by 58 % ($p < 0.01$) at 5 CH_D Cham; a dose of 10 mg / kg weight *per os* of amphetamine (Amph) has produced a greater increase in spontaneous motility of 154.06 % ($p < 0.001$);

- increasing the evasion-investigation reflex by 24.18 % at 5 CH_D Cham; to the contrary, a dose of 10 mg / kg weight *per os* of Amph has produced a great decrease of evasion-investigation reflex of 121.36 %;

- increasing the redressement reflex by 30 % at 5 CH_D Cham; Amph had no effect;

- increasing the nimbleness by 50 % at 5 CH_D Cham; Amph had no effect.

The low degree dilution has also induced in "sensitive" healthy mice with spontaneous hypomotility, a state of excitement, reported in the group toxicity test, without Amph, resulting in group lethality of 43.2 % at 5 CH_D Cham, compared to 5 CH_D alcohol; Amph has induced a 100 % lethality at $1^1/_2$ hours. The *exciting effect on CNS*, shown by potencing the Amph induced group toxicity, has been produced by low degree dilution, in mice with hypomotility namely:

- the 5 CH_D Cham, in association with Amph, has produced an increase of 40 % of lethality, compared to 5 CH_D alcohol and has increased by 20 % the amphetamine-induced lethality at $1^1/_2$ hours.

The effect on hexobarbital induced narcose was of :

- antagonizing, while using low degree dilution, 5 CH, in "sensitive" healthy mice with hypomotility, as follows: the 5 CH_D Cham has decreased by 42.9 % the percentage of narcosed mice, increased by 79.06 % the latence of narcose ($p < 0.001$) and decreased by 21.9 % ($p < 0.05$) the period of narcose;

- potencying, while using high degree dilution, 30 CH, in "sensitive" healthy mice with hypermotility, as follows: the 30 CH_D Cham has decreased by 21.46 % ($p < 0.01$) the latence of narcosis.

The diazepam (Diaz) has decreased by 31.51 % ($p < 0.01$) the latence of narcosis and has increased by 175.76 % ($p < 0.01$) the period of narcosis.

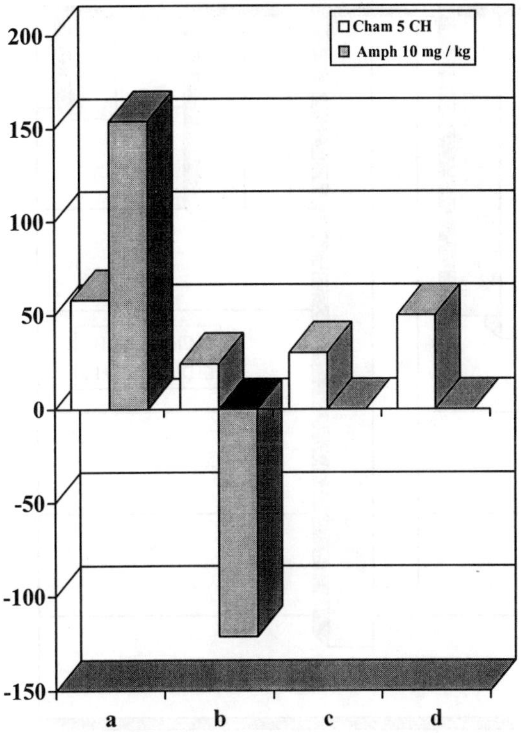

Figure 2. Stimulatory actions in healthy "sensitive" mice (with hypomotility)
a) Increase of motility (%)
b) Modification of evasion-investigation reflex (%)
c) Increase of redressement reflex (%)
d) Increase of nimbless (%)

The high degree dilution, - 30 Ch - has also induced the following pharmacotherapeutical effects, in animal organisms to which a disease syndrome has been experimentally induced:

- anticonvulsivant action in mice has been manifested by decreasing lethality by 30 % at 30 CH_D Cham; clonazepam (Clonaz) 1 mg / kg weight has induced a decrease of 50 % lethality;

- analgesic action in mice was distinguished by decreasing the number of contortions by 33.8 % at 30 CH_D Cham ($p < 0.01$); noraminophenasone (Noraminophen) 100 mg / kg, *per os*, has induced a decrease of 51.1 % contortions.

The high degree dilution, 30 CH, has the *tranquilizing effect* reported in the "sensitive" healthy mice with hypermotility by way of (Fig. 3):

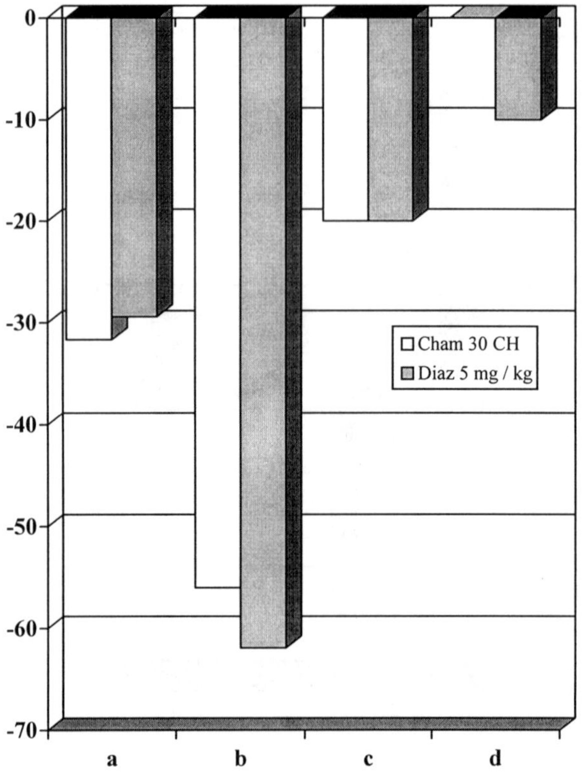

Figure 3. Tranquilizing effect in healthy "sensitive" mice (with hypermotility)
a) Decrease of motility (%)
b) Decrease of evasion-investigation reflex (%)
c) Decrease of redressement reflex (%)
d) Decrease of nimbleness (%)

- decreasing the motility by 31.68 % at 30 CH_D Cham ($p < 0.01$); a dose of 5 mg / kg weight of Diaz, *per os*, has produced a decrease of spontaneous motility of 29.41 % ($p < 0.01$);

- decreasing the evasion-investigation reflex by 56.03 % at 30 CH_D Cham; Diaz has produced a decrease of evasion-investigation reflex of 61.92 %;

- decreasing the redressement reflex by 20 % at 30 CH_D Cham; Diaz has produced a decrease of 20 %;

- the nimbleness has not been significantly modified at 30 CH_D; Diaz has decreased by 10 % the nimbleness.

No neuroleptic action has been noticed in the catalepsy test and postural test nor in the palpebral ptosis test.

No miorelaxant action has been recorded in the ipsilateral flexion reflex test and postural test. The cornean and pineal reflexes have also not been modified.

No opioid-like analgesic action has been noticed in the Hafner test. For this reason, the analgesic action, statistically-significant in chemical nociceptive stimul test, can only be of a minor analgesic type and not of a morphine-like type or can be the result of a known peripheral antispastic action.

4. Discussion

- The results of these experimental pharmacological researches indicate, for the dynamized (D) dilutions from Cham tincture, an action in the sense of stimulation of CNS, performed at low degree dilutions 5 CH, in "sensitive" healthy animal organisms with hypomotility and validate the stimulating effects on CNS described in the homeopathic patogenesis of this remedy.

Comparison with amphetamine, considered by us a substance of reference, suggests that the stimulating effect on CNS of the remedy Cham is not an amphetamine-like type.

The stimulatory action on CNS is usually related to the polinic derivatives located in the aerial part of the plant and quoted for its anticonvulsivant effect [8].

- The results also show some effects in the opposite sense, namely inhibition of CNS (tranquilizing, analgesic, anticonvulsivant), significantly performed at high degree dilutions 30 CH, in "sensitive" healthy animal organisms with hyperexcitability and hypermotility or in animal organisms submitted to experimental morbid syndrome as pains and convulsions.

- Our experimental pharmacological researches complete the other researches [9]. The cited author has noticed that, at 3 CH and 4 CH dilutions, the Cham remedy has significantly diminished the number of fights in rats become aggresive following exposure to electric shocks. Moreover, by 3 CH and 4 CH Cham or by 5 CH Coffea treatment the author has recorded decreased caffein-induced motor hyperactivity in male rats.

These pharmacotherapeutical effects prove experimentally the validity of the homeopathic therapeutical principle of similitude ("similia similibus curantur"), which consist in the disturbance regulation modality, with a similar element.

Concerning the valableness of this therapeutical principle, we discussed in another paper [10], where we explained both informationally and cybernetically, the four physiological adjustment modalities, *i.e.* iso-, homeo-, allo- and enantio-regulation, in the biocybernetical mechanisms *i.e.* feed-back and feed-forward.

- It is worth saying, that the effects noticed in our researches have had, without exceptions, a highly statistical significance ($p < 0.01$) in "sensitive", extreme individuals and have been statistically insignificant ($p > 0.05$) in "resistant" extreme individuals. Thus, we successfully demonstrated the quieting effect of the remedy Cham in high-degree dilutions of 30 CH in "sensitive" healthy mice, with spontaneous motor restfulness. We were also able to produce with the remedy Cham, an exciting effect on CNS, in "sensitive" healthy individuals, selected for their spontaneous motor hypoactivity. These results prove the objective value of experiments in healthy individuals, though "sensitive" to certain remedies, promoted by Hahnemann.

- In our opinion, the "sensitive" and "resistant" extreme individuals, within a population with normal Gauss-like biological variability, represent psycho-neuro-endocrine typologies which should be the object of future biological and pharmacological studies.

We formulate the hypothesis that the double-way (stimulating and inhibiting) effects phenomenon, made under the same physiological function, by the same drug and manifested especially in the "sensitive" and "resistant" extreme individuals, is possible as a consequence of the informational unbalance, which very low or very high concentrations of drug signals can impose on physiological balance, between both two coupled synaptic transmission systems, at the heterosynaptic level.

This asessement is based on our allopathic pharmacological studies on "sensitive" individuals [3,4,5].The experimental pharmacology researches made on type A [11] and type O [3] individuals proved that type A individuals are "sensitive" to clonidine (Clon) while type O individuals are "sensitive" to naloxone (Nalx). Thus, the 50 % efficient dosages (ED 50), determined experimentally by us, in stress-induced analgesia produced by forced swimming in mice (despair test), were for Clon lower in type A compared to type O ($p < 0.05$), while for Nalx they were lower in type O compared to type A ($p < 0.05$), either enhancing or antagonising stress-induced analgesia [4,5]. Also we emphasized on the dose-effect curves the double-way effect of Clon and Nalx, on stress analgesia, with distinct inflexion points, for the normal and the two extremes individuals [12,13].

Consequently, we consider the need for extending the allopathic pharmacology investigation from the normal healthy individuals to extreme individuals as well, within the Gauss bell-shaped curve. Ignoring or analysing the biological variability represents a real problem evolving from European pharmacological debates [6].

In conclusion, we consider that the results of these studies justify the suitableness of future researches concerning the intimate mechanism of the exciting action of Cham on CNS, an action that explains the quieting effect at high dilutions, in accordance with the therapeutical principle of similitude.

5. References

[1] Cristea, A., "Pour une pharmacologie cybernetique et informationnelle", *Cybernetica* 2, 1991, pp. 85-93.

[2] Cristea, A., "Highlighting the healthy individual type "sensitive" to a drug-Methodology ", *8th GIRI Meeting*, Jerusalem, Israel, 1994.

[3] Cristea, A., "Opioid and adrenergic types of behaviour to stress", *VIth Congres of ECNP*, Budapest, Hungary; *Europ. J. Neuropsychopharmacol.* 3, 1993, pp. 446-447.

[4] Cristea, A., "Mechanisms of stress analgesia in opioid and adrenergic types of behaviour", *VIIth Congress of ECNP*, Jerusalem, Israel; *Europ. J. Neuropsychopharmacol.* 3, 1994.

[5] Cristea, A., S. Negres, L. Bucur, E. Podoleanu, "Reactivitatea individuala la Naloxon si Clonidina, in analgezia de stress (Individual reactivity to Naloxon and Clonidine in stress analgesia)", *Farmacia* (Bucharest, Romania) 5-6, 1994, pp. 27-30.

[6] Black, J., "Biological variability: ignore, reduce or analyse?" (Plenary lecture), *1st European Congress of Pharmacology* , Milano, Italy, 1995.

[7] Cristea, A., S. Teodorescu-Negres, L. Paraschiv, V. Darie, "Experimental pharmacological researches, concerning the action of Chamomilla Homeopathic Dilutions on Central Nervous System", *7th GIRI Meeting*, Montpellier, France, 1993.

[8] Guermonprez, M., M. Pinkas, M. Torck, *Matière Médicale Homéopathique*, Ed. Boiron, France, 1989, pp. 188-193.
[9] Guillemain, J., "La recherche pharmacologique appliquée à l'Homéopathie, sa grandeur, ses faiblesses", *Homéop. Franc.* 71, 1983, pp. 191-200.
[10] Cristea, A., "Elemente similare, complementare si contrarii in reglarea sistemelor biocibernetice", *Farmacia* (Bucharest Romania) 1-4, 1991, pp. 46-51.
[11] Rosenmann, R.H., "Personality type A behaviour pattern and coronary heart disease", *R. Soc. Med.* 14, 1980, pp. 65-70.
[12] Cristea, A., S. Negres, A. Ciobanu, "The double-way effects of Naloxon and Clonidine at heterosynaptic level ", Abstr. *CINP Regional Conference*, Vienne - Prague, 1995.
[13] Cristea, A., S. Negres, D. Joean, "The double-way effects of naloxon and clonidine on experimental stress analgesia", *Rom. J. Physiol.* (Bucharest Romania) 1-4, 1995, pp. 137-143.

ENHANCED INVERSION EFFECT OF THYROXINE LOG 10^{-13} BY AGITATION. CURATIVE EFFECT FOLLOWING HYPERSTIMULATION IN FROGS

EGON LAUPPERT
Research Site for Low Energy Bio-Information, Graz, Austria

P. CHRISTIAN ENDLER
Zoological Institute of the University of Graz, Austria

ABSTRACT

The effect of diluted thyroxine was tested on the transition from the 2-legged to the 4-legged stage in the grassfrog rana temporaria. (A) Throxine dilutions *log 4* and *log 5* prepared with a minimum of agitation (final concentration in the basin water log 9-10) exerted a stimulatory effect; the dilutions *log 6* and *log 8* prepared in a process of stepwise dilution by pipetting (final concentration log 11-13) clearly inhibited metamorphosis. This is discussed as a hormetic effect. (B) When the latter dilutions were prepared in a process of stepwise dilution and vigorous agitation, this inhibitory effect became even more marked; it was even found at *log 5*. This is discussed with regard to the 'potency rule' of homoeopathy. (C) The stimulatory effect of the thyroxine dilution *log 5* was seen to decrease when an agitated dilution *log 8* was applied simultaneously. This is discussed with regard to the 'similarity rule' of homoeopathy.

KEY WORDS

Amphibia, thyroxine, reverse effect, hormesis, potency rule, similarity rule, homoeopathy

1. Introduction

1.1 Enhancement of amphibian metamorphosis by thyroxine

Thyroxine (tetra-iodo-thyronine, T_4, a thyroid hormone) plays an important role in the regulation of the speed of metamorphosis in amphibia. When thyroxine is added to the water of an aquarium with a final concentration of log 8 (10^{-8}) parts by weight, it induces or accelerates, respectively, the metamorphosis in amphibia [1,2,3]. In previous experiments with tadpoles of the grassfrog Rana temporaria at the two-legged stage (see methods), also L-thyroxine sodium pentahydrate at a concentration of log 9 parts by weight in the basin water caused an acceleration of metamorphosis of 10-17% compared to no treatment [4].

1.2 Inhibition by thyroxine

Extremely high concentration: Thyroxine, however, is not only able to enhance metamorphosis. A thyroxine concentration of log 7 parts by weight or below causes an acceleration of development to such an extent that deformities appear in the animals [5]. When L-thyroxine sodium pentahydrate was applied at molecular concentrations of log 6 and log 7 parts by weight in the basin water at the two-legged stage of rana temporaria, in both cases it blocked the development of the animals through deformation and even death (data that were undeliberately found in previous experiments, see [4]).

Extremely low concentration: In previous studies it was found that L-thyroxine sodium pentahydrate at an extremely high dilution, when prepared according to a special process of stepwise dilution and agitation, is able under certain experimental conditions to inhibit the metamorphosis from the two- to the four-legged larval stage of Rana temporaria. This was found independently in several laboratories. A total of about 3000 frogs was observed [4].

1.3 Aim of the study

This inhibiting effect of an extremely high dilution led to the idea that also at a high or moderate concentration of the hormone an inhibitive effect of thyroxine may occur [6].

(A) We wanted to observe the influence of 'conventionally' prepared test dilutions thyroxine log 4 - log 8. In other words, the final concentrations in the basin water were log 9 /8 (see Methods) - log 13 /12. This experiment was done in the context of research on hormesis [survey in 7].
(B) We wanted to investigate whether or not a special preparation process including vigorous agitation influences the effects observed in (A). This was done in the context of research on the 'potency rule' described in homoeopathy [8,9].
(C) We intended to combine a metamorphosis stimulating high molecular dilution and an agitated moderate dilution of thyroxine. This was done in the context of research on the 'similarity rule' described in homoeopathy [8,9].

2. Methods

Animals and Staging: Experiments were performed in summer (June, July, August) using rana temporaria from different Austrian pools. Early stage: For the main experiments, we exclusively chose those two-legged tadpoles which had just started to develop hind legs and where the hindlegs were slightly straddled to the extent that one could just see an open triangle between thigh, shank and tail, but without being able to look through it easily (in the region of stage 31, Gosner's staging table [10]). Depending on the experiment, the tadpoles were observed altogether 10 - 20 times at intervals of 8 h. Within this period, the forelegs, preformed under the skin, broke through within a few minutes. Thus, four-leggedness seems adequate to define a final stage. Late stage: For additional experiments, tadpoles with already straddled hindlegs were chosen.

Further conditions: The transitions were observed in disposable basins. The room temperature was 23±1°C. Indirect natural light was used. One set of basins was always used for the treatment with one of the testing solutions (see below). The amount of tap water per animal was 300 ml. The number of animals per basin was the same (5 - 20) in each experiment.

Preparation of testing solutions and exposition to probes: The tadpoles were observed under the influence of the following testing solutions.

Simply diluted thyroxine:
These solutions were prepared by gently dissolving 10 mg of L-tetra-iodo-thyronine sodium pentahydrate (Sigma) in 100 ml of 40% ethanol (= thyroxine log 4), or 1 mg in 100 ml (log 5), or 0.5 mg in 500 ml (log 6).

Step by step diluted (pipetted) thyroxine: The solutions were prepared by first gently submerging 10 mg of L-tetra-iodo-thyronine sodium pentahydrate in 100 ml of 40% ethanol (= thyroxine log 4) and then adding 1 ml of this solution to 9 ml of bidistilled water using a disposable pipette (1 : 10). The follow-up dilution log 5 was given a rest of 10 min to allow even diffusion (no agitation). Starting from the stock solution log 5, solutions up to log 8 were prepared in an identical (1 : 10, dilution by pipetting). Every dilution was given a rest of 10 min to allow even diffusion (no agitation). These test dilutions were called DP5 - DP8.

Step by step diluted and vigorously agitated thyroxine: Starting from the dilution log 5 as the stock solution, solutions up to log 10 were prepared as above, except that at every step the solution was agitated according to standardized instructions (Homöopathisches Arzneibuch 1991): at every step, a sterile 20 ml bottle was filled with 10 ml of the dilution, and then was banged down 30 times at short regular intervals against a rubber impediment to create mechanical shocks. This was always followed by a rest of 1 min (handling time). These test solutions were called DH5 - DH8.

Dilutions of ethanol in bidistilled water were prepared analogously for control. They were called W5 (4% ethanol) - W10.

Furthermore, a combination of dilution thyroxine log 4 and step by step diluted and vigorously agitated thyroxine DH8 was tested versus dilution thyroxine log 4. For blinding purposes, water as a testing substance was combined with thyroxine log 4 for treatment of one group and thyroxine DH8 was combined with tyroxine log 4 for treatment of the other group. Furthermore, a water control group was observed.

In each case 3 µl of testing solution(s) per 1 animal and 300 ml tap water were added to the basin water 20 times at intervals of 8 h. This means that, e.g., a solution DH5 was diluted by the factor of 10^5 to log 10 at the first application and reached log 9 concentration by the 10th application. In one experiment, 24µl of the testing solutions per animal were added to the basin water once in the beginning of the experiment.

Evaluation of the data: Cumulative frequencies (Fa) of four-legged tadpoles were totalled and compared graphically to those of two- or three-legged animals at corresponding times. With regard to the small numbers and to possible distortions due to different group sizes, no statistical evaluation was made for this first communication.

Laboratory and blinding: The study was organized by P.C.E. and W. Pongratz. Blinding of the laboratory worker was done by P.C.E. Laboratory work was performed by E.L., except for study 3.1.4 which was performed by C. Heckmann.

3. Results

3.1 Explorative part of the study

3.1.1 Experiments with the solutions thyroxine log 4 - 6

In these experiments, thyroxine solutions log 4, log 5 and log 6 (neither pipetted nor agitated, see Methods) were compared to a water control group (50 - 75 animals per group). During the treatment, the Fa of 4-legged frogs was 10 - 35% (log 4) or 15% (log 5) greater than that of the water control group. In other words, the chance of entering the final stage was greater in the groups treated with the thyroxine dilutions log 4 and log 5, but not log 6. This was true both for the repeated and for the single application of the substances (see Methods).

3.1.2 Step by step diluted and pipetted thyroxine

The test substances DP5, DP6 and DP8 (30 animals in each of the 3 groups) were compared to a water control group (30 animals) (Fig. 1).

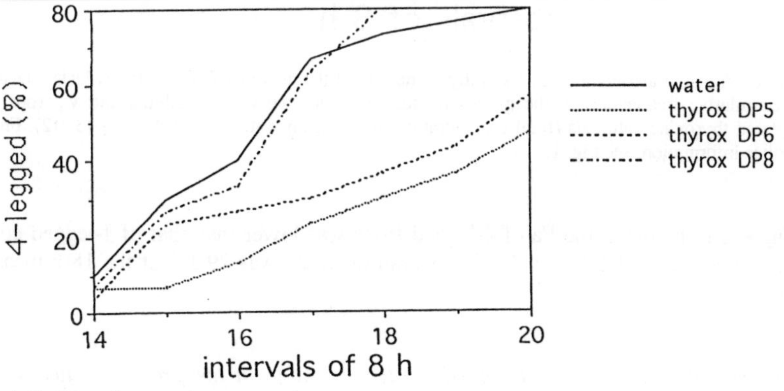

Fig. 1: For legend, see next page

Fig. 1: Pipetted moderate dilutions of thyroxine and the metamorphosis of frogs. Ordinate = cumulative frequency (Fa, in %) of animals entering the 4-legged stage . Abscissa = course of time in units of 8h. DP5 - DP8, W = cumulative frequencies of the groups treated with the step by step diluted, but not vigorously agitated thyroxine solutions (final concentration in the basin water log 10 /9 - log 13 /12) or water, respectively.

During the treatment with DP5 and DP6, but not with DP8, the Fa of the 4-legged frogs was lower than that of the water control group. The overall decrease (DP5 + DP6 + DP 8) was 23.3% at the 18th measuring point.

3.1.3 Step by step diluted and vigorously agitated thyroxine

The test substances DH5, DH6 and DH8 (21 animals in each of the 4 groups) were compared to a water control group (36 animals) (Fig. 2)

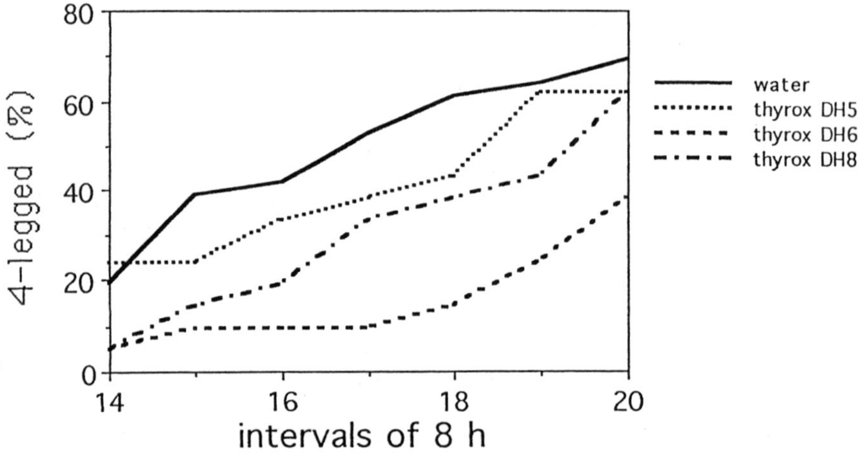

Fig. 2: Agitated moderate dilutions of thyroxine inhibit the metamorphosis of frogs. DH5 - DH8 = cumulative frequencies of the groups treated with the step by step diluted and vigorously agitated thyroxine solutions (final concentration in the basin water log 10 /9 - log 13 /12). For further information, see Fig. 1.

During each treatment, the Fa of 4-legged frogs was lower than that of 4-legged controls in *all* of the groups DH5 - DH8. The overall decrease was 29.3% at the 18th measuring point.

In general, the chance of entering the 4-legged stage in the groups treated with the agitated DH-dilutions was also smaller than in the groups treated with the DP-solutions.

3.1.4 Combination of dilution thyroxine log 4 and step by step diluted and vigorously agitated thyroxine DH8

A combination of the (stimulatory, see above) test substance thyroxine log 4 and the (inhibiting, see above) test substance thyroxine DH8 was compared both to thyroxine log4 as well as to a water control (20 animals in each of the 3 groups (Fig. 3).

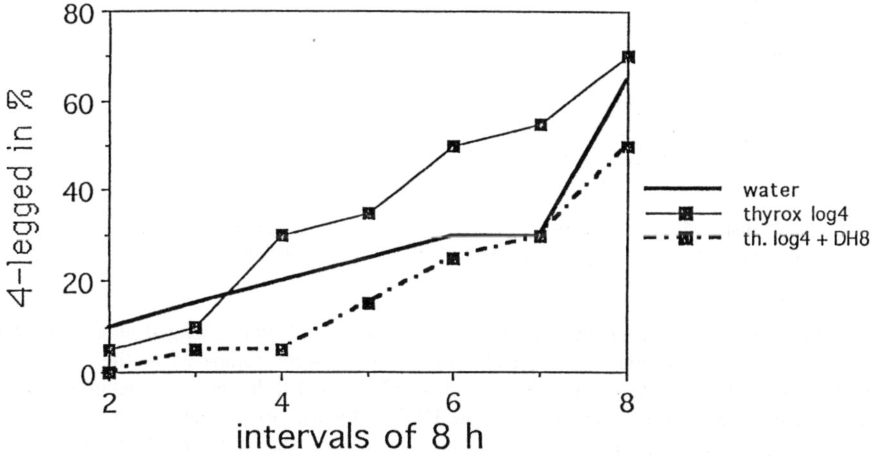

Fig. 3: An agitated dilution of thyroxine (DH8) inhibits metamorphosis of frogs stimulated by thyroxine log 4. For further information, see Fig. 1 and text.

During treatment, the group given thyroxine log 4 showed an acceleration of metamorphosis compared to water control while the group treated with thyroxine log 4 + thyroxine DH8 showed an inhibition of metamorphosis. The difference of Fa between the two groups was 20 - 25%.

3.2 Confirmatory part of the study

The test substances DH 6 and DH 8 (25 and 78 animals at the early 2-legged stage) were compared to the water controls W6 and W8 (25 and 78 animals) (Fig. 4). During treatment, the Fa of 4-legged frogs in the DH-groups was lower than that of the water control group.

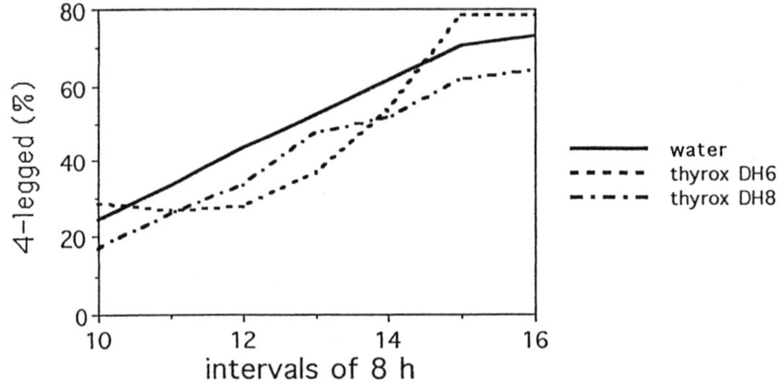

Fig. 4: Agitated dilutions of thyroxine (DH6, DH8) inhibit the metamorphosis of frogs. In the diagram, the water control for DH6 is normalized. For further information, see Fig. 1.

In a first series of experiments starting with animals at the *late* 2-legged stage, the test substances DH 6 (10 animals) and DH 8 (24 animals) were compared to the respective water controls. Only the group treated with DH8 showed a decrease of metamorphosis. Thus, it was decided to concentrate on the DH8 in further experiments using animals at the late 2-legged stage. In a second series, the test substance DH8 (75 animals) was compared to the water control WP8 (75 animals) (Fig. 5).

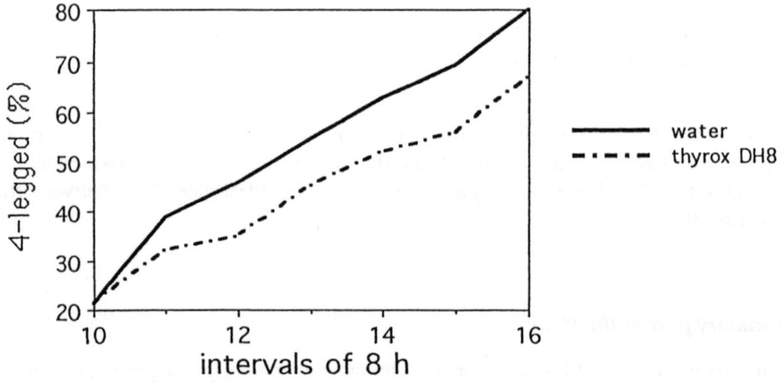

Fig. 5: For legend, see next page.

Fig. 5: An agitated dilution of thyroxine and the metamorphosis of frogs treated since the late 2-legged stage. For details, see Fig. 1.

During treatment, the Fa of 4-legged frogs in the DH-group was lower than that of the water control group.

4. Discussion

4.1 Stimulatory effect

Thyroxine at a concentration of log 8 or log 9 (not specially prepared in a step by step dilution process) is known to *stimulate* amphibian metamorphosis [1-4]. Stimulation is found regardless of whether the final concentration is reached by a single application of an appropriate thyroxine dilution at the beginning of the experiment (see Introduction) or whether it is reached by consecutive applications at 8 h intervals (see Results). When a dilution log 4 or log 5 (no step by step dilution process) was dissolved in the basin water, giving a final concentration of log 8 - 10, it exerted a stimulatory effect (10 - 35%) compared to the water control (see Results, 3.1.1). A dilution log 6 did not show any effect.

4.2 Inversion of effect: hormesis and the homoeopathic 'potency rule'

Both in the explorative as well as in the confirmative experiments presented here, in general a *decrease* in speed of metamorphosis in rana temporaria tadpoles was observed under the influence of step by step prepared dilutions of thyroxine log 5, log 6 and log 8. Thus, final concentrations in the basin water (after 10 applications at intervals of 8 h) were log 10 /9 - log 13 /12 of hormone. One set of the test dilutions 'D*P*5 - D*P*8' was prepared in a step by step dilution process by pipetting only and one set 'D*H*5 - D*H*8' was prepared in a step by step dilution process with vigorous agitation between steps. The inhibitory effect was found for both DP and DH dilutions, exept for DP8. The effect was generally more marked when vigorously agitated DH dilutions were used than when pipetted DP dilutions were used. It was shown that the experiment should best be started with animals at a defined early 2-legged stage. One experiment on DP5 and DP6 (20 animals in each group) starting at an earlier stage in development did not confirm the result (unpublished). Experiments with animals at a later stage only confirmed the effect of DH8 (see Results).

These findings are consistent with the results of C. Heckmann, Tübingen, FRG, who spent a brief study period at the Zoological Institute of Graz University in preparation of his thesis at the Medical Faculty of Tübingen University.

- Again, as in the original study (see Results 3.1.2), during treatment with the *non-agitated* thyroxine dilutions DP5 and DP6, but not with DP8, the cumulative frequency of the 4-legged frogs Fa was lower than that of the water control group (20 animals in each group). The overall decrease (DP5 + DP6 + DP8) was 10.0% at the 16th measuring point.

- Likewise (see Results 3.1.3), during treatment with the *agitated* dilutions DH5, DH6 and DH8, the Fa of 4-legged frogs was lower than that of the water control group in *all* of the DH-groups (20 animals in each group). The overall decrease, again, was more marked if the thyroxine dilutions were agitated, reaching 21.7% at the 16th measuring point. In the case of DH8, if animals at the later 2-legged stage were used, the decrease in Fa was 7% (300 + 300 animals).

These findings can be related to general assumptions of *hormesis* and *homoeopathy* research. The concept of hormesis describes effects of non-agitated moderate dilutions. These effects are the inverse of the effects of their mother substances [7]. It seems possible that research protocols in this field concern without proper distinction both ideally non-agitated dilutions and dilutions prepared in a *non-standardized* process of stepwise dilution and agitation by stirring, pipetting etc.

The 'potency rule' of homoeopathy [8] states that such inversion effects are more marked if the dilution is prepared according to a standardized protocol of stepwise dilution and vigorous agitation (succussion) [7,8,9].

4.3 The 'similarity rule'

The observed reduction of the speed of thyroxine-controlled metamorphosis in frogs by dilutions of this agent (thyroxine) [4,11,12] is consistent with the biomedical 'similarity rule' [8,9,13,14]. In its most elementary form, this rule concerns the curative effect on a living system of an agent at low concentration after intoxication with the same agent at high concentration [14].

4.3.1 Non stimulated animals

In the paper presented here (E.L. and P.C.E.) as well as in subsequent work (Heckmann, soon to be published) it was found that a thyroxine dilution DH8, compared to water control, is able to inhibit metamorphosis in frogs. If animals at the standard early 2-legged stage were used to start the experiment, the decrease in Fa of 4-legged animals was about 25% (50 + 50 animals). If animals at the late 2-legged stage were used, the decrease in Fa was less than 10% (375 + 375 animals). These differences are statistically significant.

4.3.2 Artificially stimulated animals

Furthermore, it was found that the same dilution DH8 is also able to inhibit metamorphosis in frogs treated simultaneously with a high concentration of thyroxine (log 4). When animals at the standard early 2-legged stage were used to start the experiment, the decrease in Fa of 4-legged animals was 25% (20 + 20 animals, see Fig. 3) whereas with animals at the late 2-legged stage, the decrease in Fa was 17% (100 + 100 animals). These differences are statistically significant.

These results show that in general, the relative inhibitory effect of thyroxine DH8 was more marked if the animals (both test- and control group) were simultaneously treated with a relatively high, stimulatory, concentration of thyroxine.

One further attempt to repeat this finding (100 animals in each group) failed since the positive control group thyroxine log 4 showed no stimulation of metamorphosis.

Inhibition of frog metamorphosis by information from thyroxine is not unknown in research on homoeopathy. In multicentered studies with a total of about 3000 frogs, under certain experimental conditions, a significant decrease in development speed (5 - 20%) was found when step by step diluted and vigorously agitated thyroxine (DH10 - 16, DH30) was applied [4,11,12].

From these data it can be concluded that the effect of step by step diluted thyroxine is linked to the natural or experimental thyroxine level of the tadpoles. Previous experiments grouped according to the natural length in time of the observed transition showed that the inhibitory effect of a thyroxine dilution is the more marked, the slower natural development is [4, p.62]. At that time, it was speculated that the effect of thyroxine DH30 increases, with a *decreasing* thyroxine level. This assumption may now have to be revised. Experiments with tadpoles at different defined thyroxine levels, including hyper- and hypothyroxic animals, may be of interest.

With regard to the effect patterns of thyroxine dilutions DP5 - DP8 and DH5 - DH8, respectively, the idea might arise that the inhibitory effect found at DH8, but not at DP8, is due to a higher content of thyroxine molecules caused by a better solution through the stepwise agitation process. The finding, however, that a combination of the dilutions thyroxine log 4 + DH8 slows down development compared to the dilution log 4 alone proves that there exists an inhibitory effect of dilution DH8 that is independent of the overall thyroxine concentration in the basin (which is even *higher* for the log 4 + DH8 group than for the log 4 group).

Recent physics research has revealed that water dipoles may develop phase coherent oscillations through radiation coupling [15,16]. It is proposed that these could be modulated as a time-ordered pattern of signals and induce coherent wave propagation (electron propagation, [17]). Further theory suggests that the phase coherent oscillations may originate information pattern formation through isotopicity effects in the dilutions [18,19].

5. Annotation

This study was financed by the Jubiläumsfonds der Österreichischen Nationalbank, Vienna, Austria.

6. References

(For the convenience of the reader, many references refer to one textbook only [20]. Previous publications of the respective authors are listed there.)

[1] Giersberg H. Hormone. Berlin: Springer 1972
[2] Ganong WF. Lehrbuch der Medizinischen Physiologie. Berlin: Springer 1974
[3] Wehner R and Gehring G. Zoologie. Stuttgart: Thieme 1990
[4] Endler PC et al. A zoological example on ultra high dilution research. In [20]
[5] Pitt-Rivers R and Trotter WR. The thyroid gland. London: Butterworth 1964
[6] Benveniste J, Paris. Personal communication
[7] Oberbaum M and Cambar J. Hormesis: dose-dependent reverse effects of low and very low doses. In [20]
[8] European Committee for Homoeopathy. Homoeopathy in Europe. Developing standards for professional research on homoeopathy in the European Union (working title). Brussels, Belgium, chée de Bruxelles 132; in preparation for publication
[9] Popp FA. Some biophysical elements of homoeopathy. In [20]
[10] Gosner KL. A simplified table for staging anuran embryos and larvae with notes on identification. Herpetologica 1960; 16: 183
[11] Endler PC et al. The effect of highly diluted agitated thyroxine on the climbing activity of frogs. J Vet & Human Toxicol 1994; 36 /1: 56
[12] Endler PC et al. Non-molecular information transfer from thyroxine to frogs with regard to homoeopathic toxicology. J Vet & Human Toxicol 1995; 37 /3: 259
[13] Andersch P and Endler PC. Glossary on homoeopathy. In [20]
[14] Van Wijk R and Wiegant FAC. Cultured mammalian cells in homoeopathy research. The similia pinciple in self-recovery. Utrecht: Universiteit Utrecht 1994
[15] Del Giudice E et al. Physical Review Letters 1988; 61: 1085
[16] Del Giudice E. Is the memory of water a physical impossibility? In [20]
[17] Smith CW, Salford. Personal communication.
[18] Berezin AA. Ultra high dilution effect and isotopic self-organisation. In [20]
[19] Schulte J. Conservation of structure in aqueous ultra high dilutions. In [20]
[20] Endler PC and Schulte J (eds).Ultra high dilution. Physiology and physics. Dordrecht: Kluwer 1994

STIMULATION OF CELLULAR SELF-RECOVERY BY APPLICATION OF THE SIMILIA-PRINCIPLE

R. VAN WIJK and F.A.C. WIEGANT
Department of Molecular Cell Biology, Utrecht University
P.O Box 80056, 3508 TB Utrecht, The Netherlands

ABSTRACT

The aim of the present experiments is to study the specificity of the similia-principle. Previously it has been shown that the recovery of cells that were damaged by a specific stressor, is stimulated by the application of a low dose of the same stressor. The process of recovery is evaluated in terms of enhanced synthesis of heat shock proteins, which are involved in repair-processes, and the development of tolerance. With respect to the specificity of the similia-principle, our working hypothesis is that the most effective low-dose stimulation of heat shock protein induction in damaged cells is obtained with the same stressor or a stressor which has an identical pattern of induction of the different stress proteins. With less similarity between the damaged condition and the low-dose stimulating condition, the low-dose stimulation will be less or even absent. The present paper deals with the relationship between degree of similarity and efficiency of low dose stimulation. To this end, arsenite, heat shock and cadmium are used as damaging conditions. The same three conditions, at a low dose, are applied afterwards. The results confirm the aspect of specificity of the similia principle, which can be verified further by extending the number of damaging conditions.

1. Introduction

The stimulation of self-recovery by applying the similia-principle is considered to be the essence of homeopathy. The similia-principle indicates that a substance which causes certain symptoms in a healthy organism, may be applied in a curative sense to a diseased organism with identical symptoms.

Previously, it has been shown that the recovery of cells that were damaged by a certain stressor, are stimulated by the application of a low dose of the same stressor. In this respect, the stimulatory effect of low doses could be observed after a heat shock (Van Wijk et alp 1994) as well as after a treatment with arsenite (Ovelgönne et alp 1995a).

As parameters for stimulation of self-recovery we used specific molecular processes underlying recovery. In this respect a particular set of proteins, indicated as heat shock proteins (hsps) or stress proteins are supposed to play a crucial role (Welch 1992; Parsell and Lindquist 1994). This set of proteins includes different protein families with approximate molecular masses (in kD) of 28 (hsp28), 60-65 (hsp60), 68-73 (hsp70), 84-90 (hsp84) and 100-110 (hsp100). Each hsp induced after the treatment with a stressor is supposed to have a characteristic function with respect to the recovery of specific cellular damage induced by this stressor.

Although not all the effects of the various stressors have been identified at the molecular level, it can be assumed that cellular damage in general requires a differential activity of hsps and that specific cellular damage requires the presence of specific hsps.

In this paper we focus on the specificity of low doses of stressors with respect to the enhancement of hsp-induction.

2. Hypothesis

With respect to the specificity of the similia-principle, our working hypothesis is that the most effective low-dose stimulation of hsp-induction in damaged cells is obtained with the same stressor or a stressor which has an identical pattern of induction of the different stress proteins. With less similarity the low-dose stimulation of hsp-induction will be less or even absent. If for instance arsenite is applied as the disturbing agent, it can be hypothesized that stimulation of sub-optimal hsp-induction will only occur by the application of a small dose of arsenite and not, for instance, by a small dose of cadmium, unless cadmium, applied as the disturbing agent, shows the induction of a similar pattern of hsps.

The present study deals with the relationship between degree of similarity and efficiency of low dose stimulation. To this end, arsenite, heat shock and cadmium are used as damaging conditions, whereas the same three conditions at a low dose are applied afterwards.

3. Stressor-specificity with respect to induction of gene expression

In our study arsenite, heat shock and cadmium were used in order to identify a possible stressor-specific induction in the pattern of heat shock proteins in Reuber H35 rat hepatoma cells. If a non-specific expression is observed, i.e. when all stressors induce the same set of hsps to an identical degree, then it would be *im*possible to study the specificity of the similia-principle.

3.1. Stressor-specific induction of hsps

A reliable method to compare the effects stressors have on the stimulation of hsp synthesis, is to establish their concentration dependency and then to compare the hsp patterns which were induced in these concentration ranges. To this end, Reuber H35 rat hepatoma cells were exposed to different concentrations of the various stressors. For a careful comparison of these stressor conditions, the stressors were all applied in the same experiment in order to exclude slight differences in culturing conditions.

The effect on the induction of hsps was studied as well as the effect of these stressors on the inhibition and recovery of protein synthesis and on cell survival

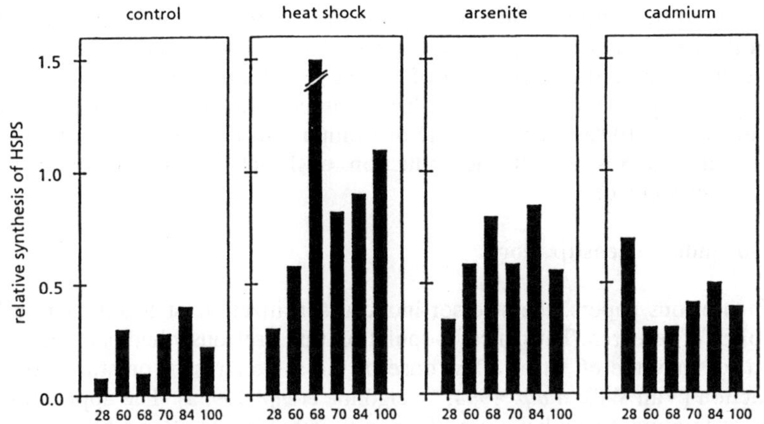

Figure 1 - The effect of three different stressors (heat shock at 42°C for 60 minutes, 200 μM arsenite for 2 hours and 100 μM cadmium for 2 hours) on the synthesis of the major hsps is shown (a-hsp28; b-hsp60; c-hsp68; d-hsp70; e-hsp84; f-hspl00). In the control situation several hsps are synthesized at a low level. Synthesis of the major stress proteins is expressed relative to the synthesis of actin.

(Wiegant et alp 1994). A comparison of the stressor-specific induction of the major hsps was made under conditions of identical impact on cellular physiology; i.e. those stressor conditions were chosen that resulted in approximately equal percentages of cell death (iso-survival dose: 40-50%). These iso-survival doses were: 1 hour at 42°C; 2 hour of 200 μM arsenite and 2 hour of 100 μM cadmium. The pattern of hsp synthesis of stressed cells consists of the hsp28, hsp60, hsp68, hsp70, hsp84 and hsp100 (figure 1). Cells exposed to cadmium show the highest relative synthesis of hsp28 if compared with cell cultures treated with heat shocks or arsenite. No induction of hsp60 is observed after cadmium treatment, whereas a large induction of hsp60 is observed after heat shocks and arsenite. Hsp68 and hsp70 are both induced by heat shocks as well as arsenite and to a lesser extent by cadmium. More or less the same pattern is observed with respect to hsp84 and hsp100. (Wiegant et alp 1994 for a more detailed analysis). Based on these results it was concluded that a stressor-specific induction of the hsps has been demonstrated.

It can be proposed that the differential induction of the group of hsps induced by a specific stressor is attuned to the need of the cell and that it can be taken as an indicator for the type of cellular damage which is under repair.

3.2. Stressor-specific induction of hsp mRNA

In order to determine whether the stressor-specific enhanced expression of HSP-genes is caused by a differential gene expression, experiments were performed in which both the levels of specific hsps and of hsp mRNAs were determined in parallel cultures after the application of a heat shock, arsenite or cadmium. The stressor-specificity of the synthesis of hsps was confirmed by hsp MRNA data (Ovelgönne *et alp.* 1995b). The inability of cadmium to induce hsp60 appeared to correlate with an absence of the induction of hsp60 mRNA at the various concentrations studied.

4. Stressor-induced sensitization

In previous papers, the stressor-induced sensitization of hsp-induction has been subject of research. Thus, a cell population that previously had been damaged was used to study the effect of a low dose of the same stressor on stimulation of hsp-induction (Van Wijk *et alp.* 1994; Ovelgönne *et alp.* 1995a). The application of a mild heat treatment or a low dose of arsenite immediately after a heat shock or an intoxication with arsenite respectively, stimulated the synthesis of hsps as well as the development of tolerance. Cells were highly sensitized; i.e. they responded to relatively low doses, whereas these low doses of stressors were without any effect in control cells during the time period studied.

5. Cross-sensitization; experimental approach

We argued that with a study of the stressor-induced sensitization of hsp-induction, the similia-principle can be studied in its most elementary form. In such a study it became of interest to determine to what extent hsp-induction in sensitized cells can also be stimulated by low doses of other (non)-related stressors. Are sensitized cells also stimulated by low doses of non-related stressors?

In this respect it was hypothesized that the specificity of the similia-principle reveals itself in the fact that for the stimulation of hsps by low doses identical or strongly similar conditions must be applied. If condition 'A' is applied as the disturbing agent, the stimulation of a suboptimal hsp-induction will only result from the application of a small dose of condition 'A' and not by a small dose of condition 'B' or 'C', unless B or C has a similar effect on the cellular physiology as condition 'A' has.

If, however, the stimulation of self-recovery by a low dose is an a-specific phenomenon, then a low dose of stressor 'B' or 'C' can also be expected to be able to stimulate the hsp-induction. Such a phenomenon, indicated as cross-sensitization, has been studied with Reuber H35 cells exposed to heat shock, arsenite or cadmium.

A number of experiments were performed in which a pretreatment was given with either a heat shock (30 minutes at 41.5°C), sodium arsenite (100 μM for 1 hour) or cadmium chloride (100 μM for 1 hour). After the pretreatment the

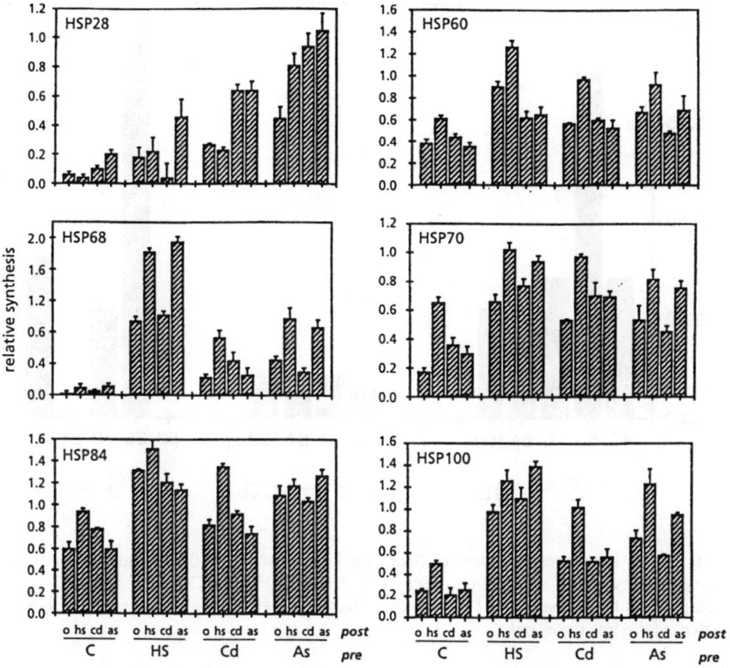

Figure 2 - Cross-sensitization. The effect of a low doses of a stressor on the synthesis of the major hsps, which was applied immediately following a pretreatment with a high doses of either a heat shock (HS: 42°C for 30 minutes), cadmium (Cd: 100 μM for 1 hour) or arsenite (As: 100 μM for 1 hour). The incorporation of [^{35}S]-labelled amino acids into the major hsps over an 8-hour period was expressed relative to the total incorporation into cellular protein. The results for each hsp are grouped per effect of the high-doses stressor. Of each group of 4 histograms, the first bar shows the induced synthesis of the indicated hsp in control conditions (0) or by a pretreatment with the high-doses stressor (HS, As or Cd), whereas the other three show the effect on the synthesis of hsps of the post-incubation in a low doses of heat shock (hs: 8 hours in 40.5°C), cadmium (cd: 8 hours 10 μM) or arsenite (as: 8 hours 10 μM). Results are the mean of three experiments.

cultures were exposed to low, mostly subliminal conditions of heat (40°C or 40.5°C), arsenite (3 or 10 μM), cadmium (3 or 10 μM) or by an incubation at control conditions (37°C) for a period of up to 8 hours. In this period the stimulating effect on the synthesis of various hsps was studied. The results are shown in figure 2. This rather complicated pattern requires further analysis.

6. Analysis of the specificity of low dose stimulation

The first step in the analysis of the results of these experiments was to determine the extent of low dose stimulation by the stressor that previously had sensitized the cell.

- A small increase in temperature after a more severe heat shock enhances the

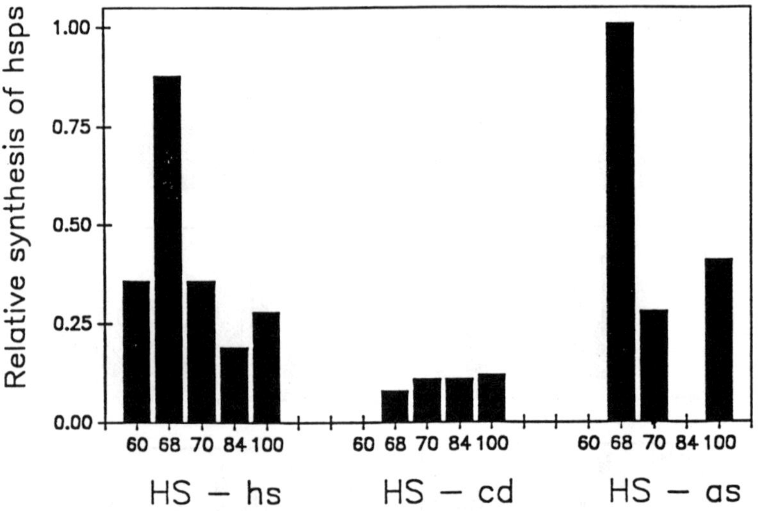

Figure 3 - The effect of a low doses of a stressor (heat shock at 40.5° (hs); 10 μM cadmium (cd); or 10 μM arsenite (as)), immediately applied after a high doses of heat shock (HS: 42°C for 30 minutes), on the extra synthesis of the major hsps over a period of 8 hours. The effect of the HS has been subtracted in order to show the stimulatory effect of the low doses only.

induced synthesis of all studied hsps, but mainly hsps 60, 68, 70 and 100 and to a lesser extent hsp84 (figure 2).

- Of the hsps induced by a treatment of arsenite, mainly the subgroup of hsps, consisting of hsp28, hsp68, hsp70 and hsp100, was further stimulated by arsenite at a subliminal concentration (figure 2).
- Of the hsps induced by a pretreatment with cadmium, a subsequently added subliminal concentration of cadmium, stimulated the synthesis of a subgroup of hsps which consisted of hsp28, hsp68 and hsp70 (figure 2). These subgroups of hsps thus seem to be indicative for a particular role in the stimulation of self-recovery.

As an example of further analysis, this extra synthesis of these hsps, induced by low dose stimulation after heat shock, is also presented in figure 3 (left panel).

The second step in the analysis involved the stimulation of synthesis of the selected hsps by subliminal concentrations of other stressors than the ones used for the pretreatment. The results of this comparison is shown for heat-shock treated cells (figure 3).

If heat shock was used as the disturbing pretreatment, hardly any extra stimulation of hsp-synthesis is observed upon the subsequent application of a low dose of cadmium (figure 2 and figure 3 -middle panel-). In contrast when a low dose of arsenite is applied to cells which were previously heat shocked, a large extra synthesis of hsp68, 70 and 100 occurs. Interestingly no extra enhancement at all of hsp60 and 84 is observed (figure 2 and figure 3 -right panel-). This demonstrates a rather specific effect of low doses of the used stressors on heat-treated cell cultures.

Specificity of low dose stimulation was also derived from experiments with arsenite- and cadmium-preterated cells.
- If arsenite was the first disturbing agent, the synthesis of hsp28, hsp68, hsp70 and hsp100 was also stimulated when a mild heat treatment was applied. Application of a subliminal cadmium concentration only stimulated hsp28, whereas no effect of cadmium was found on the synthesis of hsp 68, hsp70 and hsp100 (figure 2).
- If cadmium was used as the disturbing agent for the pretreatment, stimulation of hsp28, hsp68, hsp70 occurred after a successive application of a subliminal cadmium concentration. The synthesis of these hsp protein species was also stimulated by a very mild heat shock when applied as a second treatment. The application of a subliminal arsenite concentration stimulated only hsp28 and hsp70.

This analysis of the experiments on cross-sensitization is very elementary and takes no account of the quantitative differences between the stimulatory effect of small doses of the various stressors. Nevertheless, from the data it can be observed that even this elementary analysis demonstrates a stressor-specificity in the synergistic action between a pretreatment and a subliminal condition on hsp-induction. A schematic presentation of the cellular capacity to enhance synthesis of specific hsps upon low dose stimulation is given in Table 1.

Table 1: Cross-sensitization

	First stressor		
	Heat shock	Arsenite	Cadmium
Second stressor			
Heat shock	++++	++++	++++
Arsenite	+++	++++	++
Cadmium	-	+	++++

The range -, +, ++, +++, ++++ is indicative for the degree of extra synthesis of hsps by a low dose of the second stressor.

7. Relationship between stressors; evidence for the similia-principle

Based on a visual comparison of the hsp-induction pattern of the three stressors, it was concluded that the heat shock- and the arsenite pattern resemble each other more than the heat shock- and the cadmium-pattern. As a quantitative measure for similarity between the inductions of the various hsps after exposure to the stressors, we calculated the ratio of the different induction values and expressed it as the percentage similarity for each hsp. Subsequently, the percentage of similarity between stressor A and B was calculated as the mean of the values found for the individual hsps. Based on this calculation, the hsp-inducibility by heat shock and by arsenite show a high degree of similarity (77.2%), while the degree of similarity between arsenite and cadmium (64.0%) or between heat shock and cadmium (53.8%) were less. This higher degree of similarity is now also reflected in a higher degree of extra synthesis of hsps upon low dose stimulation and thus confirms the aspect of specificity of the similia-principle.

8. Discussion

The aim of our research program was to study the process of self-recovery and its stimulation by low doses of compounds applied to diseased biological systems according to the similia-principle. For the study of the similia-principle it is for the researcher of importance to be able to concentrate on the most elementary level where changes manifests themselves. Therefore, we have proposed the following question in our research program:

> Which molecular processes and interactions are of importance in the process of cellular self-recovery and which molecular parameters should be used to evaluate (stimulation of) cellular self-recovery?

In our study we have focused on the synthesis of hsps as a measure of cellular self-recovery. This choice is based on our own experimental results (Van Wijk and Wiegant 1994) as well as on data from the literature (for recent review see: Nover 1991; Welch 1992; Parsell and Lindquist 1994). The pattern of synthesis of the individual hsps can be considered as the pattern of symptoms specifically linked to the recovery process of the cell after damage.

With respect to the substantiation of the similia-principle we have provided evidence in which the specificity of the similia-principle is substantiated. Further studies applying detailed quantitative analysis of hsp patterns will be presented elsewhere.

Finally, for the interest of homeopathy it is of importance to ask the following question:

9. Do studies on the similia-principle at the cellular level have any significance for clinical practice?

It has been argued that the similia-principle has been formulated in homeopathy for the organismal level. There is, of course, a clear difference between research on the similia-principle as described here and the clinical practice. The first difference is the systemic level or level of organization which is studied or treated (cells versus organisms). Secondly, in our experiments cells are brought into a pathological condition, whereas patients who come to a doctor already are in this condition. Thirdly, cells are grown in cultures and treated as a group. Their behaviour and response with respect to the process of self-recovery is evaluated as the mean of the total population. There is of course a variability in self-recovery. Thus, not only cells showing a fast and spontaneous recovery, but also cells showing a slow and hampered self-recovery are evaluated. In contrast to research with patients, a single case approach or a time-series design for studies on the behaviour of individual cells is not possible or is extremely complex.

However, we expect that the assumption of the similia-principle as being a general biological phenomenon will gain weight by studies aimed at the analysis of regulatory processes at the cellular level underlying this principle. In our view, the demonstration of the stimulation of self-recovery by compounds applied according to the similia-principle, will substantiate the fundamental basis of homeopathy and will improve the methodology of research on the similia-principle in bio-medicine.

Acknowledgement
The financial support of the HomInt organisation is gratefully acknowledged.

10. References

Nover L. (1991). *Heat shock response*. CRC Press, Boca Raton.

Ovelgönne J.H., Souren J.E.M., Van Rijn J., Van Wijk R. and Wiegant F.A.C. (1995a). Enhancement of the stress response by low concentrations of arsenite in arsenite-pretreated H35 hepatoma cells. *Toxicol.Appl.Pharmacol.* **132:** 146-155.

Ovelgönne J.H., Souren J.E.M., Wiegant F.A.C. and Van Wijk R. (1995b). Relationship between cadmium-induced cell killing, inhibition of total protein synthesis and increased expression of heat shock genes. *Toxicology* **99:** 19-30.

Parsell D.A. and Lindquist S. (1994). Heat shock proteins and stress tolerance. In: *The biology of heat shock proteins and molecular chaperones*, edited by Morimoto R.I., Tissières A. and Georgopoulos C. CSHL Press, New York. pp.457-494.

Van Wijk R., Ovelgönne J.H., de Koning E., Jaarsveld K., Van Rijn J. and Wiegant F.A.C. (1994). Mild step-down heating causes increased transcription levels of hsp68 and hsp84 mRNA and enhances thermotolerance development in Reuber H35 hepatoma cells. *Int.J.Hyperthermia* **10**: 115-125.

Van Wijk R. and Wiegant F.A.C. (1994). *Cultured mammalian cells in homeopathy research; the similia principle in self-recovery*. Utrecht University, Utrecht.

Welch W.J. (1992). Mammalian stress response: cell physiology, structure/function of stress proteins, and implications for medicine and disease. *Physiol.Rev.* **72**: 1063-1081.

Wiegant F.A.C., Souren J.E.M., Van Rijn J. and Van Wijk R. (1994). Stressor-specific induction of heat shock proteins in rat hepatoma cells. *Toxicology* **94**: 143-159.

SPONTANEOUS VARIABILITY AND INDUCED INCREASE OF OLFACTORY SENSITIVITY TO DILUTED ODORANTS

THOMAS RADIL
Institute of Physiology, Academy of Sciences, Prague
Monell Chemical Senses Center, Philadelphia

and

C.J. WYSOCKI
Monell Chemical Senses Center, Philadelphia

ABSTRACT

The remarkably high variability of olfactory psychophysical thresholds could be interpreted as a transitory ability to detect, under certain, still undefined conditions, highly diluted odorants. One of the factors contributing to this phenomenon might be represented by central modulation of detection thresholds.The present experiments deal with a particular case of tuning the threshold toward low concentrations of an odorant (amyl acetate, resembling banana smell). It is characterized by a decrease by about one or two orders of magnitude of ascending limit thresholds (estimated by gradually increasing stimulus concentrations from subthreshold values in a sequence of trials) in comparison with descending trials (estimated in the opposite way). Two factors seem to be relevant:unidirectional stimulus ordering from sub-threshold toward supra-threshold values, probably activating physiological processes underlying stimulus expectation;and the lack of additional references concerning stimulus intensity (other than the ordered sequence of preceding ascending concentrations).

1. Spontaneous olfactory threshold variability

Among normal individuals olfactory threshold measures exhibit considerable variation and fluctuate from moment to moment. The difference typically amounts several orders of magnitude (1, 2, 3, 4, 5, 6). Repeated threshold measurments over time might be as variable for a given subject as are single threshold measurments for a group of subjects (5). Very little is still known about the physiological significance of factors underlying this variability in olfactory thresholds. Nevertheless, these differences might be interpreted as a real, albeit transitory, ability of unknown nature to perceive , under crtain conditions, highly diluted odorants.

The basic condition of olfactory perception is undoubtely the detection of odors by membrane receptors on the surface of olfactory neurons, a highly sensitive and selective process underlying sensory thresholds for different odorants. It holds that „Nihil est in intellectu quod non fuerit in sensu". Perceptual thresholds for detecting and recognizing odors, requiring at least partial participation of conscious processes, seem to be, however, considerably higher than actual sensory events. Thus there seems to exist a relatively broad range between both types of thresholds within which "spontaneous" and perhaps regulated shifts of olfactory sensitivity could take place (Fig. 1).

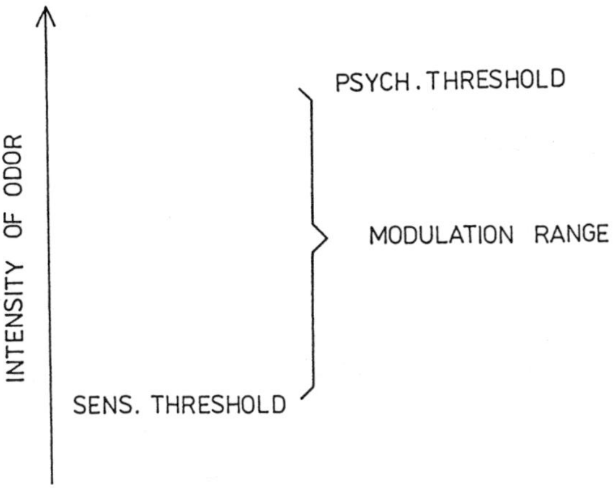

Fig. 1. Schematic representation of the sensory and psychic detection threshold, and of the modulation range within which subjective olfactory sensitivity might vary.

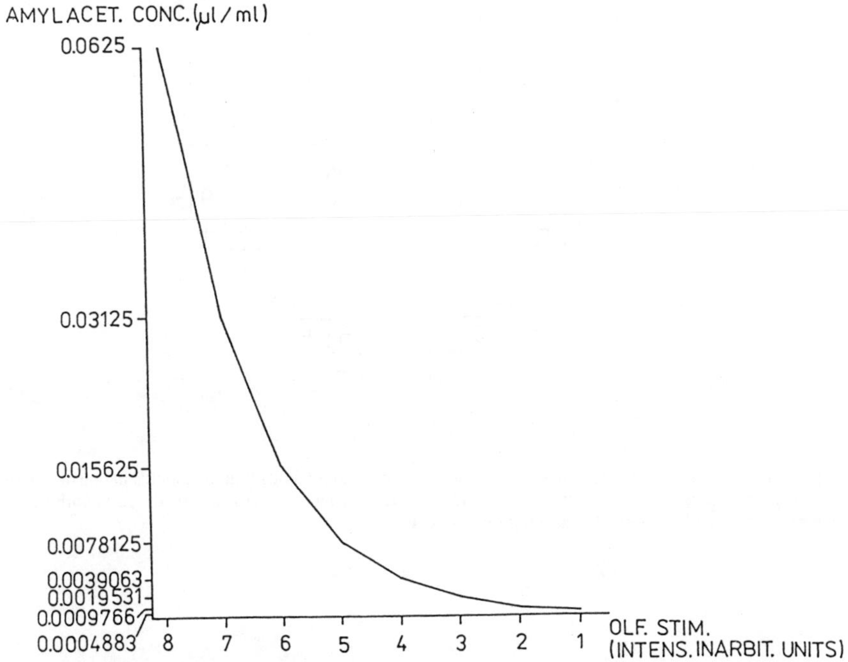

Fig. 2. Vertical axis -- microliters of amyl acetate per one milliliter of odorless solvent (light, white, mineral oil). Horizontal axis -- concentration expressed in arbitrary units.

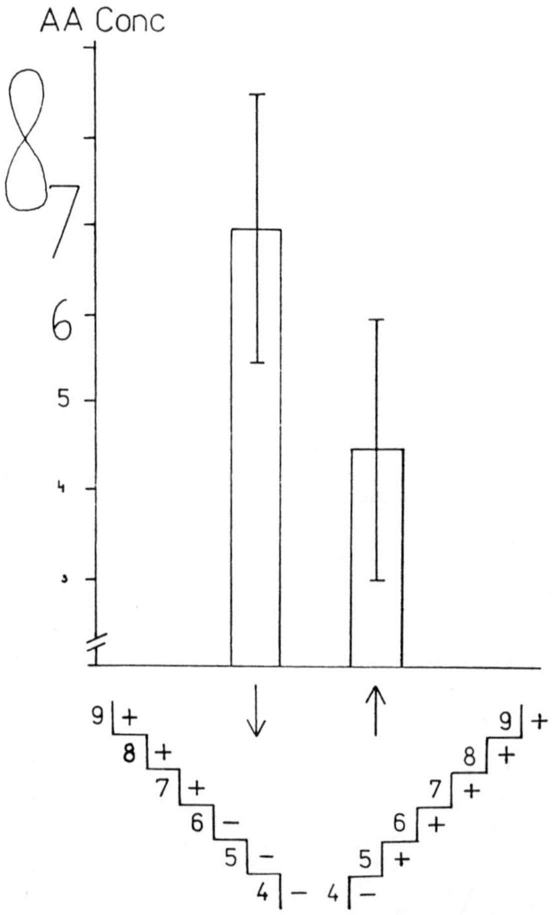

Fig. 3. Average difference between descending (left) and ascending (right) limit thresholds in one of the groups of experimental subjects (n=10, average values with standard deviation). Vertical axis; the size of arbitrary numbers corresponds to odorant concentration. Fort details see text.

2. Threshold tuning

One of the factors contributing to the extreme "spontaneous" variability of sensitivity to odorants might be linked to central modulation of detection thresholds. Our present experiments have focused on a particular case of tuning psychophysical detection thresholds toward highly diluted stimuli, based mostly upon the difference between the thresholds obtained via the ascending limit method (estimated by gradually increasing stimulus concentrations from sub-threshold values in a sequence of trials) and those obtained by a descending limit (estimated in the opposite way) (7, 8).

Physiological changes appearing at the psychological level as shifts of observers' judgment criteria (9) linked to adjustments of thresholds and of subjective scaling of stimulus intensities, also occur in taste perception (1), and might be a phenomenon of general significance (11, 12).

Specific experiments revealed that the difference in ascending and descending thresholds for amyl acetate (having a banana-like smell) in the present experiments was about one to two orders of magnitude in individual cases (Figs. 2, 3). Furthermore, other results demonstrate that this difference is not caused by olfactory adaptation (13) induced by successive administration of high stimulus concentrations during descending limit threshold measurements. The results of other experiments pointed to the significance of two factors in increasing detection sensitivity during ascending threshold measurements. The first one, delineated by means of experiments manipulating the sequence of stimulation trials, was appropriate unidirectional stimulus ordering from sub-threshold to supra-threshold levels, apparently triggering neurophysiological processes underlying stimulus expectation. The second factor was represented by the lack of additional references concerning stimulus intensity (other than that of the ordered sequence in the ascending series).

The results of these experiments suggest that as individuals are exposed to an olfactory gradient, e.g., approaching the source of an odorant, their ability to detect the odorant should be maximal, reflecting, perhaps, a physiological mechanism of increasing sensitivity of olfaction to highly diluted odorants.

3. References

[1] J. E. Ammore, D. Venstrom and A. R. Davis, "Measurement of specific anosmia", Perceptual Motor Skills, vol. 26, pp.143-164, 1968.

[2] K. S. Brown, C. M. Maclean and R. R. Robinette, "The distribution of the sensitivity

to chemical odors in man", Hum. Biol., vol. 40, pp. 456-472, 1968.

[3] P. H. Punter, "Measurements of human olfactory thresholds for several groups of structurally related compounds", Chemical Senses, vol. 7, pp. 215-235, 1968.

[4] R. L. Doty, T. Gregor and R. G. Settle, "Influences of intertrial interval and sniff bottle volume on the phenyl ethyl alcohol olfactory threshold", Chemical Senses, vol. 11, pp. 259-264, 1986.

[5] J. C. Stevens, W. S. Cain and R. J. Burke, "Variability of olfactory thresholds", Chemical Senses, vol. 13, pp 643-653, 1988.

[6] R. L. Doty, "Psychophysical measurement of odor perception in humans", in D. G. Laing, R. L. Doty and W. Breipohl (Eds.) *A Human Sense of Smell*, pp. 93-134, Springer Berlin 1992.

[7] R. M. Pangborn, H. W. Berg, E. B. Roesler, and A. D. Webb, "Influence of methodology on olfactory response", Perceptual and Motor Skills, vol. 91-103, 1964.

[8] W. S. Cain, J. E. Cometto-Muniz and R. A. De Wijk, "Techniques in the quantitative study of human olfaction", in *Science of Olfaction*, M. J. Serby and K. L. Chobor (Eds.) pp. 279-308, 1992.

[9] D. M. Green and J. A. Swets, *Signal Detection Theory and Psychophysics*, New York, Wiley 1966.

[10] R. M. Mattes and H. T. Lawless, "An adjustment error in optimization of taste intensity", Appetite, vol. 6, pp. 103-114, 1985.

[11] J. G. Beebe-Center, *The Psychology of Pleasantness and unpleasantness*, Russell and Russell, New York, 1965.

[12] R. G. Christman, *Sensory Experience* (2nd Edition) pp. 46-59, Harper and Row 1979.

[13] E. P. Köster, "Human psychophysics in olfaction", in D. G. Moulton, A. Turk and J. W. Johnston (Eds.) *Methods in Olfactory Research*, pp. 343-374 Academic, New York, 1975.

Supported, in part, by NIH grants TW04839, DC00298 and DC00214

THE STATE OF ENHANCED SENSITIVITY AS OPTIMAL CONDITION TO STUDY LOW DOSE EFFECTS

F.A.C. WIEGANT and R. VAN WIJK
Department of Molecular Cell Biology, Utrecht University
P.O Box 80056, 3508 TB Utrecht, The Netherlands

Introduction

The essence of homeopathy consists of a stimulation of self-recovery by application of remedies according to the similia-principle. The aim of our research is to contribute to the understanding of the regulatory processes which underly self-recovery and to explain in what way low doses of compounds, which were selected according to the similia-principle, are able to stimulate these processes. In general, (self-)recovery processes are initiated when a living system is disturbed or when a diseased state is induced. Directly following the application of a high dose of a stressor, biological systems show a transient state of enhanced sensitivity followed by the development of tolerance (reduced sensitivity). The transient period of enhanced sensitivity, which reflects the disturbed or diseased state of the living system, is one of the essential characteristics of our model. It offers the opportunity to study the stimulation of self-recovery processes by low doses of a stressor, which under normal conditions do not have any effect on cellular physiology. For our studies we used cell cultures as model system. The advantage of cell cultures is the possibility of high standardisation. In this paper the occurrence, origin and duration of the state of enhanced sensitivity is reviewed. Furthermore, it will be shown that the initial degree of sensitization is related to the degree to which self-recovery, as expressed by a tolerant state, is developed later on. The stimulatory effect of low doses on self-recovery processes is described elsewhere in this volume.

The state of enhanced sensitivity; its occurrence

The change in sensitivity of cells after a heat shock has been discussed extensively in the literature (for reviews: Hahn and Li, 1990, Nover, 1991). An exposure to a heat shock may either result in an increase or in a decrease in sensitivity for a second heat treatment (i.e. sensitization or tolerance respectively).

- Sensitization is commonly observed by exposing cells to an initial heat treatment, immediately followed by a second treatment at a lower hyperthermic temperature. A sensitizing effect has been observed in all *in vitro* and *in vivo* studies.
- The development of tolerance is observed when a fractionated treatment protocol is used. In this protocol, an initial exposure to heat is followed by an incubation at a normal temperature (37°C) and subsequently by an incubation at the same or a higher temperature, in order to determine the degree of tolerance. More cells survive the higher temperature when pretreated in comparison with non-pretreated cells. Examination of the kinetics of induction and decay of tolerance showed that an

interval of about 4-8 hours at normal temperature was required before the maximum thermotolerance manifested itself. The tolerance persists for at least 24 hours.
An exposure to damaging conditions other than heat shock also result in (bi-phasic) alterations in the sensitivity to a second exposure with the damaging condition. In this respect, we studied the induction of sensitization as well as of tolerance by a treatment with sodium arsenite. For this purpose both the step-down protocol and the fractionated treatment protocol were used. As in heat shock, arsenite damages cellular proteins, which triggers the induction of stress proteins. These proteins are involved in repair processes (Van Wijk and Wiegant, 1994).
A short incubation of the cell culture with sublethal concentrations of arsenite (100 μM for 1 hour) induces a transient increase in sensitivity to a second arsenite treatment with low doses of arsenite (1 and 3 μM). When these low concentrations, which are without effect in control conditions, are applied immediately after the first treatment, an increase in sensitivity is observed as reflected by a decrease in cell survival. This enhanced sensitivity is transient, since it is followed by an enhanced tolerance towards arsenite later on in time (Wiegant *et al.,* 1993).

The state of enhanced sensitivity; its origin
The cause of the change in sensitivity which occurs after the application of harmful circumstances to cells is related to the pool of free stress proteins.
Between the time the damage induced by a stress condition has consumed most of the free stress proteins and the time that the subsequent production of new stress proteins has taken place, the cell is in a period of increased sensitivity for damage. When the increased production of newly produced stress proteins has replenished the free pool, a reduced sensitivity (i.e. a tolerance) to the damaging circumstances develops (Van Wijk and Wiegant, 1994).

The state of enhanced sensitivity; its duration
The conditions which determine the duration of enhanced sensitivity are: 1) the nutritional condition, 2) the naturally occurring individual variability, and 3) the dose of the applied damaging principle.

ad 1. *Nutritional status.* The response of cells to environmental insults depends essentially on the ability to mobilize energetic compounds. In order to maintain an adequate energy level, metabolism and protein synthesis, optimal nutrition is required. When the whole cell population is placed under restricting nutritional conditions, self-recovery processes are hampered or they develop in a suboptimal way, because of a diminished capacity or ability to react to harmful conditions. This may even lead to chronically disturbed self-recovery processes. The importance of the nutritional status for stimulation of stress proteins and development of tolerance has been demonstrated (Van Rijn *et al.,* 1992). When potentially reversible damage remains present beyond a certain time limit, the probability of cell death increases.

ad 2. *Individual variability.* Although a population of genetically identical cells, cultured under similar conditions, is supposed to react in an identical way to the application of a similar damaging circumstance, individual cells appear to show differences in the self-recovery response. The variability is explained by the occurrence of variability in the amount of the stress proteins at the time of stressor application, and/or in the activity of other regulatory elements that are essentially needed to activate the synthesis of these proteins. With a large amount of free accessible stress proteins, damage can be directly reversed without the need to initiate their synthesis. With respect to the aim of our study we are especially interested in the slowly recovering subpopulation and whether the recovery of this subgroup can be stimulated by the application of low stressor doses. An experimental limitation of our model, however, is that this stimulating effect is hardly observed in the mean of the total population.

ad 3. *Stressor dose.* Both the time and the temperature of an initial heat treatment determine the state of enhanced sensitivity (Van Wijk and Wiegant 1994). In this respect, a lower initial heat treatment reduces both the degree of thermo-sensitization being developed as well as the period during which thermosensitization is present. Cells thus showed a faster development of tolerance and consequently a shorter period of enhanced sensitivity after 15 minutes at 42.5°C, if compared with a treatment of 30 minutes at the same temperature.

With respect to other stressors than heat, and in order to determine the duration of the sensitization of cells the effect of a fractionated exposure of cells to arsenite was studied. Cell cultures were pre-treated with various arsenite concentrations. At different times after this first treatment, the arsenite sensitivity was determined and expressed as the LD50 value (figure 1). As can be observed in this figure, immediately following the first treatment, a lower concentration of arsenite is required to kill 50% of the remaining cell population. The maximal sensitization was observed at 1 hour after the pretreatment, whereas sensitization disappeared 3 hours after pretreatment. At that time, the LD50-value was more or less similar to the value found in non-pre-treated cells. Six hours after an arsenite treatment, cells are significantly more tolerant. These data demonstrate the development of arsenite induced arsenite tolerance after a period of enhanced sensitivity (Wiegant *et al.*, 1993). The effect of 300 μM arsenite resulted in a longer period and a higher degree of sensitivity as compared with the effect of 100 μM arsenite. In summary, the period of enhanced sensitivity depends on the severity of the initial stress condition.

Is the degree of sensitization related to the level of tolerance developed later on?

As shown in figure 1, it can be observed that the strength of the initial arsenite treatment not only determines the sensitivity of cells to a subsequent arsenite treatment, but also the extent to which tolerance is able to develop in the remaining cell population (Wiegant *et al.*, 1993, Van Wijk and Wiegant, 1994). Apparently, the degree of tolerance is related to the degree of the initial sensitization.

Stimulation of self-recovery is related to the occurrence of enhanced sensitivity

Finally, we analyzed whether the self-recovery processes in cell cultures, which were

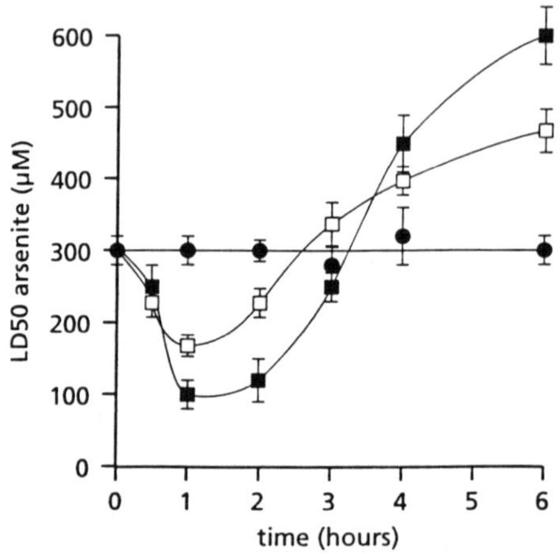

Figure 1: The concentration of sodium arsenite required to kill 50% of the cell population (LD50) at various times after pretreatment during 1 hour with 0 (●), 100 (□) or 300 μM (■) sodium arsenite. After pretreatment, cells were rinsed three times and incubated in arsenite-free medium for the indicated time (0-6 hours). Then, the medium was replaced by serum-containig medium with various arsenite concentrations (0, 30, 50, 100, 175, 300 or 600 μM) for 1 hour. After rinsing, cells were further incubated for colony formation. In order to determine the LD50-value, linear regression analysis was used.

either or not in their transient state of enhanced sensitivity, could be stimulated by low doses of stressors. It has been shown that these low doses, which in control conditions are without effect, are effective when applied immediately following the high dose stressor treatment; i.e. when cell cultures are in their period of enhanced sensitivity (Van Wijk and Wiegant, 1994). However, low dose stimulation of self-recovery processes did not take place when most cells of the cell culture were not anymore in the period of enhanced sensitivity (Ovelgönne *et al.,* 1995).

References

Hahn G.M. and Li G.C. (1990). Thermotolerance, thermoresistance and thermosensitization. In: *Stress proteins in biology and medicine*, edited by Morimoto R.I., Tissieres A. and Georgopoulos C. Cold Spring Harbor Laboratory Press, pp.79-100.

Nover L. (1991). *Heat shock response*. CRC Press, Boca Raton.

Ovelgönne H.H., Wiegant F.A.C., Souren J.E.M., Van Rijn, H. and Van Wijk, R. (1995). Enhancement of the stress response by low concentrations of arsenite in arsenite-pretreated Reuber H35 hepatoma cells. *Toxicol.Appl.Pharmacol.* **132**: 146-155.

Van Rijn J., Van den Berg J., Van Aken J.M. and Van Wijk R. (1992). Induction of thermotolerance and heat shock protein synthesis during nutritional deprivation. *Int.J.Hyperthermia* **8**: 377-394.

Van Wijk R. and Wiegant F.A.C. (1994). *Cultured mammalian cells in homeopathy research; the similia-principle in self-recovery.* Utrecht University, Utrecht, The Netherlands.

Wiegant F.A.C., Souren J.E.M., Van Rijn J. and Van Wijk R. (1993). Arsenite-induced sensitization and self-tolerance of Reuber H35 hepatoma cells. *Cell Biol.Toxicol.* **9**: 49-59.

CLINICAL EVIDENCES

HIGH DILUTIONS AS A TOOL OF IMMUNOMODULATION

MADELEINE BASTIDE
Laboratory of Immunology, Faculty of Pharmacy,
University of Montpellier 1, France

and

FREDERIC BOUDARD
Laboratory of Immunology, Faculty of Pharmacy,
University of Montpellier 1, France

ABSTRACT

The novel concept of immunomodulation proposed herein does not involve a new molecule or extract but rather a new approach to the use of conventional substances to modulate the immune response. The families of molecules we studied have different kinds of activity: thymic hormones which induce T lymphocyte maturation, bursin which participates in B lymphocyte maturation and cytokines which take part in cellular interaction. Our investigations were carried out with thymulin and natural mouse ap interferon or bursin, we sometimes tested other thymic hormones or interleukin 2. To transmit this information, we studied the effect of a series of highly diluted test substances according to the homeopathic procedure. Our results indicate that serial dilutions present immunomodulatory activity. Furthermore, the activity is specific and the higher the dilution, the greater the capacity of the test substance to transmit its information to the target cell or organism as shown in bursectomized chickens receiving highly diluted bursın. We suggest that the regulation of biological systems involves the notion of "information" rather than the quantity of molecules it receives. Therefore, it is possible that besides the regulatory mechanisms in living organisms reacting in response to hormones or cytokines and obeying the laws of classical pharmacokinetics, other mechanisms reacting in response to "information" might exist. These regulatory effects could be expressed at another level than those of the classical effector mechanisms of the cells.

1. Introduction

Immunopharmacologists are studying molecules able to modulate the immune system in the context of modern pathologies including i.mmunodepressive syndromes.

These substances are intended either to improve the anti-infectious and anti tumoral defence mechanisms or to prevent or treat autoimmune diseases. Various compounds of chemical or biological origin, termed immunomodulators, have been studied by classical pharmacological methods which are based on the toxicological and pharmacokinetic data of the test compounds. Since the effect of these drugs is a function of their half-life and serum level, they have to be used at particularly high doses to attain high blood levels and tissue concentrations. The case of cytokine α interferon or interleukin 2 exemplifies this point. These substances are used in human therapy at doses equivalent to several million units. As expected, such a treatment is associated with severe side effects. Administration of such high doses also contradicts the fact that these cytokines are secreted in a microenvironment and act according to a paracrine mechanism.

The novel concept of immunomodulation proposed herein does not involve a new molecule or extract but rather a new approach to the use of conventional substances to modulate the immune response. Given the fact that the immune system is subjected to a multiplicity of interwoven regulatory mechanisms characteristic of systemic organization, we suggest that the regulation of biological systems involves the notion of "information" rather than the quantity of molecules it receives. Indeed, the immune response is under the control of the idiotypic network, which is a reservoir of information on antigenic structures. Therefore, it is possible that besides the regulatory mechanisms in living organisms reacting in response to hormones or cytokines and obeying the laws of classical pharmacokinetics, other mechanisms reacting in response to "information" might exist. These regulatory effects could be expressed at another level than those of the classical effector mechanisms of the cells[1].

As in Homeopathy which can be considered as a model of "informative regulation" transmitted by the successed dilutions, we decided to use serially homeopathic diluted solutions of endogenous molecules active on the immune system. To transmit this information, we studied the effect of a series of highly diluted test substances. Our results indicate that serial dilutions have immunomodulatory activity. Furthermore, the activity is specific and the higher the dilution, the greater the capacity of the test substance to transmit its information to the target cell or organism. The families of molecules we studied have two different kinds of activity: thymic hormones which induce T lymphocyte maturation, bursin which participates in B lymphocyte maturation and cytokines which take part in cellular interactions.

2. Immunomodulatory products tested

The molecules studied belonged to endogenous molecules. As such, they were genetically "recognized" by the organism whatever their administration, molecular or non molecular: as the immunomodulating agents were prepared according to the homeopathic process, the "high dilutions" may be diluted in such a way that the molarity was less than 1 molecule / liter and the solution employed did not theoretically contain any molecule.

We have tested molecules isolated from the immune system: thymic hormones which induce intra or extra-thymic maturation of the T lymphocytes in humans or mammals, bursin which participates in B lymphocyte maturation in the bursa of Fabricius of aves [2] and cytokines which take part in cellular interactions. Although most of our investigations were carried out with thymulin [3] and natural mouse αβ interferon which contained β2 interferon identified now as interleukin 6, we sometimes tested other thymic hormones (thymosin and thymopoïetin), interleukin 2 (IL-2) and bursin, a tripeptide released from probursin [2]. Investigating the activity of such different kinds of molecules, we were able to verify the specificity of the information they transmitted.

The immunomodulating agents were prepared as follows: the mother solutions (1mg/ml) were prepared in distilled water. Next, successive 100-fold dilutions vortexed 30 s or succussed vertically on a special machine for 2 minutes were carried out using an alcohol-water-glycerine solvent that had been filtered through a 0.45-μm Sartorius membrane or distilled water; the dilutions used for injections or in "in vitro" tests were prepared in sterile 0.15 M NaCl. The following solutions were prepared: thymulin (Serva, Le Perray en Yvelines, France) at 10^{-x} mg/ml (where x=5, 6, 7, 8 , 9, 11, 13, 15, 17, 19, 23); thymopoietin II (Serva) at 10^{-x} mg/ml (where x=9, 15, 19, 23); thymosin 5 (kindly supplied by Hoffman-LaRoche) at 10^{-x} mg/ml (where x=9, 15, 19, 23); tuftsin (Serva) at 10^{-x} mg/ml (where x=9, 19); natural murine αβ interferon, 2 x 10^7 IU/ mg of protein (kindly provided by Dr.Michael G.Tovey (Paris) or Interchim, Montluçon France) at 10 IU/ml in 0.15M NaCl and 10^{-9} IU /ml; natural semi-purified human IL2, 330 IU/mg of protein (kindly provided by D Fradelizi, Paris) at 10 IU/ml and 10^{-9} IU/ml; bursin at 100μg, 100fg, or a pool 10^{-30}-10^{-40} M. Aliquots of all test solutions were stored at -20°C or 4°C until use and vortexed or succussed again before experimentation. The control mice or cultured macrophages received unsuccussed solvent andcontrol chicken embryos received the solvent diluted in the same manner as the test substances .

3. Immunomodulation experimentation in mice

Classical immunopharmacological ex vivo models were used to evaluate the effect of the diluted immunomodulators. The specific humoral immune response was studied by the Jerne plaque forming cell technique (PFC) on Swiss (OF1) or New Zealand Black (NZB) mice challenged with sheep red blood cells (SRBC). The specific cellular immune response of C57BL/6 or NZB mice against P 815 mastocytoma cells or EL4 cells, respectively, was evaluated using the ^{51}Cr-release technique after in vivo immunization. The non-specific natural killer response of C57BL/6 mice against YAC-1 cells was also evaluated using the ^{51}Cr-release technique. The in vitro techniques used were the lymphocyte transformation test (LTT) and chemiluminescence evaluation of adherent peritoneal cells from BALB/c mice.

3.1. Ex vivo pharmacological studies in healthy mice

Effect of thymulin: Swiss mice pretreated with 10^{-9}, 10^{-15}, 10^{-19} or 10^{-23} mg/ml of thymulin showed a significantly depressed specific humoral immune response to SRBC [4, 5, 6, 7, 8]. All control mice received the unsuccussed solvent intraperitoneally under the same conditions as the test mice (four injections 0.2 ml i.p.). We studied the effect of pretreatment of different groups of C57BL/6 mice treated at days -15, -13, -11 and -8 before sacrifice with 0.2 ml of thymulin solutions ranging from 10^{-5} mg/ml to 10^{-19} mg /ml ; we observed a significantly depressed specific cellular immune response with the highest dilutions as shown in Figure 1 [6, 7, 9]. The concentration of 10^{-19} mg/ml had an effect that was able to be reproduced. In other experiments, thymulin significantly stimulated the NK response ($P < 0.001$)[1, 9].

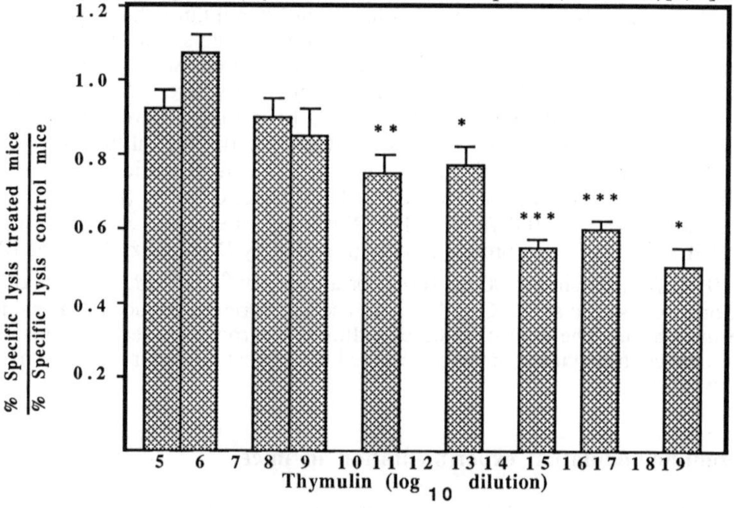

Effector cells/target cell=200.
Newman-Keuls Multiple Comparison Test.
* $P < 0.05$ ** $P < 0.025$ *** $P < 0.001$

Figure 1: Allospecific cellular immune reponse of mice treated with different dilutions of thymulin.

Effect of other thymic hormones: the immunopharmacological effects observed in healthy mice treated with thymic hormones show that they did not affect the synchonizers. We studied thymosin 5 at doses of 10^{-9} and 10^{-19} mg/ml for two years and thymopoïetin II peptide 29-41 at the doses of 10^{-9}, 10^{-15}, 10^{-19} and 10^{-23}mg/ml for one year. In both cases, we observed a decrease in PFC of treated mice compared with the control mice in June and an increase in PFC during the cold season (November-December). These seasonal variations were observed over the period of two consecutive years in Swiss mice treated with thymosin 5 and over the period of one year in thymopoïetin-II treated mice. The period of the physiological circannual rhythms was not changed but the amplitude was increased [4, 5, 10].

Table I: Specific humoral immune response of Swiss OF1 mice treated with very low doses of natural αβ mice interferon

IFN / injection[a]	Exp.I	Exp.II	Exp.III	Exp.IV
0	55±5[b]	209±23	167±13	11±2
2 IU	140±39 $P<0.05$[c]	328±32 $P<0.01$	193±24	17±4
2×10^{-10}IU	164±16 $P<0.05$	353±14 $P<0.01$	256±25 $P<0.05$	58±8 $P<0.001$

[a] Mice were treated with 4 .i.p.injections 0.2ml.
[b] Number of PFC/10^6 cells ± s.e.m.
[c] Statistical significance (Newman-Keuls Multiple Comparison Test).

Effect of leucocytary interferon : we also studied the effect of pretreatment with natural murine αβ interferon (including β 2 interferon now identified as interleukin 6) on the specific humoral response of Swiss (OF1) mice by evaluating plaque forming cells (PFC). The mice received four injections of either 2 IU or 2 x 10^{-10} IU before challenge with SRBC. The lowest concentration significantly increased the specific humoral immune response in the four experiments as shown in Table I [4, 8, 10]. We evaluated the effect of the same dilutions of interferon on the specific cellular secondary response in C57BL/6 mice immunized with P 815 cells; the mice who had received a dose of 16 x 10^{-10} IU interferon before and after the first injection of P 815 cells showed an increase ($P < 0.001$) in the cellular response [1, 9, 11]. The non-specific natural killer (NK) cellular response to YAC-1 cells was tested in C57BL/6 mice treated with leucocytary interferon. The results showed the effect of interferon at 2 IU and 2 x 10^{-10} IU/injection [1, 9, 11].

Effect of IL-2: we also studied the effect of IL-2 on cellular and humoral immune responses of mice using doses of 2 IU et 2 x 10^{-10}/injection[1, 8]. Using the same treatment protocole as for the C57BL/6 mice to evaluate the cellular immune response (pretreatment during three weeks before and after the two immunizations with P815 cells and treatment during the last day before the test), we obtained a significant stimulation of the cytotoxicity test in treated mice compared with the control mice. Concerning the humoral immune response, we obtained different results according to the treatment and immunization schedule (1, 8). We did not observe any effect when the mice were pretreated before SRBC immunization; if the animals were treated before and after the immunization, we observed a significant increase in PFC (primary response). If the treatment was performed after two immunizations (secondary response), we observed a significant suppression of the response. These results can be explained by the fact that modulation of the immune response by IL-2 is dependent on T-lymphocyte activation.

3.2. *In vitro studies in healthy mice*

Mouse peritoneal macrophages from female BALB/c mice aged 4 to 12 weeks were incubated with natural murine αβ interferon at 1.6 IU and 1.6 x 10^{-10} IU/ml. The production of reactive oxygen intermediates was evaluated by the chemiluminescence test. The treatment was able to modulate the PMA-induced chemiluminescence of macrophages. The two concentrations tested caused a decrease in the chemiluminescence of adherent cells from mice aged 4 to 5 and 8 to 12 weeks and an increase in the chemiluminescence of adherent cells from mice aged 6 weeks [12].

3.3. *Experimental ex vivo studies in immunocompromised mice*

Mutant mice such as New Zealand Black (NZB) whose thymus undergoes degeneration and who show a decrease in the thymulin secretion at the age of four weeks were chosen for the study. Female NZB mice aged 6 weeks were treated with 10^{-9}, 10^{-15}, 10^{-19} and 10^{-23} mg/ml of thymulin or with 2 IU and 2 x 10^{-10} IU/mouse of natural interferon. All four dilutions of thymulin significantly stimulated the specific humoral immune response. The cellular immune response was evaluated after treatment with 10^{-9} and 10^{-19} mg/ml of the hormone; the lowest concentration of thymulin significantly stimulated the cellular response against EL4 cells [1, 4, 9]. When the NZB mice were treated with interferon, we obtained a significant decrease in the cellular immune response. We always observed an opposite effect when we compared the results obtained in healthy mice with those obtained in immunocompromised mice. In addition, a crude extract of thymus at 10^{-9}, 10^{-15},10^{-19} and 10^{-23} mg/ml also showed opposite effects on the immune response when administered in healthy and NZB mice[1, 4, 9]. We also tested ex vivo the mitogenic response to ConA and PHA of lymphocytes from NZB mice treated with thymulin ; these mice are high responders to these mitogens as we verified in control mice. The female NZB mice treated with 10^{-15} mg/ml thymulin showed a significant decrease in lymphocyte stimulation [4]. Taken together, we conclude that treatment with very low doses of thymulin caused an improvement in the immune response of NZB mice. However, these results have to be considered very carefully since unpublished results of thymulin-treated MRL mice showed an aggravation of their autoimmune pathology .

4. Effect of in ovo administration of high dilutions of bursin in bursectomized chickens

We also tested the effect of bursin on the humoral immune response of bursectomized chickens[13]. Chick embryos were surgically bursectomized at 80h of incubation and then given saline or bursin in ovo, a tripeptide isolated from the bursa of Fabricius, at days 6 and 9. Four groups of chickens were tested: one untreated sham-operated control group (N); one saline treated bursectomized group (BX+S); one 100 fg bursin treated group (BX+fg); one 10^{-27}g bursin treated group (BX+ 10^{-27}). The 10^{-27}g bursin solution was prepared by 1/100 serial dilutions according to the homeopathic preparations. The four groups were repeatedly immunized with porcin thyroglobulin (Tg) at days 21, 30, 39 after hatching. Plasma corticosterone levels and serum titers antibodics against thyroglobulin were evaluated on days 20, 29, 38 and 47 after hatching. The immunized N group had high concentrations of corticosterone and ACTH on day 38 but not the bursectomized BX+S group. This level was completely restored in the two treated groups. Specific antibodies to Tg remained at very low background in the BX+S group in spite of repeated stimulations, whereas N group produced a strong specific antibody response. In ovo administration of very low doses of bursin (100 fg and 10^{-27}g) led to high antibody production[13]. These results were repeated in another experiment including a randomized tripeptide control (Trp-Leu-Leu-NH_2) at the concentration of 100 fg. Melatonin was evaluated in addition to the corticosterone, ACTH and antibodies anti-thyroglobulin. The results were able to be repeated and melatonin also showed a peak on day 38 in normal and treated chickens while the controls (bursectomized BX, solvent treated bursectomized BX+S and peptide control treated bursectomized BX+Pf) showed no increase in melatonin secretion (fig 2).

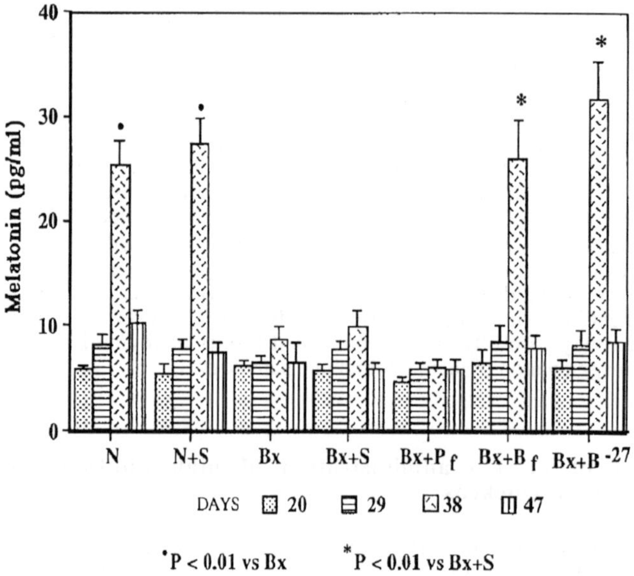

Figure 3: Effect of highly diluted bursin on the release of melatonin in bursectomized chickens.

These results were emphasized by a study on the circadian rhythm of these hormones. The animals underwent a 12L-12D light-dark schedule for 12 weeks. Then, we measured the circadian rhythms of plasma ACTH, corticosterone (CORT), melatonin (MLT) and the pineal enzymatic activity of N-acetyl-transferase (NAT) and hydroxyindole-o-methyl-transferase (HIOMT). The data showed that intact and bursin-supplied bursaless chickens exhibited pronounced diurnal rhythms of plasma ACTH and CORT, with low light-time levels and significantly higher night-time values reaching a sharp peak at midscotophase. The waveform rhythms of both hormones were severed after bursectomy . In all groups and regardless of the treatment, a daily rhythm of MLT with middark crest was recorded and persisted in Bx birds despite 50% reduction in height. This waveform profile was in keeping with the rhythm of NAT, but HIOMT remained arhythmic. Neither the saline nor the control compound (Trp-Leu-Leu-NH_2) could reverse the effects of bursectomy.

The corrective effect of bursin has been demonstrated in independent double-blind experiments and involved numerous physiological parameters measured in distinct laboratories. Compellingly, these arguments warrant that infinitesimal doses of bursin specifically induced recovery immune and neuroendocrine performances in our Bx recipients. Nevertheless, possible involvement of porcin thyroglobulin (Tg) in either control or bursin-treated birds is not precluded, since Tg is related to thyreotropic activity. Therefore, due to its precocious, time-limited, dose-dependent, pleiotropic and long-term action, bursin occupies a privileged position in the hierarchical scale of biologically active compounds.

At this point, it is worth noting that highly dilute bursin was as efficient as 100 fg of the same compound in cancelling the multiple effects of bursectomy. Nowadays, a growing set of experiments aimed at substantiating the biological effects of high dilutions are under way, and some reports are compelling [14, 15, 16]. Nevertheless, whether underlying mechanisms fit into a molecularist theory remains a central question, because highly dilute solutions are theoretically devoid of active material compounds. Several theories are put forward, but a deciding factor in the efficiency of high dilutions seems to lie in the way such solutions are prepared, including sequential centesimal dilution steps with potent vertical stirring between two successive steps. We demonstrated that in serially succussed diluted ^{3}H-thymidine solutions, radioactivity decayed linearly, without showing any difference between succussed and unsuccussed solutions, which precludes the assumption of a different molecular distribution due to succussion. As a working hypothesis, it can be proposed that non-molecular specific information corresponding to the original molecule (bursin) is conveyed to the recipient organism which deals with this special cue to restore normal immune and neuroendocrine functions. Finally, using a relevant experimental model, we have provided evidence that minute amounts of a compound originating in the bursa of Fabricius can functionally mimic *in vivo* the biological effects of this organ. Further investigations based on the electromagnetic properties of high dilutions may shed more light on the possible means by which the recipient organism receives the message delivered by highly dilute bursin.

However, the route (s) by means of which the bursal information influences neuroendocrine or immune functions and underlying mechanisms have yet to be unraveled.

5. Discussion

In previous research, there are several reports describing the effects of high dilutions (pico- and fento-molar concentrations) in *in vitro* models of biological systems. Protein kinase C isolated from fresh rat brain was found to be activated by picomolar concentrations of lead [17]. Leung-Tack et al. [18] demonstrated an inhibitory effect of a synthetic peptide (postin) at a concentration of 10^{-14}M on phagocyte functions. Since these experiments were carried out in vitro, no pharmacokinetic or tissue diffusion data was reported.

However, recent data [19] has shown a better stimulating effect of very low doses of thymulin (1ng/kg) compared to a higher concentration (1µg/kg) in restraint stressed mice.

The experimental results reported here demonstrate the efficiency of highly diluted solutions of hormones, peptides and cytokines in biological systems. They raise the question of the nature of the signal given by thymulin, for example at a concentration lower than 10^{-23} M, or bursin used at the theoretical concentration of 10^{-27}g.This solution contains theoretically no active molecule.

We conclude, therefore, that the succussed preparations obtained by successive 100-fold test dilutions of immunomodulators transmitted "information" specific to the molecules used to prepare the solution, even though no active substance was present in the solution.

Many problems remain to be solved: among them, what is the nature of this information? Physicists are now studying the possibility of a physical electromagnetic carrier of information. We are now in a new paradigm of information, different from the mechanistic one [1, 14, 15, 16].

References

[1] M.Bastide and F.Boudard, " A novel concept of immunomodulation", in *Forum sur l'Immunomodulation*, M.Guenounou Ed, John Libbey Publishers Paris, 1995, pp 303-316.

[2] T. Audhya, D.Kroon, G.Heavner, G.Viamontes, G.Goldstein, " Tripeptide structure of bursin, a selective B-cell differentiation hormone of the Bursa of Fabricius", *Science,* 231, pp 997-998, 1986.

[3] J.F.Bach, M.Dardenne, "Studies on thymus products. II: Demonstration and characterization of a circulating thymic hormone? ", *Immunology*, 2, pp 353-366, 1973.

[4] M.Bastide, "Immunological examples on ultra high dilution research", pp 27-33, in *Ultra High Dilution, Physiology and Physics,* Endler and Schulte Eds, Kluwer Academic Publisher, Dordrecht, 1994.

[5] M.Bastide, M.Doucet-Jaboeuf, V.Daurat, "Activity and chronopharmacology of very low doses of physiological immune inducers", *Immunol. Today,* 6, pp 234-235, 1985.

[6] M.Bastide, V.Daurat, M.Doucet-Jaboeuf, A.Pelegrin, P.Dorfman. "Immunomodulatory activity of very low doses of thymulin in mice", *Int.J.Immunotherapy*, 3, pp 191-200, 1987.

[7] M.Doucet-Jaboeuf, J.Guillemain, M.Piechaczyk, Y.Karouby, M.Bastide, "Evaluation de la dose limite d'activité du Facteur Thymique Sérique", *C.R.Acad.Sci.*, 295, III 1982.

[8] M.Doucet-Jaboeuf, "Activité de très faibles doses d'hormones thymiques sur la réponse immunitaire humorale de la souris: incidence sur le rythme circannuel", *Thèse Docteur es Sciences Pharm.& Biol., Université de Montpellier I* , 1986.

[9] V.Daurat ," Modulation de la réponse immunitaire de la souris par des doses infra-pharmacologiques d'immunomédiateurs", *Thèse Docteur ès Sciences Pharm. & Biol., Université de Montpellier I* , 1988.

[10] M.Doucet-Jaboeuf, A.Pélegrin, M.C.Cot., J.Guillemain, M.Bastide, "Seasonal variations in the humoral immune response in mice following administration of thymic hormones", *Ann.Rev.Chronopharmacol.*, 1, pp 231-234, 1984.

[11] V.Daurat, P.Dorfman, M.Bastide," Immunomodulatory activity of low doses of interferon α,β in mice" , *Biomed.& Pharmacother.*, 42 , pp 197-206, 1988.

[12] V.Carriere, M.Bastide, "Influence of mouse age on PMA-induced chemiluminescence of peritoneal cells incubated with α,β interferon at very low and moderate doses". *Intern.J. Immunotherap.*, 6, pp 211-214, 1990.

[13] B.J.Youbicier-Simo, F.Boudard., M.Mekaouche, M.Bastide., J.D.Baylé. "Effects of embryonic bursectomy and *in ovo* administration of highly diluted bursin on adrenocorticotropic and immune response of chickens", *Int.J.Immunotherapy,* 9, pp 169-180, 1993.

[14] M.Bastide ,A.Lagache, *The paradigm of Signifiers,* Alpha Bleue Publishers., Paris, 1992.

[15] M.Bastide, A.Lagache, C.Lemaire-Misonne, "Le paradigme des signifiants: schème d'information applicable en Immunologie et en Homéopathie", Revue Internationale de Systémique, 9, pp 237-249, 1995.

[16] M.Bastide, A.Lagache, "A new paradigm applied to high dilution effects on the living body", see the corresponding chapter in this book.

[17] J.Leung-Tack,J.C.Martinez,J.L.Sansot, Y.Manuel, A.Colle," Inhibition of phagocyte functions by a synthetic peptide lys-pro-pro-arg (postin)", *Protides Biol.Fluids Proc.Colloq.*, 34, pp 205-208, 1986.

[18] J.Markovac, G.W.Goldstein," Picomolar concentrations of lead stimulate brain protein kinase C", *Nature*, 334, pp 71-73, 1988.

[19] O.Masahiro, M.Masatara, S.Chikako, N.Katsuhiko. "Restorative effect of short term administration of thymulin on thymus-dependent antibody production in restraint-stressed mice." *Int.J.Immunopharmac*,15, pp 757-762, 1993.

HEALING CHRONIC EAR WOUNDS IN A MURINE MODEL USING SILICA (SiO_2) AS A HOMEOPATHIC REMEDY

M. OBERBAUM, Z. WEISMANN & Z. BENTWICH

R. Ben-Ari Institute for Clinical Immunology, Kaplan Hospital, Rehovot, Israel

ABSTRACT

The therapeutic effect of homeopathic dilutions of silica was investigated for two years in an experimental murine model of chronic wound induction. In conventional science, silica is used in pharmacological concentrations as a substance that is both inhibitory and cytotoxic to macrophages. In homeopathy, silica is used, among other things, as an anti-inflammatory remedy, especially for treating chronic wounds, ulcers, abscesses, etc. Silica is also regarded as an agent that causes rejection of foreign bodies from living organisms. In this study, each experiment consisted of three to four groups of ten mice. Holes were made in the ears of mice by perforation with a dental wire, which was then left hanging in the ear so that the constant friction would cause a persistent mechanical irritation. For 4 to 20 days, the following homeopathic dilutions of silica: 10^{-10} (C-5), 10^{-60} (C-30), 10^{-400} (C-200) of the mother tincture and of saline , "10^{-10}" ("C-5"), were added to the drinking water of the mice and the size of the hole was measured daily. In the first and second experiments, the size was graded from 0 to 4, and then we used an image analysis system. In 7 of 13 experiments, the size of the holes of the silica-treated animals was significantly smaller ($P<0.05$-0.001) and healed faster than those of the saline-treated control animals. Moreover, the therapeutic effect increased as the concentration of the silica decreased. The results of this study demonstrate that homeopathic dilutions of silica (even beyond Avogadro's number) have a therapeutic effect on chronic wounds. The results cannot be attributed to a placebo effect because experiments 3-6 were conducted as a double-blind study in which an objective measurement method was used. The mechanism of action is not clear. Although the effect of the low dilutions is probably due to the action of silica in conventional systems, the effect of the high dilutions remains an enigma because the C-30 and C-200 dilutions do not contain even one molecule. One explanation of this result could be an "imprint" of the "information" of the silica on the diluent. The successive augmentation of the therapeutic action of silica from low through medium to high dilutions suggests that the mode of action of all dilutions have the same foundation.

1. Introduction

Potencies of silica are commonly used as homeopathic remedies for abscesses, discharging fistulas, chronic ulcers, discharging wounds, and keloids [1]. Silica is a deep-acting constitutional remedy as well and promotes rejection of foreign bodies [2]. Contrary to the basic law of individualization in homeopathy, potencies of silica can be applied indicatively without the need to individualize, which is a major advantage for experimental studies. Another exception to this law is that silica can be prescribed in most cases that require its "anti-chronic inflammation" properties.

In conventional scientific research, unpotentized silica is used as a substance that is inhibitory and cytotoxic for macrophages [3–6]. Both in homeopathy and in conventional science, silica has been shown to have a strong influence on inflammatory processes, particularly in the post-acute phase of the inflammation.

The aim of the present study was to investigate the ability of various dilutions of silica to facilitate wound healing. To establish a condition similar to that where silica would be used in homeopathy, we developed a murine model of chronic wound induction.

2. Materials and Methods

Animals and wound induction. C57Bl/6 female mice, 12-to-16 weeks old, were used in this study. After a 2-week adaptation period, we perforated the lobe of one ear of each mouse with dental wire weighing 110±10 mg and then twisted the ends of the wire together to form an "earring." The constant friction of the wire along edges of the wound caused a chronic irritation. In the ears of untreated mice, holes of various sizes developed, depending on the properties of the tissue, the weight of the hanging wire, and the detrimental effect of the constant irritation caused by the earring. In some cases, the wire eventually "cut" the ear, which then fell off. In most cases, however, the wound healed, and the hole closed around the wire, first producing a granulation that was later replaced by scar tissue.

The animals were randomly assigned to three (experiments 2–9) or four (experiments 10–13) groups of ten animals each. (Note: the preliminary experiment, No. 1, had only four animals per group). In the first 9 experiments, the mice received silica dilutions of C-5, C-30, and saline, whereas in experiments 10–13, an additional dilution of C-200 was given. In only the first two experiments, the mice received 3 ∞l intraperitoneal injections daily, and thereafter, the silica was added to the drinking water.

Preparation of Silica Dilutions. Decimal dilutions (100^{-n}) of silica were prepared by Dolisos (France), according to the official French pharmacopoeia. The first two steps were (a) triturating silicium oxide (SiO_2) with lactose powder at a 1:100 relation (C-1 and C-2 dilutions), and (b) dissolving one part of the triturated powder in 99 parts of 70% alcohol (C-3 dilution). The C-3 dilution was vigorously sucussed using a dynamizator. The same technique was used for all subsequent dilutions. In the last two dilutions, saline rather than alcohol was used as the diluent; the last two dilutions were renewed once a week and kept in a refrigerator at 4_C. For the saline control, identical dilutions were prepared by first "diluting" an aliquot of alcohol in alcohol and then using saline as the diluent for the last two dilutions.

Method of evaluation. In the first three experiments, two independent observers made a subjective evaluation of the size of the hole, using a scale of 0 = a closed hole that was tight around the wire, to 4 = a very large hole. In subsequent experiments, an objective evaluation of the size of the hole by an image-analysis system (Galai, Haifa, Israel) was added on the last day of the experiment. We then photographed the hole with an electronic camera and transferred the data to an IBM/AT computer, which calculated the hole size. Evaluation of the hole size of an ear that had been "cut" by the wire was calculated by using the border of the hole as the edge of the ear.

3. Results

Preliminary experiment. To discover whether silica could elicit any reaction at all when used for treating a chronic wound, we performed a preliminary experiment using only four animals per group. All mice were treated for 32 days with three daily intraperitoneal (i.p.) injections (0.5 ml each) of the silica dilutions. On day 14, the earrings were inserted into the mouse ears. Five days later, a clear difference between the hole size was seen in the ears of treated and untreated mice. The difference in hole size was maintained until the end of the experiment on day 32. In the treated group, granulation tissue developed that closed the hole until it completely disappeared (Fig. 1)

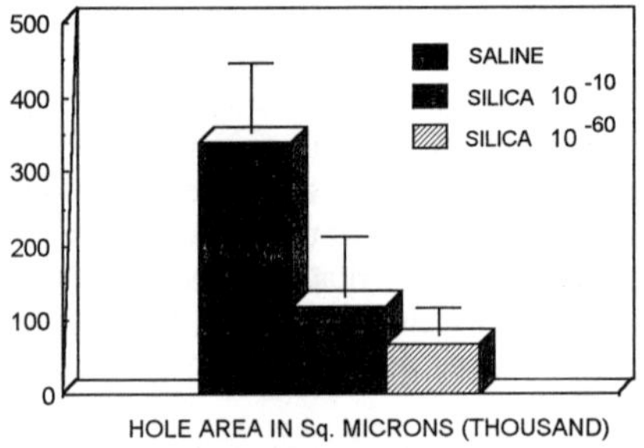

Figure 1. Evaluation of hole size in wounded ears receiving intraperitoneal silica administration. 14 days after daily i.p. injection of silica dilutions or saline one ear of each mouse was perforated with dental wire, which was twisted to form an "earring." 18 days after insertion of the earrings, the ears with holes were photographed, and the hole size in the photograph was evaluated using an image analysis system.

No sign of hole closure around the wire was observed in any animal in the control group. Image-analysis evaluation of the hole size revealed a significant difference between the treated and untreated groups ($p<0.01$ in one-way ANOVA), as well as between each treated group and its parallel control group ($p<0.01$, Student t-test).

Kinetics of the therapeutic effect. The results of the preliminary study prompted us to study the kinetics of the therapeutic effect. We evaluated the size of the hole daily, beginning on day 3 of the experiment and continuing until day 20. After the first experiment, the dilutions were administered via the drinking water to avoid the stress of holding the animals while injecting them thrice daily. Fig. 2 shows the difference between the hole sizes in the treated and the untreated groups. In untreated mice, the hole size increased, reaching its maximum on day 9 ($p<0.05$ in the Kruskal-Wallis one way ANOVA and the Pearson chi square test). Comparing the hole sizes of mice treated with the lower dilution (C-5) and those of mice treated with the higher dilution (C-30) indicated that the healing process was stronger at the higher dilution. After day 10, however, the hole size was the same in both the treated and untreated groups. Contrary to the results obtained in the first experiment, the beneficial therapeutic effect did not continue until the end of the experiment.

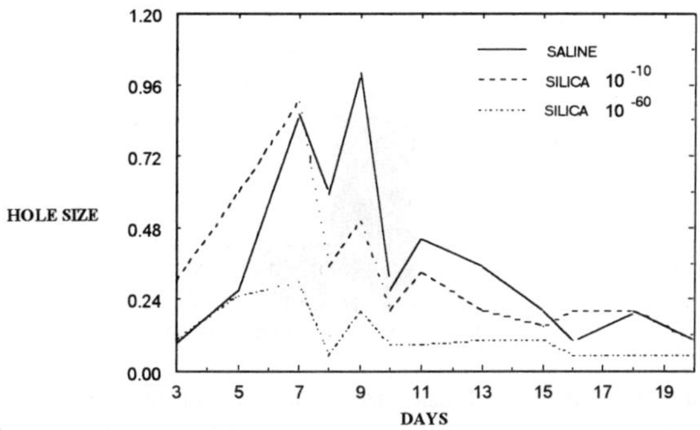

Figure 2. Evaluation of hole sizes in wounded ears of mice receiving oral silica administration. Dental wire was inserted into the ears of mice as described in Fig. 1. In parallel, either silica dilutions or saline was added to the drinking water throughout the experiment. The hole size was subjectively evaluated daily, beginning from the third day and continuing until the twentieth day of the experiment.

Influence of the "earrings" on the therapeutic effect. We examined the question of whether either an earring or chronic irritation is necessary for achieving successful therapy. In this experiment, we made a hole in the second ear as well, using an instrument that makes holes of a constant size. Earrings were not inserted into the second ear. To improve the evaluation of the hole size, we scanned the hole with an image analysis system for the measurement. Table 1 shows a good correlation when the subjective evaluation and the objective measurement of the hole sizes are compared.

Table 1. Comparison between objective and subjective evaluations of hole sizes in the wounded ears.

Treatment Group	Method of Evaluation	
	Subjective*	**Image Analyzer****
Saline	1.59	37269
SiO_2 10^{-10}	0.5	23644
SiO_2 10^{-60}	0.4	16469

*Mean of the subjective evaluation using a scale of 0 = a hole closed tight around the wire, to 4 = a very large hole. **Mean hole size (square microns)

In this experiment, which was terminated after 5 days, we also observed a significant effect of the silica dilutions on wound healing in the ears with earrings. Wound healing in animals receiving higher dilutions was significantly increased in comparison with the healing in animals receiving low dilutions ($p < 0.05$, one-way ANOVA) (Fig. 3). In perforated ears that did not bear an earring, however, no significant difference was observed in the hole sizes of the treated and the untreated groups (Fig. 4).

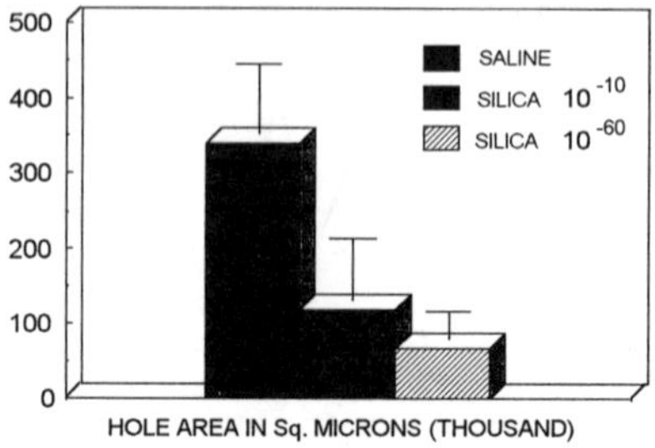

Figure 3. Evaluation of hole size, with and without dental wire. The protocol was the same as that in Fig. 2. in addition, a hole was made in the other ear without inserting wire earrings. The experiment continued for 5 days,. Fig. 3 shows the hole size in ears with dental wire earrings.

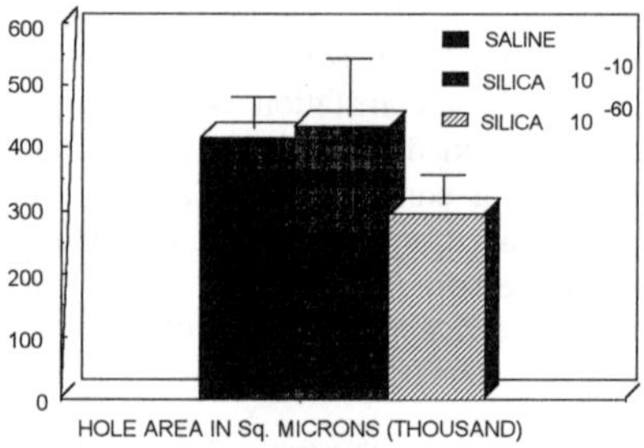

Figure 4. See legend of Figure 3 for protocol. Hole size in ears without dental wire earrings: no significant difference was seen between the treated and untreated groups.

The therapeutic effect of very high dilutions. To determine whether the observation that higher dilutions have stronger therapeutic properties is consistent, we added another dilution to the experiment, 200-C. Similar to the trend observed in the previous experiments, a stronger effect occurred on day 7 of the experiment in animals treated with the 200-C dilution. In the ears of the saline-treated mice, the average area of the holes was 72799 square microns, whereas the comparable average areas of the holes in mice treated with Silica C-5, C-30, and C-200 were significantly lower(56195 square microns, 31484 square microns, and 28222 square microns (Fig. 5) ($p<0.001$ in one-way ANOVA).

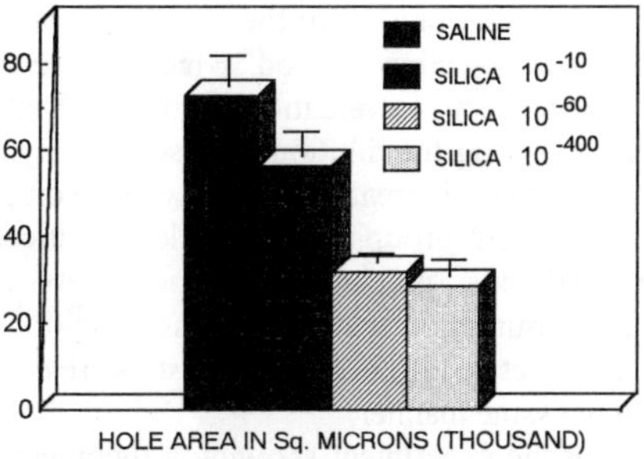

Figure 5. The protocol was similar to the previous experiment except that a group treated with a higher dilution of silica, (10^{-400}) was added. The experiment continued for 7 days.

Summary of repetitions. In three of the nine remaining experiments, the silica dilutions produced a clear therapeutic effect. In the other six experiments, no significant therapeutic effect was obtained. Technical difficulties probably led to the failure of the first three experiments. At first we tried to hang silver earrings to evoke a stronger inflammatory effect, but the silver caused a very strong irritation, causing very large holes that did not tend to close. In the second experiment, we used very thin dental wire, which was not heavy enough to cause sufficient chronic irritation. Thus, the wounds remained very small and closed rapidly and tightly around the wire. In the third experiment, we inserted earrings into both ears of each animal to determine whether different hole sizes would develop in each ear. No significant difference was seen between the groups, however, nor between two ears of an individual mouse. In the remaining three experiments, the holes did not develop and closed immediately around the wire; for which no clear explanation could be found.

4. Discussion

The experiments described in this study cover a period of almost 2 years. The results demonstrate that oral ingestion of homeopathic dilutions of silica (SiO_2) enhanced the healing of chronic irritated ear wounds, which resulted from inserting a metal earring in the ears of mice. The results were reproducible and statistically highly significant. The process of wound healing reached its peak after a period of 3 to 10 days, after which the treated and untreated groups became similar (experiment b, Fig. 2). In experiment b, the wound healing appeared to proceed in two phases (Fig. 2): in the first phase, the chronic irritation and the weight of the earrings led to an enlargement of the holes (the peak in Fig. 2 occurred between days 7 and 9). In the second phase, the reparative process began: the wound started to heal and decrease in size. The second phase peaked in all groups simultaneously (on day 10, Fig. 2). This observation suggests that the main action of the silica occurred during the first phase, either halting or slowing down the process of wound enlargement but not diminishing the eventual wound size.

The results of this study could not be due to mistakes in the evaluation methods because (a) we evaluated the hole size subjectively in the first two experiments; (b) we added an objective method using an image analysis system, in the third experiment; and

(c) we used the objective method only from the fourth experiment onward. It should be noted that the results of the objective method, which is free from observer bias, were similar to the results of the subjective evaluation (Table 1). That the healing effects could be due to "mistakes" in preparing the dilutions is also unlikely because the same effects consistently occurred in almost all treated groups when compared with the untreated control groups. Moreover, both groups were handled identically, kept in similar cages and under the same conditions, precluding the possibility that these factors could influence the results. The assumption that the "stress condition" of the injected animals could cause the difference between the groups must be rejected as well because both groups were injected in the same manner.

The positive results of the experiment showing a therapeutic effect only in ears that bore earrings provide additional support to the idea that active dilutions of silica may have a curative effect in healing chronic wounds. No difference between the groups was seen in hole sizes of ears that had been perforated but did not bear earrings (experiment c, Figs. 3 and 4).

The idea that some molecules of SiO_2 remained in the solution by adhesion to the tube walls must be rejected because higher dilutions produced a stronger therapeutic effect ($10^{-400}>10^{-60}>10^{-10}$). This result reflects the basic concept of homeopathic medicine in which higher dilutions produce deeper, stronger, and more persistent therapeutic activity.

The question of the reproducibility of homeopathic experiments is one of the big problems of homeopathic research. Thus, explaining the "failures" of some experiments (6 out of 13 experiments failed) is problematic. The failure of first three experiments was probably due to technical difficulties during our search for optimal conditions for the study, but the reason for the failure of the remaining three experiments is unknown. It must be stressed, however, that the significance of the successful experiments was more than enough to draw the conclusion that the effect was more than would be expected at random. Moreover, a "therapeutic effect" was never obtained when using any control dilution.

The "conventional" influence of SiO_2 on biological systems is an interesting aspect of these experiments. Many scientists have reported the effect of SiO_2 on macrophages, which in turn influences the immune response [3-6]. Reports have also been published about the curative effects of silica in such experimentally induced autoimmune diseases as myasthenia gravis and diabetes mellitus [7,8]. To our knowledge, only one controlled study examining the influence of very low doses of SiO_2 has been reported in an animal model. In that study, Davenas, Poitevin & Benveniste [9] found that the activity of macrophages obtained from C57Bl/6 mice receiving 10^{-19M} SiO_2 in their drinking water was significantly increased relative to the activity of macrophages obtained from an untreated control group. Although the concentrations used in that study were several orders of magnitude lower than those used in conventional studies, the silica dilutions still contained some molecules. In experiments 10–13 of the present study, we diluted the silica to beyond the Avogadro number; such dilutions are not considered to contain even one molecule of SiO_2.

We are aware of the problematic nature of the results described here. Not only did we find that the lower the concentration, the stronger the reaction, but also that "dilutions" not containing a single molecule performed a biological activity. The reason for this phenomenon remains an enigma. Recently several attempts have been made to explain the activity of such "dilutions" [10–12], almost all of which claim that the "drug information" is transferred to the diluent, which then becomes the carrier of such "drug information." According to this theory, if the drug/remedy is the "key" and the diluent is the "lock", the therapeutic effect must be attained with the "lock" rather than with the "key."

Acknowledgement - We thank Dr. Virginia Buchner for editorial assistance during the preparation of this manuscript.

References

[1] Jouanny J., *The essentials of homeopathic materia medica*, pp. 86–91. Laboratoires Boiron, Ste-Foy-les-Lyon, France, 1985.

[2] Metzger J., *Gesichtete Homoeopathische Arzneimittellehre,* 1320–1332. Karl F. Haug Verlag-Heidelberg, 5te Auflage 1981.

[3] Moseley P.L., Monick M., Hunninghake G.W., "Divergent effects of silica on lymphocyte proliferation and immunoglobulin production." *J. Appl. Physiol.* 65:350–357, 1988.

[4] Gennari M., Bouthillier Y., Ibanez O.M. *et al.,* "Effect of silica on the genetic regulation of antibody responsiveness." *Ann Inst Pasteur Immunol.* 138:359–370, 1987.

[5] Souvannavong V., Adam A., "Macrophages from C3H/HeJ mice require an additional step to produce monokines: synergistic effects of silica and poly(I:C) in the release of interleukin 1." *J. Leukocyte Biol.* 48:183–192, 1990.

[6] Bowers W.E., Ruhoff M.S., Goodell E.M., Stoltenborg J.K., "The effect of silica treat-ment on accessory-cell-dependent rat T-lymphocyte proliferation." *Immuno-biology* 176:179–194, 1988.

[7] Kinoshita I., Nakamura T., Satoh A., *et al.*, "Role of the macrophage in the pathogenesis of experimental autoimmune myasthenia gravis." *J Neurol Science* 87:49–59, 1988.

[8] Charlton B., Bacelj A., Mandel T.E., "Administration of silica particles or anti-Lyt2 antibody prevents beta-cell destruction in NOD mice given cyclophosphamide." *Diabetes* 37:930–935, 1988.

[9] Davenas E., Poitevin B., Benveniste J., "Effect on mouse peritoneal macrophages of orally administered very high dilutions of silica." *Eur. J. Pharmacol.* 135:313–319, 1987.

[10] Anagnostatos G.S., Vithoulkas G., Garzonis P., Tavouxoglou C., "A working hypothesis for homoeopathic microdiluted remedies." *Thr. Berlin J. on Research in Homoeopathy* 1:141–147, 1991.

[11] Bernard G.P., "'Microdose paradox–a new concept." *J. Am. Inst. Homeopathy,* 58:205–212, 1965.

[12] Kumar A., Jussal R.A., "A hypothesis on the nature of homoeopathic potencies." *Br. Homoeopathic J.*, 68:197-204, 1979.

WATER AND THE DIAGNOSIS AND TREATMENT OF ELECTROMAGNETIC HYPERSENSITIVITY

CYRIL W. SMITH
Department of Electronic and Electrical Engineering, University of Salford
Salford, M5 4WT, England

ABSTRACT

The clinical effects of coherence relate particularly to electrically hypersensitive patients. These are persons who usually have some chemical damage, and in whom one or more regulatory systems are on the verge of instability. Very weak EM fields or water 'imprinted' with their appropriate frequency can take them from stability to instability or vice versa. Water "memory" can be imprinted with an alternating magnetic field, an alternating magnetic vector potential or, by a person holding and/or succussing a glass tube of water. The latter forms the basis of a technique for determining the stabilising and de-stabilising frequencies for patients too sensitive to be exposed to any coherent frequency.

1. Coherence and EM Hypersensitivity

This clinical work has been going on continuously since 1982, when Dr. Jean Monro asked whether I could do anything to help her with her electrically sensitive allergy patients [1,2]. Initially, we found that we needed to test these patients in a chemical and particulate clean environment, with coherent alternating magnetic fields at strengths of the order of 30 nT in the ELF, typical of those to be found in the environment. Eventually we found that we needed to cover frequencies ranging from millihertz to gigahertz. We began by observing the various patients' reactions and noted down their observed and reported symptoms. These turned out to be the same as the symptoms produced in them by chemical, food, or inhalant challenges and were in general characteristic of the individual patients rather than some particular electrical frequency. We sought those frequencies which had a therapeutic effect, in that when they were present in the environment the patients' symptoms disappeared.

When a biological system is measured over a wide range of frequencies, the coherent frequencies found are not regularly spaced like the harmonics of a musical instrument or an inductance-capacitance circuit but, fit better to a fractal equation. There are stimulating frequencies which are therapeutic and these seem to be able to speed up and synchronise the activity of biological systems. They alternate with depressing frequencies, the persistant presence of which paralyses the small frequency jumps in the coherence frequencies which seem to be naturally present in living systems. Exposure to a depressing frequency is most unpleasant and quickly leads to a strong urge to "get-away" from it.

2. Diagnosis

2.1. Problems in Diagnosis

Difficulties were soon encountered. Many patients would have severe reactions at some frequency during test which for clinical reasons these must be avoided if at all possible. These might occur at any coherent frequency to which a patient happened to be sensitive, but, the frequency at which this might occur was not apparent until the patient started to react during testing. One had some indication of likely levels of patient sensitivity from the results of chemical and particulate testing which, had usually taken place before electrical testing was considered necessary. In particular, the tester needs advance warning if the patient had any tendency to anaphylaxis, or cardiac or respiratory problems, so that appropriate preventative measures can be taken. Testing electrically hypersensitive patients is a clinical procedure, it is not to be undertaken without medically qualified assistance and the ready availability of resuscitation facilities.

2.2. Water in Diagnosis

Some patients develop delayed reactions that come on hours, or even days, after the initial electrical challenge. A technique was needed that was not restricted to those patients who were either not very ill, or to those who reacted to an electrical challenge within seconds. To be able to help these other patients, I borrowed from the techniques of radiesthesia [3].

Furthermore, some patients did not have a body field strong enough to be detected. In this case, there is usually at least one coherent frequency in the theta (4-7 Hz) or delta (below 4 Hz) brain-wave region which will re-establish a "normal" body field pattern when the patient is exposed to it. This is conveniently done by having the patient hold a tube of water which has been 'imprinted' with the required frequency. The patient should then develop a strong enough body field for testing to be possible. This may be done by direct challenge with coherent frequencies, or by surrogate testing having the patient imprint a tube of water by succussion and subsequently measuring it.

2.3. Water in Therapy

As a result of clinical necessity, water was found to be capable of retaining the frequency information of an alternating magnetic field in a form which could be used therapeutically. This came about when Dr. Monro and I first encountered a patient who needed a therapeutic frequency in the microwave region; it was not practicable to supply the patient with a microwave oscillator. The empirical rule for therapy of electromagnetic hypersensitivity is to use an oscillator at one of the highest stimulatory frequencies found or, to imprint some or all of the stimulating frequencies into water and to use this as a medicament. The imprinting is easily done by succussing a thick and tough glass tube half filled from any source of water that the patient can tolerate, on a wooden surface next to a coil fed from an oscillator at the required frequency. The strength of the oscillation is not critical over a very wide range. To succuss the tube, it should be held above the water level so that the tester's body fields are not imprinted along with the frequency imprint, and given sharp rap on the wooden surface but, not hard enough to break the glass.

If a sample of 'imprinted' water is placed on one side of a toroid excited by an alternating current, and a sample of 'clean' water is placed equidistant on the other side and succussed, the latter acquires a phase-inverted imprint of the original water. A practical application for this is in allergy or surrogate electromagnetic hypersensitivity therapy since, it provides a way of inverting the phase of frequency components of allergens outside the available bandwidths of electrotherapy or electroacupuncture apparatus. That is, it can convert depressive frequency imprints into stimulatory ones.

It seems that the above techniques act by stimulating the autonomic nervous system. This is the first body system to be affected in chemical and electrical sensitivities. Work with electromagnetically hypersensitive patients led me to the investigation of endogenous body frequencies and to the development of therapies involving the stimulation of patient specific frequencies by applied electromagnetic fields or potentised water.

3. Nervous and Endocrine Systems

There is cooperation between the nervous and endocrine systems in the integration of the higher body functions. The nervous system is designed to transmit bio-information directly to and from restricted regions of the body through nerve fibres. The central nervous system, composed of the brain and spinal cord, is well established in vertebrates and has specialised cells (neurons) connected with other neurons and receptor or effector cells. In the higher organisms, the close and complementary relationship between the nervous and endocrine systems form the control networks of the neuroendocrine system [4] which make a higher regulatory performance possible.

There are neuroendocrine system effects following on from electromagnetic field exposure [5]. Originally, neurons seem to have been both excitatory and secretory in nature, but the majority of neurons have developed their excitatory role. They can generate electric spikes as a result of an excitation and very rapidly transmit a signal through nerve fibres to other cells. Thus, the nervous and endocrine systems may not always be independent of each other. Hormonal information can be transformed into neural information, and neural spikes can generate chemical signals. Furthermore, hormones may modify the excitability of neurons, while in turn the secretion of hormones may sometimes be controlled by neural inputs to endocrine organs. The activity of the nervous and endocrine systems may be modified by genetic information and by the internal and external environment.

The pancreas, is both an endocrine gland in respect of the production of insulin but also an exocrine gland in respect of its production of pancreatic juice, has parasympathetic innervation and increased parasympathetic activity increases the secretion of pancreatic enzymes.

4. The A.N.S. and Electromagnetic Hypersensitivity

4.1. Clinical Evidence

On re-examining the case notes of electromagnetically hypersensitive patients I found that half of the most commonly reported symptoms could be attributed to inappropriate responses by the diencephalon and endocrine integrated autonomic nervous system (A.N.S.). The A.N.S. related effects were selected from data obtained by the provokation and neutralization of the symptoms of 200 patients tested in England during the period 1984-1992 with frequencies ranging from 1.8 Hz to 4.2 GHz. These included the following A.N.S. systems: eye and visual, heart, blood vessels, respiration, stomach, intestine, sweat glands, saliva, muscle, thermo-regulation, syncope, yawning, sleep/wake, emotional state.

The results of tests on 120 patients included a total of 661 frequencies between 1.8 Hz and 300 Hz each corresponding to the onset of symptoms that could be regarded as being under the control of the hypothalamus-mediated autonomic nervous system. The Kolmogorov-Smirnov statistical test showed that this data was not randomly distributed. The frequency bands containing 2 Hz and 50 Hz had significantly higher counts than those containing the other frequencies. The symptoms at these two frequencies were consistent with the stimulation of the hypothalamus-mediated autonomic nervous system. Furthermore, the counts at the 25 Hz subharmonic and the higher harmonics of 50 Hz (the UK power supply frequency) were significantly above average. No single symptom predominated at either frequency. The symptoms most frequent at 2 Hz involved: thermal regulation, eyes, heart, respiration. At 50 Hz they were predominantly musculo-skeletal, involving pain and involuntary movements.

Challenging patients living in the USA at their 60 Hz power supply frequency gives a higher score than challenging them at the 50 Hz of Europe's power supplies [6]. This may be an example of an acquired sensitivity. The 2 Hz reactions were common to patients on both sides of the Atlantic and may represent a fundamental resonance in the autonomic nervous system.

A set of data was obtained within a two week period during a visit to the Environmental Health Center, Dallas, in November 1992. Then, 55 patients were seen and tested under environmentally controlled and consistent conditions. The results for their neutralizing frequencies showed a remarkable spread of frequencies, ranging from brain-wave frequencies, to microwave frequencies. Three significant peaks showed up on 'Monte-Carlo Simulation' (Dr. R.D. Baker, personal communication), these were at 0.7-0.8 Hz ($p < 0.0001$), 7-8 Hz ($p = 0.0002$) and 800-1000 MHz ($p < 0.0001$). It is well documented that 7.8 Hz stimulates the hippocampus and several commercial stimulating devices use this frequency.

Of these 55 patients, 32 had abnormal body fields when checked before testing commenced. The results showed three well-defined peaks: 0.6-1.0 Hz ($p = 0.0045$), 4.4 Hz ($p = 0.0019$), 6.2 Hz ($p = 0.0004$). The peak, 0.6-1.0 Hz, is in the band of EEG delta waves. The peaks at 4.4 Hz and 6.2 Hz span the band of EEG theta waves. These frequency bands were found to be useful for stimulating a normal body field in electrically depressed patients. They were also effective in giving relief from pain in the limb stump of amputees. Surprisingly they restored the "normal" body field into the region of space where the leg would have been before amputation. No effects on body fields were found for frequencies within the alpha wave band (8-13 Hz).

This work with electromagnetically hypersensitive patients [7] led the writer to the investigation of endogenous body frequencies and the development of therapies involving the stimulation of patient specific frequencies by applied electromagnetic fields or potentised water and, because of the extreme sensitivity of some patients has led to the development of frequency measurement techniques based on radiesthesia.

Quboa [8] developed an a.c. electric field strength and magnetic flux density meter for environmental measurements. The (f.s.d.) measuring range was 0.1 - 1000 V/m and 100 nT - 100 μT in steps of 5 Hz from 10 - 95 Hz; 50 Hz from 100 -950 Hz; 500 Hz from 1000 - 19000 Hz. This was applied to the measurement of the homes and work-places of electrically hypersensitive patients and the field and frequency distributions of electrical apparatus such as fluorescent lighting, televisions and computers, to determine the fields that could trigger reactions in these patients. Attempts to measure any fields and frequencies in the vicinity of reacting patients with this apparatus were not successful. From clinical case records, it is clear that certain patients have emitted significant radiation when reacting, enough to be incompatible with electronic apparatus.

The autonomic nervous system is part of a control network involving the cerebral cortex, the diencephalon and the endocrine system. There are electrical pathways, there is the secretion of a chemical substance in response to an electrical stimulus, and there are interaction sites at which the absorption of a chemical substance gives rise to electrical signals. In tests involving electromagnetically hypersensitive patients, the environmental electromagnetic fields appear to be affecting a feedback loop comprising: cerebral cortex

- autonomic nervous system - hypothalamus - cerebral cortex. Such a neural pathway is consistent with the rapid responses (within seconds of challenging) which these patients show. Direct patient testing does not detect slow responses which may involve the endocrine system and occur hours or days later. In such cases the surrogate method must be used for investigations.

4.2. Theoretical Implications

The A.N.S. integrated neuro-endocrine system gives a biological system the advantages of the rapidity of a neural network with the stabilisation provided by a chemical feedback path. If the phase shift around any feedback loop reaches 360° when the gain exceeds unity, then self-sustaining oscillations result at the frequency for which this condition applies. The provision of a single very long time-constant is one way of preventing the occurence of a 360° phase shift. This the endocrine system provides.

Very high electrical amplification is needed to trigger a nerve impulse from a single electronic charge or photo-event. This may amount to a gain of 10^9 to 10^{10}. It is very difficult to prevent $1{:}10^{-9}$ to 10^{-10} parts of electrical feedback being present, enough to give unity gain around the loop. That is all that is necessary for instabilities and oscillations to occur at whatever frequency gives a 360° phase shift. With the chemical feedback of the endocrine system, reverse reactions determined by chemical pathways limit unwanted feedback. The rate of unwanted chemical reaction pathways is either the reverse reaction rate of an enzyme reaction or the rate of spontaneous bond breaking under thermal agitation which is negligible for stable compounds.

What is needed from the clinical and environmental point of view, is a battery of tests to assess dysfunction in specific parts of the autonomic nervous system following electromagnetic or chemical challenge.

5. Chemical Effects and EM Hypersensitivity

5.1. Free Radicals

McLauchlan [9] has produced evidence to support a theory of how electromagnetic fields may cause physical damage to biological systems. Weak magnetic fields may affect cells by allowing free radicals to survive in the body in larger numbers and thereby damage the DNA.

In chemical reactions, bonds break and then re-form. Most of these bonds consist of pairs of electrons with opposite spins. When a reaction occurs, the bond breaks and then both partners 'reclaim' their electron from the bond and search for a new partner. Free radicals are these 'ex-partners', which have become free-roving agents. In order to reform a bond, two radicals that meet must have electrons with opposite spins (a 'singlet pair'). Often, 'same-spin' radicals meet each other, with the result that no coupling occurs, and the pair quickly separate to search for an 'opposite-spin' radical with which to form a singlet pair.

Unsuccessful pairings are more likely than successful ones because, there is only one way that an opposite-spin singlet pair can form but, three ways in which like-spin electrons can unite. Triplet pairs, when exposed to a very low static magnetic field, easily break and re-form into a singlet pair. But, as the field is increased, two of the three triplet states become decoupled from the singlet state completely. This stops two-thirds of the radical pairs from reacting and could occur in a magnetic field as low as 8 mT. The magnetic field does not affect the nature of the product, only the amount of it formed.

5.2. Oxidative Stress

In the present environment, the body's oxidative defence mechanisms are placed under stress due to environmental pollutants and nutritional deficiencies [10]. Of importance in this context are the essential dietary constituents, the anti-oxidant vitamins, A, C and E, together with selenium, an essential component of the per-oxydolytic antioxidant, glutathione peroxidase.

Evolution has not provided for the increased antioxidant protection required to counter the oxidative stresses placed upon the human body by the modern environment. Atmospheric pollutants, such as oxides of nitrogen, sulphur dioxide and ozone, and carbon-based pollutants such as polycyclic hydrocarbons, all compromise the natural oxidative defences of the cell, as do solvents, alcohol and even the polyunsaturated fatty esters in modern diet. Electromagnetic fields can increase oxidative stress by increasing free radical concentrations and this extra oxidative stress occurs in an already stressed cell. It provides the ideal conditions for damage to the DNA.

To simulate realistically the effects of electromagnetic fields, it is essential either to carry out experiments under conditions of oxidative stress or under conditions of antioxidant depletion. The electromagnetically hypersensitive patients already have a body load which includes a wide range of chemical stressors when their electromagnetically triggered symptoms appear.

5.3 Nuclear Magnetic Resonance

Nuclear magnetic resonance is a phenomena associated with atomic nuclei having an odd number of either protons or neutrons. The observation of NMR effects depends upon having a system which can detect a net absorption of energy by a small excess population of nuclei. Because of this, technological applications use very strong magnetic fields and high frequencies. but, biological systems do not have this limitation [11]. The nuclear magnetic resonance frequency is proportional to the magnetic field strength and the Larmor precession frequency. In the case of the proton, this frequency is about 2 kHz in a typical geomagnetic field. NMR effects for protons, and other nuclei, have been observed in measurements of biological cell dielectrophoresis, dielectric constant and loss measurements of biological materials; in mean-generation-time, cell size and population measurements for cells cultured under NMR conditions. Microwave induced (posterior, subcapsular) cataracts on *in vitro* bovine eye lenses are produced at very low microwave power densities when the microwaves are modulated at a frequency which satisfies proton NMR conditions. The results of satisfying proton NMR conditions during the growth of bacterial cells were unexpected and spectacular. They resulted in the same total cell mass but, with almost twice as many cells of half the size [12]. Magnetic resonance conditions may also affect singlet to triplet transitions.

6. References

[1] R.Y.S. Choy, J.A. Monro and C.W. Smith, 1987. "Electrical Sensitivities in Allergy Patients". *Clinical Ecology* 4(3), 93-102.

[2] C.W. Smith, R.Y.S. Choy and J.A. Monro, 1989. "The diagnosis and therapy of electrical hypersensitivities". *Clinical Ecolology* 6(4), 119-128.

[3] C.W. Smith, 1991. In: 9th. International Conference on "Man and His Environment in Health and Disease", Dallas, Texas, Feb 27 - Mar 3, 1991. Proccedings available on audio tape from INSTA TAPE Inc., P.O. Box 1729, Monrovia, CA 91017-5729, USA.

[4] H.A. Fischer, 1989. "Photons as Transmitters for Intra- and Intercellular Biological and Biochemical Communication - The Construction of a Hypothesis", in Popp, F-A. et al. (eds.) *Electromagnetic Bio-Information,* Munich: Urban & Schwarzenberg, pp.193-199.

[5] R.J. Reiter and M.K. Vaughan, 1991. "Dynamic organization of endocrine networks - the neuroendocrine reproductive axis and the pineal gland as examples". *Ann. NY. Acad.* 618(Feb), 11-27, 1991.

[6] W.J. Rea, Y. Pan, E.J. Fenyves, I. Sujisawa, H. Suyama, N. Samadi and G.H. Ross, 1991. "Electromagnetic field sensitivity". *J. Bioelectricity* 10 (1&2), 241-256.

[7] C.W. Smith, R.Y.S. Choy and J.A. Monro, 1989. "The diagnosis and therapy of electrical hypersensitivities". *Clin. Ecol.* 6(4), 119-128.

[8] K.M. Quboa, 1990. *Environmental electric and magnetic fields: measurements and communications implications*. Ph.D. Thesis, Salford University, England.

[9] K. McLauchlan, 1992. "Are environmental magnetic fields dangerous?". *Physics World* 5(1), 41-45.

[10] N.A. Cridland, 1993. *"Electromagnetic Fields and Cancer: A Review of Relevant Cellular Studies"*. London: HMSO.

[11] E. Aarholt, M. Jaberansari, A.H. Jafary-Asl, P.N. Marsh and C.W. Smith, 1988. *"NMR Conditions and Biological Systems"*, in Marino, A.A. (ed.) Modern Bioelectricity. New York: Marcel Dekker, pp. 75-104.

[12] Aarholt E.: *Biological effects of electromagnetic fields.* PhD Thesis, University of Salford, England, 1982.

INTERACTION OF A HIGH DILUTION OF *AGARICUS MUSCARIUS* L WITH DOPAMINE AGONISTS AND ANTAGONISTS IN MODULATING CATALEPSY OF MICE

N.C. SUKUL
Department of Zoology, Visva-Bharati University, Santiniketan 731 235, West Bengal, India

ABSTRACT

The ethanolic extract of a fungus *Agaricus muscarius* L at a dilution of 10^{-60} induces in albino mice a very mild catalepsy which is highly potentiated when the drug is followed by the mixed dopamine agonist apomorphine and selective D2 agonist bromocriptine. The high dilutin of *Agaricus* is produced in 30 steps and in each step the dilution (1:100) is mechanically agitated or sonicated. The drug thus prepared is called potentized. Pre-treatment of *Agaricus* markedly reduces haloperidol-induced catalepsy in mice. The potentized drug is effective only through the oral route and not through the intraperitoneal route. *Agaricus* also interferes with the locomotor activity of mice modified by the neurotransmitter drugs. It is suggested that potentized *Agaricus* acts on receptors in lingual epithelium and the impulse goes through afferent nerves to specific brain areas where specific G-proteins are expressed. These G-proteins associate with D2, D3 and D4 receptors and serve as their agonists resulting in the modulation of catalepsy in mice.

1. Introduction

Catalepsy is a transitory state of immobility in which the animals remain conscious but fail to correct the imposed posture. This behavioural state is thought of as an active inhibition of neural circuits, that mediate motor initiation, thus enabling akinesia with awkward postures. It is a central nervous system phenomenon involving a number of neurotransmitter systems [1-9]. Catalepsy can be induced by certain drugs and also by repeated restraint on the movement [5]. While the chloinergic system promotes catalepsy, the dopaminergic system antagonises it [5, 6]. We have observed that *Agaricus muscarius* extract at extremely high dilutions produces a very mild catalepsy in rats and mice [10, 11]. These dilutions, prepared according to the standard Pharmacopoeia [12], were used in combination with dopamine agonists and antagonist to see if the catalepsy of mice could be modified by the combined effect of the drugs. We also observed the effect of the drugs on the open field activity of mice. Catalepsy research helps in understanding the action of drugs on different neurotransmitter systems as well as relationships and interactons among the systems. The main purpose of our research was to develop a non-sacrifice, non-invasive animal model to elucidate the action of drugs at extremely high dilutions beyond the Avogadro number. This paper combines the results of two earlier publications [11, 13].

2. Materials and Methods

2.1. Subjects

All tests were conducted on outbred Swiss albino mice (30-35g). Each group, whether treatment or control, consisted of 10 mice. Mice were kept in conventional plastic cages with bedding. Most of the experiments were performed at controlled room temperature (22-23°C) with automated light and dark cycles of 12 hrs. Some of the experiments in which both *Agaricus* 30 and haloperidol were used together, were done under natural conditions at room temperature, 28-30°C. Mice were tested at least 2 weeks after shipment.

2.2. Drugs

Agaricus 30 was prepared from the ethanolic extract of the fungus, *A. muscarius*. The mother tincture was mixed with 90% ethanol in the proportion of 1:100 and then subjected to 100 manual downward strokes. This gives the 1st centesimal potency called *Agaricus* 1. All subsequent potencies were prepared by adding to one part of the preceding potency 99 pats of 90% ethanol and giving the mixture 10 sucessive strokes (12). *Agaricus* 30 obtained from Boericke and Tafel Co., Philadelphia,U.S.A.(Lot 165)was used in a major part of the study. *Agaricus* 30, used in combination with haloperidol, was prepared in the laboratory from the mother tincture (Boericke & Tafel). But here every dilution, instead of being succussed,was sonicated at 20 Hz for 30 sec using an ultrasonic homogenizer (Braun, Germany). This sonicated potency will be called here as Agaricus 30s. The drug was kept in the form of sugar globules totally soaked with the liquid drug with excess liquid removed by drawing. the drug was administered as an aqueous solution orally on mice at 0.05 ml/mouse by means of a glass pipette. The solution was prepared by mixing 20 globules in 1 ml of distilled water. Each globule weighed 7.18 mg on the average. Neurotransmitter drugs used in this study were as follows: 1) apomorphine (Sigma), a common dopamine agonist, that acts on both D1 and D2 receptors at doses of 5 mg/kg and also 0.3 mg/kg, 2) SKF 38393 (Smith kline Beckman C0.), a selective D1 agonist at 5 mg/kg, 3) bromocriptine (Sigma), a selective D2 agonist at 5 mg/kg, 4) haloperidol (Sigma), a common dopamine antagonist at 5 mg/kg. All these drugs and the saline (0.9%) control were given by i.p. injection in a volume dose of 10 ml/kg. Two doses of apomorphine were used because the drug is known to produce catalepsy at a low dose (0.3 mg/kg) and not at a high dose [14]. The test latency from the time of drug administration was 4 hr for *Agaricus* 30 or 30s, and 30 min for all the DA agonists and antagonist. The experimenter was blind to the test conditions.

2.3. Catalepsy test

Catalepsy, induced by drugs, usually manifests itself only when and experimental animal is placed in an awkward posture. Common methods for mice include placing the mouse on a vertical stringwrapped rod [15, 16], on an inclined wire grid [17], on an elevated horizontal bar[18, 19] or by "pinching" of scruff of the neck [5, 20]. In our study the one-trial inclined floor method [11] was used because it was very sensitive even for mild cataleptogens and did not involve stress or learning effect. In this method a smooth-bottom plastic (45 x 35 cm) or enamel tray (32 x 30 cm) was kept at a 70° incline. At the upper end pieces of sand paper (2 x 1 cm) were glued alternately to the floor in 3 or 4 rows in a checkerboard pattern that covered a width of 4 or 5cm. A paper-free border of 2.5 cm surrounded the checkerboard on the sides and the top. The smaller enamel tray was used to test catalepsy of mice treated with *Agaricus* 30s followed by haloperidol.

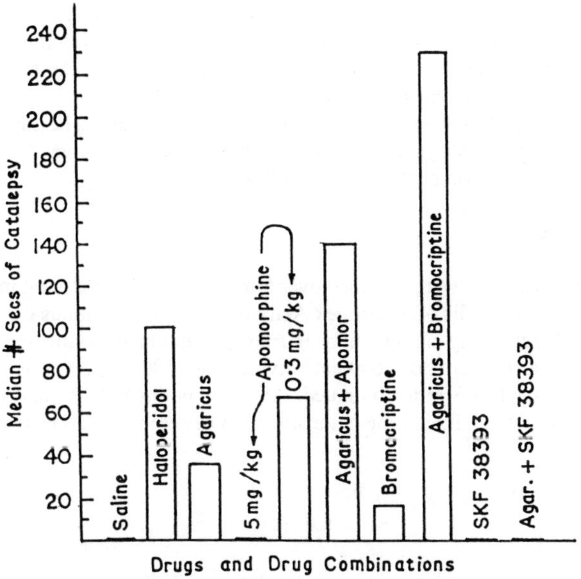

Figure 1 - Median numbers od seconds of Catalepsy in albimo mice produced by drugs and drug combinations (adapted from Sukul and Klemm, 1988).

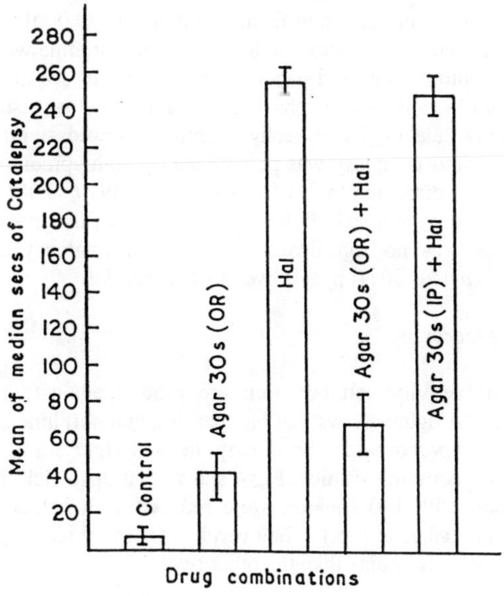

Figure 2 - Suppression of haloperidol catalepsy in albino mice by *Agaricus* 30s (oral).

A mouse was held by its tail and placed horizontally or with head upwards on the sand paper mosaic. A normal mouse initially explored the mosaic, and within about 5 to 50 sec walked or slipped head down toward the bedding material at the bottom of the tub. A cataleptic mouse sometimes explored initially but within a few seconds stopped all movement and clung to the sand paper in an awkward posture. The time a mouse remained immoblie on the strips was taken as the catalepsy duration. A convenient arbitrary cut-off time for scoring catalepsy was 300 sec.

2.4. Open field activity

Locomotor activity scores were obtained over 5 min in which horizontal movements were detected by the photocells of a Varimex Activity Monitor (Columbus C0.).

Activity scores for the drug combination *Agaricus* 30 and haloperidol were obtained in an open box with a floor area of 1600 cm^2 divided into 100 squares. The number of squares a mouse crossed in 5 min was noted. All tests, both catalepsy and open field activity, were performed in a fixed time in the aftenoon between 13.00 and 16.00 hrs.

3. Results

3.1. Catalepsy scores

Median numbers of seconds of catalepsy with different drugs and drug combinations were plotted in histograms in Fig. 1. Results with *Agaricus* 30s in combination with haloperidol were separately shown on the right side of the dotted oine. All of the treatment conditins except the 5 mg dose of apomorphine, D1 agonist SKF 38393 and *Agaricus* 30 + D1 agonist produced significant catalepsy ($P<0.0001$, Kruskal-Wallis test). The catalepsy produced by *Agaricus* and low-dose apomorphine was mild. In these treatments mice often explored the sand paper mosaic for 15-20 sec and this catalepsy could be easily disrupted by noise and by chasing the animals before testing. Post-hoc t-test comparisons of drug data vs the respective controls showed significant catalepsy with haloperidol or *Agaricus* catalepsy was potentiated by both apmomorphine and D2 agonist, bromocriptine and disrupted by D1 agonist SKF 38393 (<0.001). Catalepsy by haloperidol was significantly reduced ($P<0.02$) by *Agaricus* 30s given by oral route (Fig.2). However, there was no significant difference in catalepsy scores between haloperidol alone and *Agaricus* 30s (i.p) followed by haloperidol.

3.2. Open-field activity scores

Median numbers of activity counts/5 min are plotted against drugs and drug combinations in Fig.3. The figure shows that haloperidol, apomorphine at both low and high doses, *Agaricus* + apomorphine, bromocriptine and *Agaricus* + bromocriptine suppessed spontaneous locomotin of mice. Fig.4 shows that open field activity counts, as obseved in open box with 100 squares, were reduced with haloperidol, Agaricus (oral) + haloperidol and *Agaricus* (i.p.) + haloperidol. Of the three, *Agaricus* (oral) + haloperidol had higher activity counts than the other two.

4. Discussion

It is evident from the results that *Agaricus muscarius* extract at a dilution of 10^{-60}

with systematic mechanical agitation in 30 steps could produce marked cataleptic as well as anticataleptic effect in mice in combination with DA agonists and antagonist, respectively. The catalepsy test, therefore, serves as a reliable, non-invasive animal model to detect the biological effect of extremely high dilutions of *Agaricus muscarius* beyond the Avogadro number .

Dopamine receptors are G-protein coupled receptors which contain seven hydrophobic transmemebrane domains. There are 5 known sub-types of DA receptors. In the absence of ligands, G-protein coupled receptors interconvert between active and inactive conformations. These conformations are stabilized by agonists and anatgonists, respectively. G-proteins preferentially interact with the active conformation of receptors in the absence of agonists. Over-expression of a G-protein induces constitutive activation of compatible muscarinic receptors and this activity is blocked by appropriate antagonists [21].

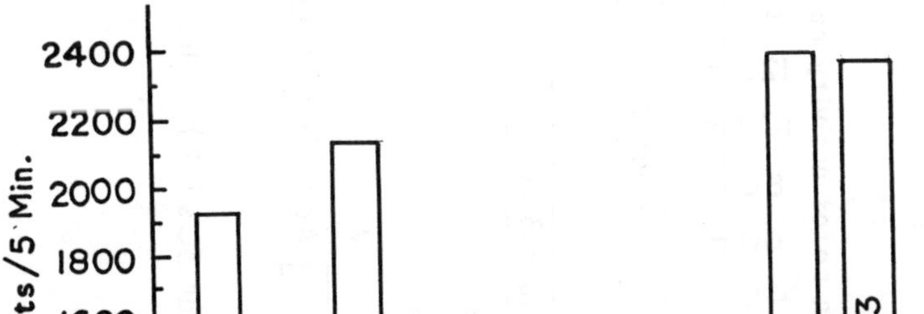

Figure 3 - Locomotor activity scores in albino mice for various drug-treatments obtained with Varimax Activity Monitor (Adapted from Sukul and Klemm, 1988).

As the potentzed *Agaricus* is put on the tongue, the receptors in the oral and paticularly lingual epithelium are stimulated and the message in the form of nerve impulse is transmitted via afferent nerves to the appropriate are of the brain. There expression of specific G-proteins occurs which selectivly couple with the DA subtypes like D1 and D4. This constitutive activation of the DA subtypes leads to inhinition of adenylyl cyclase. Activation of D1 and D5 leads to stimulation of adenylyl cyclase [22].

It seems that an optimum inhibition of adenylyl cyclase is the primary factor in the initiatin of catalepsy. D1 blockage with SCH 23390 causes a catalepsy that is potentiated by the mixed DA agonist, apomorphine and D2 agonist, bromocritine [23].

Concurrent administration of apomorphine (4 or 8 mg/kg) potentiated haloperidol catalepsy. Haloperidol in large doses reduces catalepsy [24]. The effect of potentized *Agaricus* in enhancing or reducing catalepsy with apomorphine, bromocriptine and haloperidol might have resulted from a balance of inhibition and stimulation of adenylyl cyclase through constitutive activation of D2 and D4 receptors.

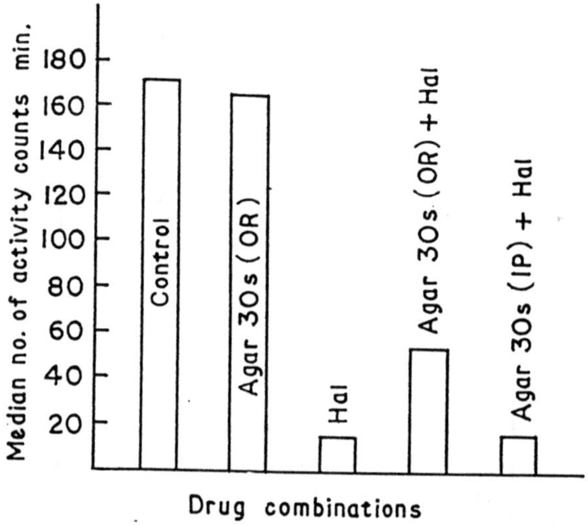

Figure 4 - Locomotor activity scores in albino mice for drug treatments obtained with an open box.

Acknowledgement

The co-authors of the two papers incorporated in the present lecuture were W.R. Klemm, S.P. Sina Babu, A. Paul and S. Ghosh.

References

[1] Zetter G., "Pharmacological differentiation of "nicotinic" and "muscarinic" catalepsy", *Neuropharmaclolgy*, vol. 10, pp. 289-296, 1971.

[2] Worms P., M.T. Willigens and K.G. Lloyd, "GABA involvement in neuroleptic-induced-catalepsy", *J. Pharm. Pharmacol.*, vol. 30, pp. 716-718, 1978.

[3] Sanberg P.R., "Haloperidol-induced catalepsy is mediated by postsynaptic dopamine receptors", *Nature*, vol. 284, pp. 472-473, 1980.

[4] Hartgraves S.L. and P.H. Kelly, "Role of mesencephalic reticular formation in cholinergic-inducced catalepsy and anticholinergic reversal of neuroleptic-induced catalepsy", *Brain Res.*, vol. 307, pp.47-54, 1984.

[5] Klemm W.R., "Experimental catalepsy: influences of cholinergic transmission in restraint-induced catalepsy", *Experientia*, vol. 39, pp. 228-230, 1983a.

[6] Klem W.R., "Cholineric-dopaminergic interaction in experimental catalepsy", *Psychopharmacology*, vol. 81, pp. 24-27, 1983b.

[7] Klemm W.R., "Experimental catalepsy is both enhanced and disrupted by apomorphine", *Psychopharmacology*, vol. 87, pp. 12-15, 1985a.

[8] Klemm W.R., "Neuroleptic-induced catalepsy: a D2 blockade phenomenon?" *Pharmacol. Biochem.* Behav., vol. 23, pp. 911-915, 1985b.

[9] Sukul N.C., L.Cherian and W.R.Klemm, "Alpha noradrenergic agonists promote catalepsy in the mouse", *Pharmacol. Biochem. Behav.* vol. 31, pp. 87-91, 198.

[10] Sukul N.C., Sujan K. Bala and Bhudev Bhattacharya, "Prolonged cataleptogenic effects of potentized homoeopathic drugs", *Psychopharmacology*, vol. 89, pp. 238-339, 1986.

[11] Sukul N.C. and W.R.Klemm, "Influence of dopamine agonists and an opiate antagonist on *Agaricus*-induced catalepsy, as tested by new method", *Arch. Int. Pharmacodyn.*, vol. 295, pp. 40-51, 1988.

[12] *The American Homoeopathic Pharmacopoeia*, 9th ed., Philadelphia: Boerick and Tafel, 1920.

[13] Paul A., S.P.Sinhababu, Souvik Ghosh and N.C.Sukul, "Suppressionnof haloperidol catalepsy by *Agaricus muscarius* through oral route", *Indian Biologist*, vol 26, pp. 14-17, 1994.

[14] Balsara J.J., T.R. Bapat, V.P.Gada and A.G.Chardorkar, "Small doses of apomorphine induce catalepsy and antagonise methampetamine sterotypy in rats", *Psychopharmacology*, vol. 78, pp. 194-197, 1982.

[15] Zetler G. and E. Moog, "Die Bulbocapnin-Katatonie, inhre Synergistern and Antagonisten", *Naunyn-Schmiediberg's Arch. Pharmacol.*, vol. 232, pp. 442-458, 1958.

[16] Zetler G., "Cataleptic state and hypothermia in mice, caused by central cholonergic stimulation and antagonized by anticholinergic and antidepressant drugs, *Int. J. Neuropharmacol.*, vol. 7, pp. 325-335, 1968.

[17] Chaillet P., H. Marcais-Collado and J. Constentin, "Catatonic or hypotonic immobility incuced in mice by intra-cerebroventricular injecton of Mu or Kappa opioid receptor agonists as well as encephalins or inhibitors of their degradation. *Life Sci.*, vol. 33, pp. 2105-2111, 1983.

[18] Wand P., K. Kuschindky and K.H. Sontag, "Morphine-induced muscular rigidity in rats," *Europ. J. Pharmacol.*, vol. 24, pp. 189-193, 1973.

[19] Snijders R., N.R. Kramarcy, W.R. Hurd, C.B. Nemeroff and A.J. Dunn, "Neurotensin induces catalepsy in mice", *Neuropharmacology*, vol. 21, 465-468, 1982.

[20] Amir S., Z.W. Brown, Z. Amit and K. Ornstein, "Body pinch induces long-lasting cataleptic-like immobilty in mice: behaviour characterization and the effect of naloxone", *Life Sci.*, vol. 28, pp. 1189-1194, 1981.

[21] Burstein Ethan S., Tracy A. Spalding, Hans Brauner-Osborne and Mark R. Brann, "Consctitutive activation of muscarinic receptors by the G-protein Gq", *FRBS Lett.* vol. 363, pp. 61-263, 1995.

[22] Civelli Olivier, "The Neuroreceptors: one key for many locks", *J. Recep. Sign. Transduc. Res.*, vol. 15, pp. 161-172, 1995.

[23] Klem W.R. and H. Block, "D-1 and D-2 receptor blockade have additive cataleptic effects in mice but receptor effects may interact in opposite ways", *Pharmacol. Biochem. Behav.* vol. 29, pp. 223-229, 1988.

[24] Toru M. and M. Tadashima, "Haloperidol in large doses reduces the cataleptic response and increases noradrenaline metabolism in the brain of the rat", *Neuropharmacology*, vol. 24, pp. 231-236, 1985.

EXPERIMENTAL PHARMACOLOGICAL RESEARCHES CONCERNING VEGETAL EXTRACTS IN HIGH DILUTIONS: II. BELLADONNA, AT VERY WIDE SCALE OF DILUTIONS, *IN VITRO*, ON THE ISOLATED RAT DUODENUM

AURELIA CRISTEA
Department of Pharmacology, Faculty of Pharmacy, Str. Traian Vuia 6, Bucharest, Romania

ABSTRACT

The Belladonna (B) tincture (containing 0.03 g % total alkaloids, expressed in atropine) was progressively diluted with water, at a geometrical ratio of 1: 100. Thus, a scale of centesimal dilutions (C) between the 1st dilution 10^{-1} (1C) and the 200th dilutions 10^{-400} (200 C), with a content of 6.023 x 10^{17} and respectively 6.023 x 10^{-381} atropine molecules/100 ml, was obtained. The dilutions were prepared in two ways: a) by simple homogenization (N dilutions) and b) by "dynamization" with manual succussion according to Hahnemann's homeopathic technique (D dilutions). The *in vitro* classical pharmacodynamic technique, on the isolated rat duodenum, mounted in the isolated organ bath, with special precautions was used. The isotonic contraction was recorded. The effects on the spasm induced by acetylcholine (ACH), at the standard dose of 10^{-6} g/ml bath, in the aqueous N solution, with a content of 33.55 x 10^{18} ACH molecules/100 ml was studied. On the scale of D dilutions 1C - 200C of B, the effect had a sinusoidal course (bidirectional multiphase effect). In the light of the author's informational theory of dosages, this double way (stimulator - inhibitor) effect is possible at different value scales of concentrations. In accordance with this theory, the information of biological and drug signals come out, may change and diversify, not only from the nature of substance and energy and also from their order and organisation in space and time, but also with their quantity, within certain limits. The definite organisation of a solvent as the specific informational substrate of very high D dilutions of B was investigated, by a calorimetric methodology.

1. Introduction

Within the framework of our conception about the informational pharmacology [1], we elaborated an informational theory of doses [2]. In accordance with this theory, the drug information comes out, may change and diversify not only from the nature of substance and energy and from their order and organisation in space and time, but also with their quantity, within certain limits.

In the light of this theory, we endeavoured to systematically develop an informational pharmacology of dosages [3]. For this purpose we initiated systematic researches on the pharmacodynamic effects of a drug, over a wide scale of doses.

The result of these researches demonstrated the double-way (bidirectional) effects of drugs, over a wide scale of doses and dilutions [4,5].

The double-way (stimulating and inhibiting) effect phenomenon made under a physiological function by the same chemical mediator or drug is now continuosly pointed out by the cell and molecular biology and pharmacology.

It is important to be mentioned that the reality of bidirectional effects of a drug, with a sinusoidal course, over a wide scale of dilutions, was known long ago in homeopathy. Therefore, the first drug chosen by us, for these studies, was Belladonna, because it is used both in allopathy and in homeopathy.

Thus in allopathy, atropina, the alkaloid extracted from *Atropa belladonna* is known to exert a spasmolytic action and is used in the treatment of colics on the basis of the therapeutic principle of contraries ("Contraria contrariis curantur"). As is known, in homeopathy, the Belladonna remedy exerts a spastic action and is also indicated at low dilutions, in the treatment of colics, but on the basis of another terapeuthic principle, that of similitude ("Similia similibus curantur").

In this lecture, we present our experimental pharmacological research, concerning the effects of Belladonna, *in vitro*, on the isolated rat duodenum, at a very wide scale of dilutions, up to the very high dilutions [4].

2. Methods

2.1. Drug and Dilution

The Belladonna (B) tincture (containing 0.03 g % total alkaloids, expressed in atropine) was progressively diluted with water, at a geometrical ratio of 1 : 100. Thus, a scale of centesimal dilutions (C) between the first dilution 10^{-2} (1C) and the 200th dilution 10^{-400} (200C), with a content of 6.023×10^{17} and respectively 6.023×10^{-381} atropine molecules / 100 ml, was obtained. The content of moles and atropine molecules / 100 ml was calculated for all the 1C - 200C dilutions, according to Avogadro's number (6.023×10^{23} molecules / 1 mol) and taking into account the molecular weight of atropine (MW = 289.41g ~ 300g) (Table 1).

The dilutions were prepared in two ways: a) by simple homogenization, without succussion (N dilutions) and b) by "dynamization" with manual "succussion" (knocking slightly, 100 times the vial against a firm substrate, rhythmically, at a frequency of 1 succussion / second for each dilution), according to Hahnemann's homeopathic technique (D dilutions).

Likewise, D dilutions 1C - 5C were prepared in water, starting from distilled water, as well as from alcohol 70°, the solvent with which the Belladonna tincture was prepared. These dilutions were used as control.

The reference substances, atropine sulphate (ATR) and acetylcholine chlorhydrate (ACH), were used in standard N aqueous solutions, 10^{-2} (~ 2C), with a content of 17.33×10^{18} and respectively 33.55×10^{18} molecules / 100 ml.

2.2. Animals

Male white Wistar rats, with an average weight of 150 g, fasting, with access only to water on the day of the experiment, were used.

2.3. Technique

The *in vitro* classical technique on the isolated rat duodenum, mounted in the isolated organ bath, with special precautions, was used.

The motor activity (motility) of the isolated duodenum, both the normal tone and the isotonique contraction, was recorded. Separate baths were used for the tests with: Belladonna, 1C - 9C dilutions, 10C - 200C dilutions, N dilutions, D dilutions, atropine.

2.4. Testing the effect on the ACH induced spasm

The intensity of the effect (decrease or increase of ACH spasm) was expressed in percentage against the contraction produced by ACH at the standard dose of 10^{-6} g / ml bath, on the same duodenum.

SUBSTANCE	DILUTIONS	MOL %	MOLECULES %
Belladonna	1	10^{-4}	6.023×10^{17}
Belladonna	5	10^{-12}	6.023×10^{9}
Belladonna	6	10^{-14}	6.023×10^{7}
Belladonna	9	10^{-20}	6.023×10^{1}
Belladonna	10	10^{-22}	6.023×10^{-1}
Belladonna	20	10^{-42}	6.023×10^{-21}
Belladonna	25	10^{-52}	6.023×10^{-31}
Belladonna	30	10^{-62}	6.023×10^{-41}
Belladonna	40	10^{-82}	6.023×10^{-61}
Belladonna	45	10^{-92}	6.023×10^{-71}
Belladonna	50	10^{-102}	6.023×10^{-81}
Belladonna	55	10^{-112}	6.023×10^{-91}
Belladonna	60	10^{-122}	6.023×10^{-101}
Belladonna	70	10^{-142}	6.023×10^{-121}
Belladonna	75	10^{-152}	6.023×10^{-131}
Belladonna	80	10^{-162}	6.023×10^{-141}
Belladonna	90	10^{-182}	6.023×10^{-161}
Belladonna	95	10^{-192}	6.023×10^{-171}
Belladonna	100	10^{-202}	6.023×10^{-181}
Belladonna	110	10^{-222}	6.023×10^{-201}
Belladonna	120	10^{-242}	6.023×10^{-221}
Belladonna	125	10^{-252}	6.023×10^{-231}
Belladonna	130	10^{-262}	6.023×10^{-241}
Belladonna	140	10^{-282}	6.023×10^{-261}
Belladonna	150	10^{-302}	6.023×10^{-281}
Belladonna	160	10^{-322}	6.023×10^{-301}
Belladonna	170	10^{-342}	6.023×10^{-321}
Belladonna	200	10^{-402}	6.023×10^{-381}
Atropine	2	2.87×10^{-3}	6.023×10^{18}
Acetylcholine	2	5.57×10^{-3}	6.023×10^{18}

Table 1. Centesimal dilutions (CH) content in moles and molecules of atropine / 100 ml, according to Avogadro's number and molecular weight.

2.5. Statistical Analysis

The statistical significance of the results was calculated by using the Student's "t" , Wilcoxon tests.

3. Results

The effect of Belladonna (B) dilutions on the spasm induced by acetylcholine (ACH)

The N water had no effect. The effect of D water consisted in a slight stimulation was statistically insignificant. The N solutions of alcohol 70^0 in water had the effect of a statistically insignificant lowering of the spasm, whereas the D aqueous dilutions of alcohol 70° had no effect.

The N solutions of B had a spasmolytic effect only up to the 9C dilution, with a content of 6.023×10^2 ATR molecules / 100 ml (unidirectional, monophasic effect) (Fig. 1).

The D dilutions of B had an effect up to the 200C, with a content of 6.023×10^{-381} ATR molecules / 100 ml. On the scale of D dilutions 1C - 200C, the effect had a sinusoidal course, *i.e.* the decrease of the contraction at the 1C - 25C, 55C - 60C and 160C - 200C dilutions and the increase of the spasm at the 30C - 50C and 70C - 150C dilutions (bidirectional, multiphasic effect) (Fig. 1).

The higher D dilutions were more efficacious. Thus, among the D dilutions which have antagonized the ACH spasm, the highest dilution studied, 200C, had the most intense effect of 71.12 %. Among the dilutions which have amplified the ACH spasm, the 75C had the higest effect of 95.45 % (Table 2).

Conversely, at the N dilutions, the efficacy decreased with the increase of the dilution, hence with the lowering of the concentration.

The statistically significant values ($p < 0.05$) are listed in Table 2.

4. Discussion

On the scale of D dilutions 1C - 200C of B, the effect had a sinusoidal course (bidirectional, multiphasic effect) (Fig.1).

In the light of informational theory of dosages [2,3], this double-way (stimulator - inhibitor) effect is possible, at different value scales of concentrations. In accordance with this theory, the drug information comes out, may change and diversify not only from the nature of the substance and energy and from their order and organisation in space and time, but also with their quantity. That means that the information carried by a certain structure quality, at various value scales of concentrations, may be different in terms of quality.

In another experimental pharmacological research, we also demonstrated the bidirectional effect of two allopathic drugs, *i.e.* naloxon and clonidine, over a wide range of doses, on stress analgesia [5].

Other researchers carried out the bidirectional effect of naloxon, at large and very small doses [6].

The statistically significant results disclosed quantitative and especially qualitative differences in the pharmacodynamics of the non-dynamized (N) and dynamized (D) preparations according to Hahnemann's technique.

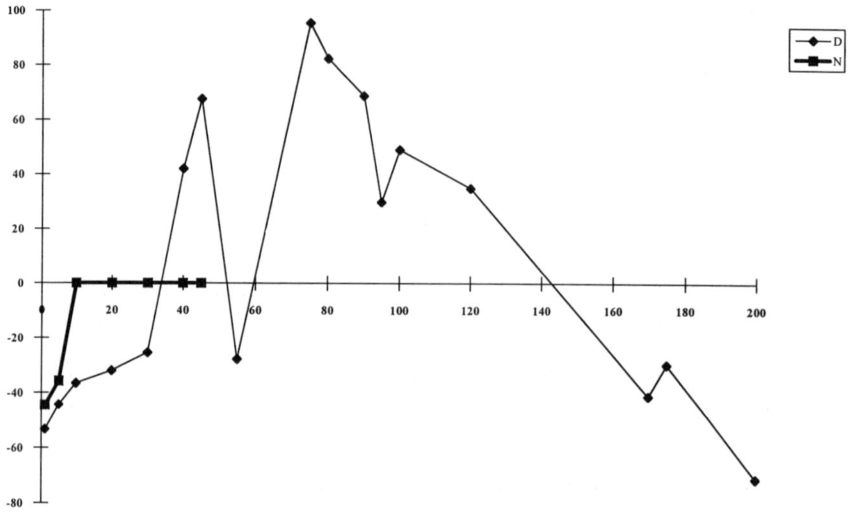

Figure 1. Statistically significant Belladonna effects ($p < 0.05$), on the spasm induced by acetylcholine (ACH)

Quantitative differences:

- At the same degree of dilution, the D dilutions had a more intense biological effect than the N dilutions.

Qualitative differences:

- The N dilutions exerted a pharmacodynamic action only up to the limit of the molecular specificity, calculated on the basis of Avogadro's number at the 9C dilution, with a content of 6.023×10^1 molecules / 100 ml.

In contrast with the above, the D dilutions exerted a statistically significant biological effect also at high dilutions, over 10C, which mathematically and according to Avogadro's number, do no longer contain any molecule of the dissolved active substance.

- The N dilutions had unidirectional monophasic pharmacodynamic action on the 1C - 9C scale.

The action of D dilutions on the spasm was bidirectional multiphasic, with a sinusoidal course on the 1C - 200C scale (Fig.1).

- Belladonna (B) in N dilutions exerted only a relaxant pharmacodynamic action on the duodenal smooth muscle, on the 1C - 9C scale, an effect known in allopathic pharmacology.

In D dilutions, on the 1C - 200C scale, B exerted also a contracting pharmacodynamic action on the duodenal smooth muscle, action which in the allopathic pharmacology is unknown for B.

- In N solutions the effect decreases with the increase of the dilution degree and rises with the increase of the concentration and dose.

In D dilutions, the efficacy rises with the increase of the dilution degree.

DILUTION C	± Effects % D	(Means ± SEM) N
1	-53.30 ± 1.10	-44.54 ± 0.10
5	-44.30 ± 1.20	-35.79 ± 1.50
10	-36.55 ± 1.30	0.00
20	-31.95 ± 1.30	0.00
30	-25.40 ± 0.50	0.00
40	+42.02 ± 1.20	0.00
45	+67.60 ± 0.10	0.00
55	-27.80 ± 0.30	-
75	+95.45 ± 1.20	-
80	+82.39 ± 0.60	-
90	+68.74 ± 1.30	-
95	+29.81 ± 0.20	-
100	+48.92 ± 1.10	-
120	+34.99 ± 0.70	-
170	-41.15 ± 1.10	-
175	-29.73 ± 1.60	-
200	-71.12 ± 2.00	-

Table 2. Statistically significant Belladonna effects ($p < 0.05$), on the spasm induced by acetylcholine (ACH).

C = centesimal dilution at the 1 : 100 ratio
D = dynamized dilutions, according to Hahnemann's technique
N = non-dynamized dilutions prepared by sample homogenization
p = probability by Student's "t" test ($n = 8$)

All these qualitative differences, detected between the N and D preparations, with respect to their biological effect, can be explained only by admitting that another form of information, qualitatively different from that existing in the N preparations, is also present in the D preparations. The appearance of this new form of information in D dilutions can be due only to dynamisation by rhythmical succussion according to Hahnemann's preparation technique.

Since in high D dilutions, exceeding 10C, the molecule of dissolved active substance is no more present as an informational substrate, according to Avogadro's number, the information in D dilutions should be looked for in another informational substrate: a submolecular structure, a certain form of energy or, merely, a specific molecular organization of the solvent.

The clearly superior potency and efficacy of the D preparations demonstrate that the information conveyed by these preparations is much more specific to the living matter. The D high dilutions, characterized by infinitesimal doses and by dynamisation according to Hahnemann's preparation technique, has a marked informational character, acting in conformity with one of the laws of information action: "Minimum quantity, maximum significance, maximum effect" [7,8].

Starting from the hypothesis that, according to the laws of information processing and storage, the structural information carried by the active atropine molecule can be conveyed, under certain condition, by the solvent, we performed - concomitantly with the pharmacological study - a research by physical methods, in order to detect in the solvent the informational substrate of high B dilutions.

For this purpose, a calorimetric methodology was used, *i.e.* the high resolution mixing calorimetry (HRMC) and the differential scanning calorimetry (DSC), the sensitivity and reproductiveness of which in the study of aqueous solutions has been demonstrated on a wide series of cases [9]. The HRMC and DSC experiments disclosed a definite organisation of water in D dilutions, in comparison with N dilutions. Two distinct structures, with a sinusoidal multiphasic course, in accordance with the bidirectional multiphasic pharmacodynamic effects, were observed in D dilutions of B [10,11].

These results, which we have obtained, in calorimetric studies, on aqueous Belladonna (B) solutions, highly diluted in a centesimal geometric progression on the 1C - 200C scale, are consistent also with those of other researchers, which were carried out on high dilutions and lead to the same conclusions [12].

A systematic study on the bidirectional rhythmic effect phenomenon has been performed [13].

5. References

[1] Cristea, A., "Pour une pharmacologie cybernetique et informationnelle", *Cybernetica* (Namur, Belgium) 2, 1991, pp. 85-93.

[2] Cristea, A., "Teoria informationala a dozelor (Informational theory of dosages)", Abstr. *6-th Nat. Conf. of Cybernetics*, Bucharest, Romania, 1994.

[3] Cristea, A., "For an informational theory of dosages. Illustration from cell and molecular biology and pharmacology", *Cybernetica* (Namur, Belgium), 1996, in press.

[4] Cristea, A., S. Nicula, V. Darie, "Pharmacodynamic effects of very high dilutions of Belladonna on the isolated rat duodenum", Abstr. *5-th GIRI Meeting*, Paris, France, 1991.

[5] Cristea, A., S. Negres, A. Ciobanu. "The double-way effects of naloxon and clonidine at heterosynaptic level", Abstr. *CINP Regional Conf.*, Wienne - Prague, 1995.

[6] Kayser, V., J.M. Benoist, A. Neil, M. Gautron and G. Guilbaud. "Behavioural and electrophysiological studies on paradoxical antinociceptive effects of an extremely low dose of naloxone in an animal model of acute and localised inflamation", *Experimental Brain Research* 73, 1988, pp. 402 - 410.

[7] Restian, A., "Informatie minima, semnificatie maxima (Minime information, maxime signification)", *Rev. de Filosofie* (Bucharest, Romania) 9, 1971, pp. 1229-

1235.
[8] Restian, A., *Homo Ciberneticus*, Ed. St. si Enc., Bucharest, Romania, 1981.
[9] Dragan, G., "Comparative study on molecular associations in solid and liquid phases of aqueous solutions. Presentation of high resolution mixing calorimetry (HRMC) and the differential scanning calorimetry (DSC)", *Acta Polymerica* 4, 1987, pp. 211-220.
[10] Cristea, A., G. Dragan, V. Darie,. "Potency of pharmacodynamic action and structure of highly diluted aqueous solutions of Belladonna", Abstr. *4-th Internat. Conf. on Water and Ions in Biological Systems*, Bucharest, Romania, 1987.
[11] Dragan, G., V. Darie, "Composite structure of homeopathic solutions of Belladonna as revealed by calorimetric measurements", Abstr. *5-th GIRI Meeting*, Paris, France, 1991.
[12] Cabaner, C., M. Bastide, "Isothermal microcalorimetric analysis of homeopathic dilutions", Abstr. GIRI Meeting, Paris, France.
[13] Husemann, K., "Rhythmusphänomene beim wirksamkeitsnachweis potenzierter heilmittel - Nachgewiesen von Kolisko (1923) bis Cristea (1991)", *Der Merkurstab* 2, 1992, pp. 73-90.

MEMBRANE ANTIOXIDANTS. MECHANISM OF ACTION OF ANTIOXIDANTS IN ULTRA-LOW DOSES.

E.B. BURLAKOVA
Institute of Chemical Physics, 117334, ul. Kosygina 4 Moscow, Russia

ABSTRACT

In 1983, staff members of our institute and colleagues at the Institute of Psychology studied the effect of antioxidants on the electrical activity of an isolated neuron of an edible snail (Helix pomatia) and obtained an unexpected result. The initial dose of the substance (10^{-3}M), was, for the neuron, not only active, but quite toxic, which made it necessary to use a weaker solution. To our surprise, a dose that was 10^4 lower than the initial dose proved not only less toxic, but more effective. Its further reduction heightened the effect, which reached a maximum (at 10^{-15}M) and then declined until the results (at 10^{-17}M) virtually coincided with the pilot results [1]. The same effects were later recorded in experiments involving animals when an antioxidant was administered, and the sensitivity of a choline mimetic arecoline was studied.

We examined these effects on animal and plant cells, at the organism level, the biomacromolecule level, etc. The action of antitumor and antimetastatic substances, radiation protection preparations, plant growth inhibitors and promoters, various classes of neurotropic preparations, hormones, adaptogens, immunomodulators, detoxifying agents, antioxidants, and physical factors (such as ionizing radiation) were subjected to analysis. Ten years of work (1983-1993) led us to believe that this is not a specific action of one individual preparation or a specific response of an individual biological agent, but a new law of interaction between a biological object and ultra-low doses of biologically active substances (BAS). Each of this substances can have a target, an enhancement mechanism and metabolic features of its own, but, when used in ultra-low doses they display a number of common properties. It was later discovered that low-intensity physical factors display analogous characteristic.

The levels of biological organization at which the effect of ultra-low doses of BAS was discovered also vary from cells, macromolecules, organs, and tissues to animal and plant organisms and even populations. This does not mean that the effect was observed, when ultra-low doses of any BAS were used on any biological object. I only wish to strss that the effect of the action of BAS concentrations of 10^{-13}-10^{-17}M and lower cannot be linked to any definite substance structure or biological organization level.

1. "Dose-Response" Dependence for Biologically Active Substances

The general laws of the effect of ultra-low doses manifest themselves most strikingly when dose dependences are examined. In some cases, this dependence is bimodal: the effect is heightened when ultra-low doses of these preparations are used and diminishes as the dose is increased, which gives way to a so-called dead space, where it is unnoticeable, and then increases again (Fig.1, curve 1). Occasionally, the dose dependence reveals a phase where the effect "changes the sign". For example, if inhibition was observed in the ultra-low dose range, it was replaced by stimulation as concentration increased , where it was succeeded by inhibition once again (Fig.1, curve 2). Cases are known when, in a very broad concentration range, the effect did not depend on the dose (Fig.1, curve 3). We studied the effect of a herbicide (a hydroperoxide) on a plant culture of cells. The preparation displayed equal activity under doses differing by six orders: 10^{-13}-10^{-7}M (Fig.2). In the space between these concentrations, a response was absent. Antioxidants have two stimulation maxima on the growth of plant cells on the dose dependence curve [2].

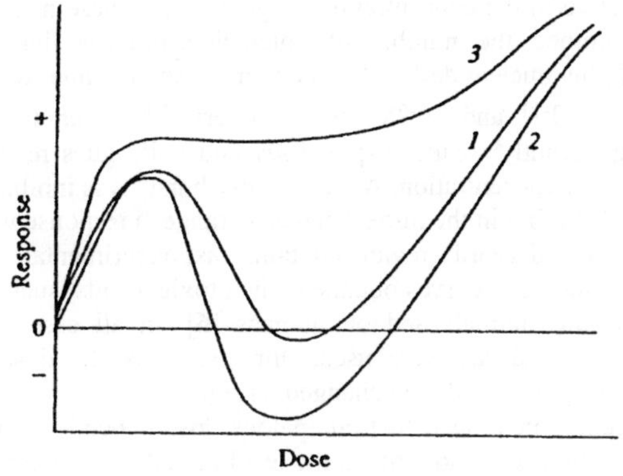

Figure 1. Types of dose-response curves.

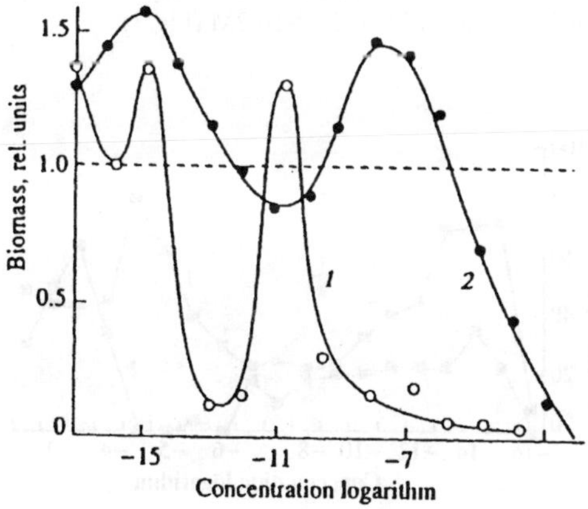

Figure 2. Effect of the peroxide of (1) succinic acid and (2) lontrel on the growth of the biomass of a callus culture of tobacco cells.

Experiments conducted in other laboratories confirmed our results. Topchieva and colleagues constructed a biphasic curve for the rate of the synthesis of antibodies by the spleen when an immunogenic polymer was administered to mice. Equal (similar) activity was displayed by doses of a polymer differing by five orders, while the polymer was virtually inactive in the concentration range between these maxima [3]. Robertson and Grutsch examined the number of colchicine mitoses that depended on the concentration of the nucleotide in the medium. The maxima were registered at a concentration of $5 \cdot 10^{-11}$ and 10^{-6}M. In the interval between these two doses, the response was significantly weaker [4]. Zaitsey and colleagues reveale the activity of enkephalin at 10^{-14}M concentration. A "respiratory burst" was inhibited when the doses equaled 10^{-14}-10^{-11}M, but in the in-between dose range, a response was absent [5].

Davis and Svendsgaard's review contains vast experimental and clinical data on the U-shaped dose-response curves obtained when toxic agents, such as lead of alcohol, were used on various animals and on humans [6]. In all cases, the response was pronounced when low doses were used, diminished as the doses were increased, returned to standard values, and then changed its sign.

I have already mentioned that anomalous dose dependence for ultra-low BAS concentrations was registered not only at the level of cell or organism response but also at the level of individual biomacromolecules. An enzyme, protein kinase C isolated from animal heart, was used to study the inhibiting effect of tocopherol. Protein kinase C is a complex, lipid-dependent, calcium-demanding enzyme, which has several active allosterically interacting centers. Writings on the subject contain data proving that within the 10^{-4}-10^{-5}M concentration range, tonopherol inhibits the activity of protein kinase C [7]. Our research into the inhibiting effect of tocopherol within a broad range of concentrations (10^{-18}-10^{-5}M) showed that there is a bimodal dependence between dose and effect (response). Tocopherol vigorously inhibits protein kinase C at the concentrations of 10^{-16}-10^{-12}M and 10^{-4}-10^{-5}M (Fig.3).

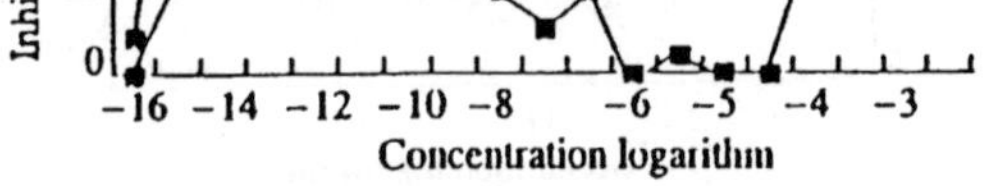

Figure 3. Effect of α-tochoferol on the activity of (1) protein kinase C and (2) protein kinase C activated with forbolic ester.

In the interval between these values, its activity is significantly lower. Analogous data has been obtained for protein kinase C activated by forbolic ester. The inhibiting effect of this enzyme was lower than that of the non-activated enzyme at low tocopherol concentration, but the bimodality of the dependence was retained [8].

2. "Dose-Response" Dependence for Ionizing Radiation

Its correct definition has a practical importance, because the choice of the model determines how accurate risk assessment will be. The study of the low-dose irradiation effect was comprised of two stages. Initially, many radio biologists maintained that low doses do not affect biological objects, that the consequences of irradiation of living systems began to make themselves felt only after a certain threshold dose had been reached, and that, therefore, the "dose-response" dependence had a threshold character.

Later, the concept of linear or quadratic-linear dependence on the dose was used to evaluate the stochastic effects of irradiation (cancer, genetic disorders, and teratogenic effects).

In the 1970s, papers on the enhancement effects of low-dose irradiation appeared, and a general notion of so-called hormesis was shaped; this notion included enhancing an organism's vital activity and raising its resistance to the action of injurious effects and its capacity for survival. In our country, the founding father of this trend was A.M. Kuzin, and in the West, it was L.F. Sagan. For a long time, this line of research attracted the attention of only a few radiation biologists, and it became a subject of scientific discussion only in the 1990s.

When measuring the risk of radiation damage to biological objects, not only well-known (threshold, linear and quadratic-linear) models were employed, but also a model that takes the process of hormesis into account. According to the latest model, under low irradiation doses, an injurious enhancement effect appears, which first increases, then diminishes, and subsequently grows monotonously as the dose is increased. I believe that, in front of the range of irradiation doses causing an enhancement (adaptive) effect, an area exists in which the irradiation dose produces the same changes in cells as higher doses. These two dose ranges (low and high), with identical effects on biological objects, are separated by the interval within which the opposite enhancement irradiation effect manifests itself. Therefore, the general "dose-response" curve is not U-shaped: it is biphasic.

The effect of low-doses of low-intensity irradiation on chromosom. aberrations in lymphocytes has been studied at seven of the world's largest laboratories (e.g., the Livermore National Laboratory United States). An insignificant peak in the growth of their number under low irradiation doses (approximately one rad) was observed; a virtually constant aberration level was observed at the dose was increased to 2-4 rads, and, only when it reached 5 rads and above, the number of chromosome aberrations began to rise monotonously.

Spitkovskii analyzed these data and compared them to the results he himself had obtained when studying the charges in the degree of chromatin compactness in lymphocyte nuclei after irradiation [9]. He established that, when the dose was only 1 rad or even lower, the diameter of lymphocyte nuclei began to grow and continued to do so until the dose reached 2.5 rads, after which it began to decrease and then increased again when the dose reached 5 rads and above.

Researchers at the Institute of Chemical Physics and the Institute of Biophysics began a study of low-intensity irradiation on DNA molecules, membranes of various organs, and tissues of mice. They discovered a bimodal dependence of the rate of alkaline elusion of DNA on the irradiation dose. In this experiment the irradiation intensity was 6 rads over 24 h, and the irradiation dose under which the first maximum was recorded was 6-12 rads. The same regularity was revealed when the values of the binding of spleen DNA by nitrocellulose filters in conditions of neutral pH were measured. Identical effects were obtained for irradiation doses differing from 10 to 20 times (Fig.4).

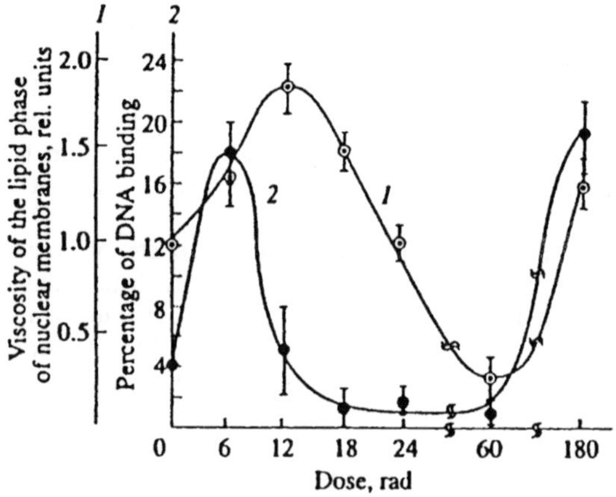

Figure 4. Measuring the viscosity of the lipid phase of nuclear membranes of (1) liver cells and (2) the percentage of binding of spleen DNA by nitrocellulose filters (pH 7) in dependence on the irradiation dose in mice.

These results can be determined by the presence of single-stranded breaks, DNA-protein cross links, and other conformational DNA changes. This explanation is, however, not true when low-dose irradiation is used on cells.

Spitkovskii proposed that a programmed subpopulation of cell exists, whose death is preceded by yet another genetically determined reaction, autoinduction, in the cells of chromosomal aberrations or other structural rearrangements in the genome. The latter lead to the emergence of new genetic variants and possible selection of those best adapted to the changed environment or to expected new effects. I believe that the mechanism of the emergence of damage in DNA under the impact of low doses may differ from that under high doses and assign the main role to the damage of membranes under the impact of irradiation. The first to change are their structural characteristics followed by the activity of membrane enzymes, including DNA repair enzymes. Indeed, the defects in the membranes of the spleen, liver, and erythrocytes (their viscosity, antioxidant activity, and lipid composition) obey the same laws: bimodal dose-response dependence and identical effect for irradiation doses differing by approximately 20 times [10].

"Paradoxical" dose-response dependencies were also discovered in the case of microwave radiation: ranges of low- and high-intensity irradiation that do equal damage to biological objects were discovered. These ranges are separated by an interval in which this effect is practically absent [11].

3. Sensitivity at Biological Objects after Ultra-Low Doses

The effect of ultra-low doses of BAS or low-intensity irradiation manifests itself not only in the anomalous dependence on the dose (although this fact may have the greatest practical significance), but in other effects as well-above all, in the modification of sensitivity of biological objects to a broad range of endogenic and exogenic influences. Thus, low-intensity irradiation causes changes not only in the physico-chemical, biophysical, and biochemical cell characteristics in irradiated animals, but also in physiological parameters. For example, the sensitivity to choline mimetics of the central nervous system of animals subjected to ultra-low doses of ionizing radiation changed, as did the sensitivity to subsequent irradiation of cell irradiated with the same dose [10].

We experimented with plants to examine the effect of herbicides. The plants were treated with two preparations derived from 2-chlorphenoxyacetic acid. The dose of one was 10^{-11} - 10^{-13}M, and of the other, 10^{-8}M. The effectiveness of each taken separately did not exceed one percent, and the expected additive effect equalled two percent. In fact, when both preparations were administered, we observed a one hundred-percent inhibition of the plants' growth. Similar data was obtained by Bonavida and colleagues. When low doses of tumor necrosis factor and adriablastine were Jointly administered to animals, a strong synergistic effect was observed [12].

Our explanation of this data allows for the fact that ultra-low doses affect the sensitivity of cells and organisms to subsequent impacts of other biologically active substances or injurious effects. Sensitivity may either increase, in which case a synergistic effect is observed, or decrease, thus revealing the protective action of these preparations.

We proposed that the channel along which the low doses of low-intensity irradiation operated could also be used by radiation protection preparations administered in ultra-low concentrations. We registered such action on the part of a 10^{-15}M dose of antioxidants during low-intensity ionizing radiation. The same doses of the same preparations revealed a protective effect against chemical toxic substances.

This explanation is also valid for the antimetastatic effect of 10^{-15}-10^{-17}M doses of lodinamine and ephasol (this was observed by N.P. Konovalova and colleagues). In the case of B-16 melanoma and Lewis's cancer, almost total suppression of the metastatic process was observed. The effect that appeared when 10^{-15}-10^{-17}M doses of ephasol were administered was comparable to the effect of 10^{-2}-10^{-4}M doses [13].

The response of biological objects to the introduction of ultra-low doses of BAS has another important feature, heterogeneity, which in determined by elementary biochemical, biophysical, physiological, and other characteristics. Goloshchapov and colleagues have established that 10^{-14}M doses of the antioxidants phenozan and anphen do not exert a uniform effect on the antioxidant activity of lymphocyte membranes [14].

In lymphocytes with a high level of peroxide oxidation of lipids, this inhibition of this process was observed when the preparations were administered, and in the lymphocytes with a low level of such oxidation, induction was observed (Fig.5).

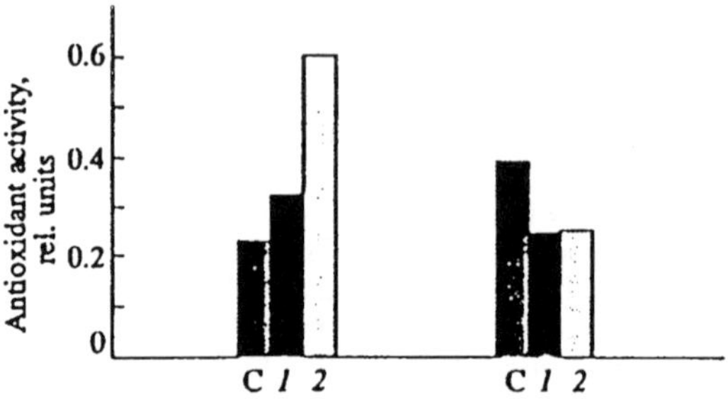

Figure 5. Effect of a 10^{-14} M dose of (1) phenosane and (2) anphen on the antioxidant activity of lymphocyte membranes (C-control)

4. Possible Mechanisms of the Action of Ultra-Low Doses

To understand how ultra-low doses operate on biological objects, the possibility of reactions of such a small number of molecules with their targets must first be explained from the kinetic viewpoint. With concentrations of 10^{-15}M and lower, the mass action law ceases to be valid and the notion of "concentration" is, in a measure, lost. The work of Blumenfeld, Grosberg, and Tikhonov contains an equation that describes the reaction between molecules and small but macroscopically linked vesicles in terms of statistical physics [15]. It is shown that the mass action law breaks down when the volume of vesicles of the reaction equilibrium constant in small enough, and the average number of free particles inside the vesicle is on the order of one or below.

Fluctuations are important, especially for biological vesicles with a size of 10^2-10^3A.

Some specialists pay attention to the kinetic paradoxes associated with the fact that many biologically active substances are effective when used in doses that are several orders lower than the constants of dissociation of these ligands from receptors.

Ashmarin and colleagues examined the action of 14 neuropeptides, four of which proved active when their doses varied between 10^{-16} and 10^{-14}M [16]. At the same time, the known constants of dissociation of neuropeptide complexes from their receptors are equal to 10^{-12}-10^{-18}M.

Similar regularities were obtained for antioxidants. The constant of their specific binding by proteins are equal to 10-8-10-9M, while the effect of antioxidants makes itself felt at concentrations of 10^{-15}-10^{-18}M. The simplest way to explain this paradox is to assume that receptors that have lower constants of dissociation of their complexes from biologically active substances exist, but they cannot be discovered using available research techniques. However, Zaitsev and Sazanov offered another original explanation

[17]. They used the model of cellular adaptation developed by Coshland to describe the chemotaxis of bacteria in response to the action of attractants (repellents). Under this model, bacteria respond not to the concentration of an attractant as such, but to the difference between the concentrations, that is, their gradient in two neighbouring points along the course of their movement. A higher (lower) concentration produces a prompt response, which fades (returns to the initial level) over a longer time than it takes to rise.

Zaitsev and Sazanov modified this model to explain the kinetic paradoxes of the ultra-low doses effect. They assumed the existence of several (in the simplest case, two) receptors or enzymes with which the ligand is bound. The receptors have differing constants of transition from the initial to the active state (and back) and operate in opposite directions. Calculating the rates of the manifestation of the general effect as the difference of the effects produced by the binding of ultra-low doses of the substance upon different receptors, the researchers demonstrated the possibility of existence of bimodal dependence with a maximum for ultra-low BAS doses. The character of the response depends, among other things, on the time of observation and the differences in the complex dissociation constants.

Using this model, it is possible to explain both the greater "distance" of effective concentrations from the ligand-receptor complex dissociation constants and the possibility to feel the effect of the action of concentrations that are lower than endogenous ones by several orders.

We response another explanation about kinetic paradoxes [18]. It is based on the idea of allosteric interaction of catalytic centers in an enzyme molecule. Let us suppose that an enzyme or a receptor contains several centers with differing affinity to the substrate, e.g., a dissociation constant of 10^{-13}M for one center and 10^{-8}M for another. When low doses of the preparation are administered, molecules are bound mainly to the high-affinity center.

When the dose is raised, the other center joins the "game". It enters into allosteric interaction with the first center, reducing its affinity with the substrate, and all of the molecules that used to be bound to the first center "leave" it. They can be bound to it again after the concentration approaches the value of the ligand complex of dissociation from the first center, which is reached under the impact of the second. This idea is used, in particular, to explain the complex response of the olfactory receptor to the changes in the substrate dose.

Blumenfeld advanced the idea of parametric resonance as a possible mechanism of action of ultra-low concentrations of biologically active substances at cellular and subcellular levels [19] . He believes that parametric resonance arises when there is coincidence between the temporal parameters of intracellular processes triggered by BAS and the typical time of the BAS approaching the target. As a result of binding of the BAS and its target, the enzyme (receptor) passes into a conformational nonequilibrium state, which at a certain relaxation phase ensures its maximum activity.

Under very high BAS concentrations, when the typical time of the approach to the target is small and the approach frequency is high, the enzyme as a whole will be in a weak final equilibrium state. At very low concentrations, when the typical time is very large, almost all of the enzyme (receptor) will remain in a low-active initial equilibrium state. Also, only for BAS doses under which the typical time of the substance's approach to the target and the temporal parameters of the intracellular processes it triggers off virtually coincide can we expect the maximum stationary concentration of the intermediate active nonequilibrium conformation.

Calculations show that the peak effect is at 10^{-11}-10^{-15}M doses. For similar but slightly higher or lower concentrations, the effect is significantly weaker. An equal or stronger effect can be attained only when the concentration is raised to 10^{-5}-10^{-4}M, when the changes follow a different pattern, due to the saturation of enzyme (receptor) centers with-ligands. According of these ideas, BAS can interact with their targets even if the dissociation constant of the BAS target complex is significantly higher (by an order) than the concentrations used. Effect weakening under higher doses can also be explained.

An analysis of the interaction of two opposite processes, for example, damage and repair. or specific activity and toxicity, also provides an opportunity to explain the bimodal nature of dose dependence. Let us suppose that, under low irradiation doses directed at a system, repair is not induced in full or at all, as a result of which we observe only the effects associated with damage to biomolecules. As the radiation dose increase, repair processes become activated and may completely "neutralize" the damage. It is then that a "dead space" in the dose-response dependence can be observed. The prevalence of repair over damage leads to the hormesis effect.

Subsequently, as the radiation dose is raised further, the damage-repair relationship intensifies, and the effect grows in a linear or quadratic-linear mode toward the dose. Thus, although thorough experimental verification of these models is necessary, quantitative mathematical and qualitative phenomenological models help explain the laws of the action of ultra-low doses.

Researchers do not agree about why the mechanism of response of biological objects to the action of ultra-low BAS doses and low-intensity physical factors has a universal character. Some believe that the common regularities in the dose-response dependence, the change in the sensitivity of biological objects to a wide ranger of factors (internal and external) and the heterogeneity of the response to ultra-low doses point only to a superficial similarity of these phenomena. In each individual case, one should look for a specific mechanism, specific targets of action, specific possibility to enhance the signal, etc. Others do not deny the specific character of the response in each individual case, but develop ideas about the universal character of the response of biological objects to ultra-low BAS doses and systemic changes in metabolism following signals from the external environment.

I support the latter viewpoint.

References

[1] Burlakova E.B., Grechenko T.N., Sokolov E.N., and Terkhova S.F., "The Effect of Inhibitors of Radical Reactions of Lipid Oxidation on Electrical Activity of the Isolated Neuron of the Edible Snail", in *Biofizika*, 1986, vol.31, no.5, pp. 921-923.

[2] Bogatyrenko T.N., Redkozubova G.P., Konradov A.A., et al., "The Effect of Organic Peroxides on the Growth of Cultivated Cells of Higher Plants", in *Biofizika*, 1989, vol.34, no.26, pp. 327-329.

[3] Topchieva I.N., Erokhin V.N., Osipova S.V., et al., "Blok Copolymers of Ethylene Oxide and Oxide (Pluronics) as lmmunomodulators and Antitumor Agents", in *Biomed. Sci.* (Tokyo), 1991, vol.2, pp. 38-44.

[4] Robertson A.D. and Grutsch I.F., "Biphasic Responses, Quantal Signal and Cellular Behavior" in *J.Theor.Biol.*, 1987, vol.25, no.1, pp. 41-47.

[5] Zaitsev S.V., Sazanov L.A., Koshkin A.A., et al., "Respiratory Burst Inhibition in Human Neutrophils by Ultra-Low Doses of [D-Ala] Methionine Enkephalin Amide", in *FEBS Lett.*, 1991, vol.291, no.1, pp. 84-86.

[6] Davis J.M. and Svendsgaard D.J., "U-Shaped Dose-Response Curves", in *J.Toxic Env. Health*, 1990, vol.30, pp. 71-83.

[7] Voskoboinic K.D., Szeveszk A., Hensey C., and Azzi A., "Inhibition of Cell Proliferation by α-Tocopherol (Role of PKC)", in *J. Biol. Chem.*, 1991, vol.226, pp. 6188-6294.

[8] Pal'mina N.P., Mal'tseva E.L., Kurnakova N.V., and Burlakova E.B., "The Effect of α-Tocopherol in a Broad Range of Concentrations (10^{-2}-10^{-16}M), on the Activity of PKC: The Link with Proliferation and Tumor Growth", in *Biokhimiya* (Moscow), in press.

[9] Spitkovskii D.M., "The Concept of the Effect of Low Doses of Ionizing Radiation on Cells, and its Possible Use for the Interpretation of Medicobiological Consequences of the Accident at the Chernobyl' Nuclear Power Station", in *Radiobiologiya*, 1992, vol. 32, no.3, pp. 382-400.

[10] Burlakova E.B., Goloschapov A.N., Zhizhina G.P., et al., "Some Specific Aspects of Low-Dose Irradiation on Membranes, Cells and Organisms", *25th Annu. Meet. Eur. Soc. Radiation Biol.*, Stockholm, 1993.

[11] Kislovskii L.D., "Response of a Biological System to Weak Low-Frequency Electromagnetic Fields Adequate to It", in *Probl. Kosm. Biol.*, 1982, vol.431, pp. 148-165.

[12] Safrit J., Tsuchitani T., Zighubolm J., and Bonavida B., "Overcoming Tumor Cell Drug Resistance by Low Doses of Recombinant Tumor Necrosis Factor and Drus", Ultra-low Doses, Doutrempuich, C., Ed. Univ. Bordeaux: France, 1991, pp. 27-43.

[13] Konovalova N.P., Franchi F., D'yachkovskaya R.F., and Volkova L.N., "Antimetastatic Activity of Low and Ultra-Low Doses of Lonidamine", in *Izv. Ross. Akad. Nauk*, Ser. Biol., in press.

[14] Goloschapov A.N., Kislova I.V., and Burlakova E.B., "The Effect of Ultra-Low Doses of Antioxidants on the State of Cell Membranes", in *Biol. Membr.*, in press.

[15] Blumenfeld L.A., Grosberg A.Ju., and Tikhonov A.N., "Fluctuations and Mass Action Law Breakdown in Statistical Thermodynamics of Small Systems", in *J. Chem. Phys.*, 1991, vol.95, no.10, pp. 7541-7544.

[16] Lelekova T.V., Romanovskii P.Ya., Aleksandrov P.N., and Ashmarin I.P., "The Action of Femto- and Picomolar Concentrations of Thyroliberin and Taphcin on the Concentration Activity of Mesentery Lymphatic Vessels in Rats", in *Byull. Eksp. Biol. Med.*, 1989, vol.108, no.7, pp. 8-10.

[17] Zaitsev S.V., and Sazanov B.A., "A Possible Interpretation of the Paradoxical Effects of Ultra-Low Doses of Biologically Active Substances", in *J. Chem. Biochem. Kinet.*, 1991, vol.1, no.3, pp. 21-26.

[18] Burlakova E.B., Konradov A.A., and Khudyakov I.V., "The Effect of Ultra-Low Doses of Chemical Agents on Biological Objects", in *Izv. Akad. Nauk SSSR*, Ser. Biol., 1990, no.2, pp. 184-193.

[19] Blumenfeld L.A., "Parametric Resonance as a Possible Mechanism of Action of Ultra-Low Concentrations of Biologically Active Substances at Cellular and Subcellular Levels", in *Biofizika*, 1993, no.1, pp. 129-132.

HIGH DILUTION EFFECTS OF *STRYCHNOS NUXVOMICA* L ON HYPOTHALAMIC NEURONS AND ADRENERGIC NERVE ENDINGS OF ALCOHOLIC RATS

N.C. SUKUL, S. GHOSH and S.P. SINHA BABU
Department of Zoology, Visva-Bharati University, Santiniketan-731 235, West Bengal, India

ABSTRACT

Rats, which are quickly exhausted and remain passive when subjected to inescapable swimming in a tub of water, are potential alcoholics. These rats were given 20% alcohol for 10 days and then both 20% alcohol and water in a two-choice bottle test. They were treated by oral route with the potentized extract *Strychnos nuxvomica* L such as *Nux vom* 1000 and 30s. The latter potency was prepared by sonication instead of succussion. Both potencies of *Nux vom* significatly reduced alcohol consumption and *Nux* 30s was more effective than *Nux vom* 1000 in this regard. *Nux vom* 30s given intraperitoneally to another batch of rats produced no effect on alcohol consumption. *Nux vom* 200 and 1000, applied on the tongue of anaesthetised alcoholic rats, increased the neuronal activity of the LHA while 20% alcohol, distilled water and *Nux vom* mother tincture decreased it. The inhibitory effect of the latter might be due to alcohol in the mother tincture rather than to the crude drug. The antialcoholic effect of potentized *Nux vom* appears to be mediated through tongue receptors and hypothalamic neurons. In another experiment rats were kept on 20% alcohol and normal food for 2 months. Then they were treated with *Nux vom* 30s, one dose every 15th day for 4 months, and kept on a two-choice bottle. Noradrenergic nerve terminals in atrio-ventricular valves degenerated considerably in untreated alcoholic rats. Treatment with *Nux vom* 30s not only reduced alcohol consumption in rats but also helped in regeneration of most of the nerve terminals.

1. Introduction

Alcoholism is a major social and public health problem in the world. There is no good remedy to help the alcoholics give up their habit. Potentized *Nux vomica* is supposed to counter the effects of alcoholism [12]. Potentized *Nux vomica* is prepared from the alcoholic extract of the poison nut [3]. The mother tincture is diluted with 90% alcohol (1:100) and shaken successively to prepare different potencies of *Nux vomica* [4, 5]. In the present study alcoholic rats were treated with different potencies of *Nux vomica*. The effect of the drug on the alcohol consumption of rats was observed.

Chronic alcoholism may lead to autonomic disfunction in patients with peripheral neuropathy [6]. Degenerative changes in the vagus and sympathetic nerves were also reported in patients with alcoholic neuropathy [6]. Prolonged consumpton of white rum (40% alcohol) produces severe degenerative lesions in the adrenergic nerve plexuses of rat atrioventricular valves [6]. In another experiment we have attempted to see whether potentized *Nux vomica* could heal up the degenerative lesions in alcoholic rats. Since the autonomic system is centrally regulated in the hypothalamus, alcoholism and the drug countering alcoholism may influence the hypothalamic neurons. We have already observed that some potentized drugs could influence hypothalamic neurons [7, 8]. In view of this the response of hypothalamic neurons to potentized *Nux vomica* has also been examined.

2. Materials and Methods

Four batches of male albino rats (190-230g), each of 15 individuals, were selected as potential alcoholics after subjecting them to an inescapable swimming test. Rats were released in a large plastic tub containing water 20cm deep and kept there for 10 min. The period of immobilization, as evidenced from passive swimming with the head seldom protruded over the surface of water, was recorded for each rat. The period of immobilization is directly proportional to the depressive-like state. The highly active rats are characterized by a period shorter than 130s and the lowly active ones by the period longer than 300s. It is the lowly active rats that are potential alcoholics.

These rats were kept individually in cages with food and drink ad lib. Initially the drink provided was 20% alcohol in a bottle. After a period of 10 days with only alcoholic drink, each rat was provided with two-choice bottles, one containing water only and the other 20% alcohol. The amount of water and of 20% alcohol consumed by each rat was measured every 24 h. One batch of rats was treated orally with *Nux vom* 1000 (obtained from Boericke and Tafel, Philadelphia, USA), one dose every day.

Another batch was given *Nux vom* 30s also orally (prepared in the laboratory), one dose every 15th day. The 3rd batch was treated intraperitoneally with *Nux vom* 30s. The drugs were in the form of sucrose globules soaked with the liquid potentized drug (in 90% alcohol). The globules were dissolved in distilled water at the rate of 1 globule (average weight 6.18 mg/ml). The solution was forced into the mouth of a rat at 0.5ml per dose by a glass pipette. The fourth batch remained as untreated control. The drug *Nux vom* 30s was prepared in the laboratory by successive dilution followed by sonication [8]. Each dilution of *Nux vom* was sonicated at 20 Hz for 30s in an ultrasonic homogenizer (Labsonic 2000, Braun, Germany). *Nux* 200 and tincture were obtained from M.Bhattacharya and Co., Calcutta.

One batch of 15 rats was given 20% alcohol by force drinking at 1.5 ml/kg of body wt. for 7 days. They were given food and water ad lib. After 7 days of alcohol consumption, neuronal activity in the lateral hypothalamic area (LHA) of these rats in response to 20% alcohol, distilled water and *Nux vom* 200 and 1000 was recorded. Rats were anaesthetized with peraldehyde (1.28 mg/kg, i.m.) and fixed on a sterotaxic table.

A hole was trefined on the skull in the LHA (A, 4.6-0.5; L, 1.5-0.5; H, 2.7- 0.7 according to the atlas of Konig and Klippel) [9]. A glass- coated silver electrode (tip 0.5 mm, resistance 0.002) was inserted into the LHA, sealed with dental acrylic on the skull and was connected through a head-stage pre-amplifier (AK 100, Digitimer, U.K.), a pre-amplifier and filter (Digitimer) to an oscilloscope (L&T Gould) and a 2-channel recorder (L&T). While gross activity in the LHA recorded in 7 rats, unit activity in the LHA was recorded in 7 rats. In case of gross activity a spike of fixed amplitude was selected and its frequency per 20s was counted. In case of unit activity, impulses per 20s were counted. Drugs like *Nux vom* 200 and 1000, 20% alcohol and distilled water were applied on the tongue of the rats and the effect was recorded. Single unit discharges in the LHA were recorded by means of glass coated silver microelectrodes (tip 3.5 μm in diam., resistance 1.5-2.5 MΩ). After each experiment, the rat was perfused with saline and 10% formalin through the left ventricle of the heart and the recording site was examined histologically.

In another experiment 20 rats were allowed to drink only 20% alcohol for 2 months after which they were given two-choice bottles, one containing water, another 20% alcohol. Simultaneously, half of these rats were treated with *Nux vom* 30s, one dose every 15th day for a period of 120 days. Then all these rats together with 10 nonalcoholic control rats were autopsied and their atrioventricular valves taken out. The valves were rinsed in cold calcium-free Tyrode's solution for 2 min, stretched, mounted whole on glass slides and allowed to dry in air at room temperature for 15 min with relative humidity maintained at 50%. The tissue was then treated with formaldehyde gas of optimal humidity (70%) generated from paraformaldehyde at 80°C for 1 h, mounted with immersion oil and examined under a microscope [6].

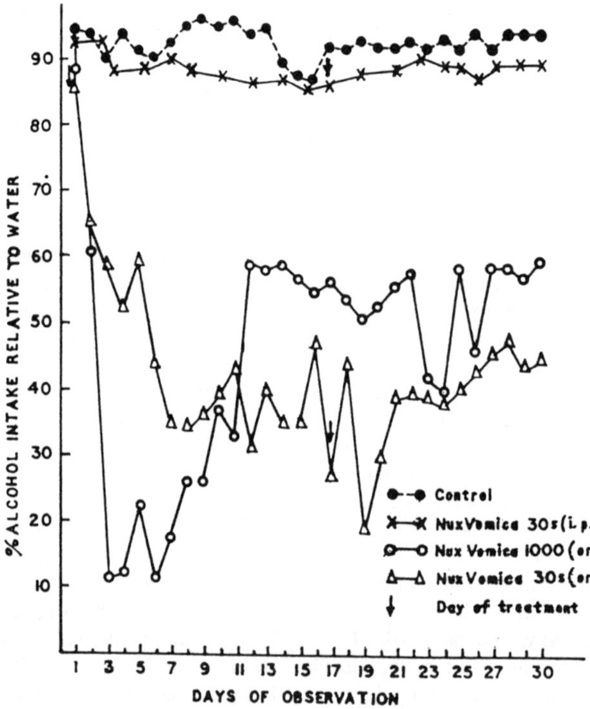

Figure 1- Graph showing reduction in the mean percentage of alcohol consumption relative to water in rats following oral administration of *Nux vom* 1000, one dose every 15th day. No marked variation in alcohol consumption in untreated control and *Nux* 30s (i.p.) treatment during the period of observation.

3. Results

The percentage of alcolhol intake relative to water was calculated everyday for each rat by the following way:

$$\frac{\text{Amount of alcohol consumed}}{\text{Total amount of water and alcohol taken}} \times 100$$

The mean percentages for each of the three baches of rats were plotted in a graph against days of observation in Fig. 1. Both *Nux vom* 1000 (oral) and *Nux vom* 30s (oral) reduced alcohol consumption significantly ($p<0.005$). *Nux* 30s (i.p.) did not produce any

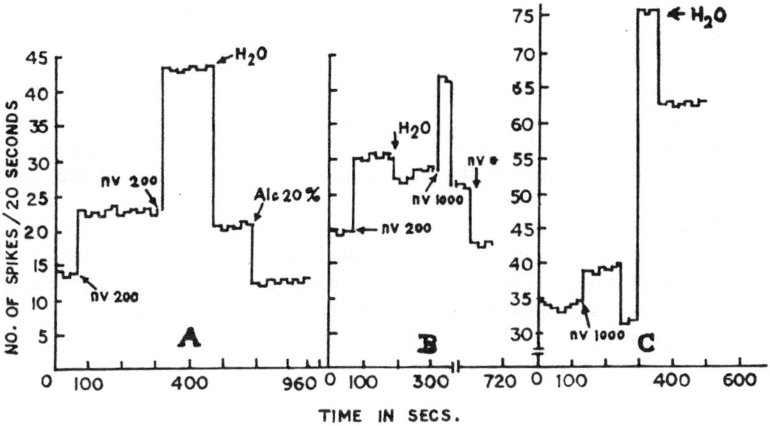

Figure 2 - Histogram showing excitatory effect of *Nux Vom* 0, 200 and 1000 and inhibitory effect of distilled water and 20% alcohol, applied on the tongue, on the LHA neurons of rats given 20% alcohol at 15 ml/kg body weight for 7 days. A, B gross and C unit activity.

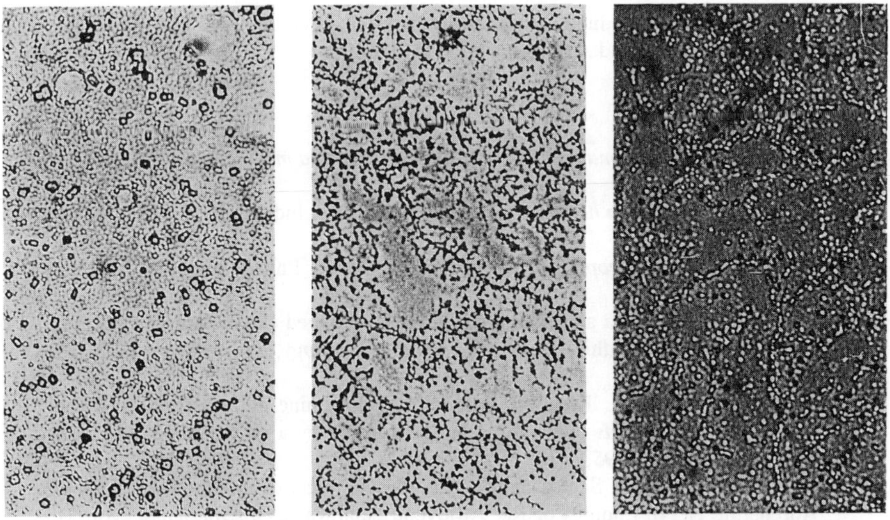

Figure 3 - Degeneration of noradrenergic nerve endings in atrio-ventricular valves of alcoholic rats (a), as compared to nonalcoholic rats (b); treatment with *Nux Vom* 30s restored the condition (c).

effect on alcohol consumption (Fig.1). *Nux vom* 200 and 1000 increased the neuronal activity in the LHA of alcoholic rats (+124.48 ± 25%, $p<0.05$ for 200 and 116 ± 20%, $p<0.05$ for 1000). The increase was more pronounced with *Nux vom* 1000 than with *Nux vom* 200 (Fig.2).

Distilled water and 20% alcohol, when applied on the tongue, reduced the neuronal activity in the LHA (52.78 ± 15%, $p<0.05$ for distilled water and 58.24 ± 5% for 20% alcohol) in comparison with *Nux* 1000. Application of *Nux vom* (mother tincture) also reduced the activity.

Consumption of 20% alcohol for 6 months led to the degeneration of adrenergic nerve plexuses (30 to 40%) in atrio-ventricular valves of rats (Fig. 3a) as compared to non-alcoholic control (Fig.3b). Treatment with *Nux vom* 30s restored the condition to a great extent (Fig. 3c) and reduced alcohol consumption (50%) ascompared to the untreated alcoholic rats.

4. Discussion

The results clearly indicate that potentized *Nux vom* at dilutions 10^{-60} and 10^{-2000} could significantly reduce alcohol consumption in rats. Potentized *Nux* could also regenerate adrenergic nerve terminals. The action of the drug appears to be mediated through tongue receptors and hypothalamic neurons. The results further confirmed our earlier observaton that sonicated drug proved more effective than succussed drug [10].

Acknowledgements

The present study partly includes our earlier publicaton co-authored by A.Paul, S.P. Sinhababu, N.P. Kurzina and A.S. Batuave [11].

References

[1] Boericke W., *Pocket mamual of homoeopathic materia medica*, Calcutta. sett Dey & Co. 1927.

[2] Kent J.T., *Lectures on homoeopathic materia medica*, Ind. ed. 1962. Calcutta, sett Dey & Co., 1911.

[3] *The American Homoeopathic pharmacopoeia*, 9th ed. Philadelphia, Boericke and Tafel, 1920.

[4] Sukul N.C., S.K. Bala and B. Bhattacharya, "Prolonged cataleptogenic effects of potentized homoeopathic drugs", *Psychopharmacology*, vol. 89, pp. 338-339, 1986.

[5] Sukul N.C. and W.R. Klemm, "Influence of dopamine agonists and an opiate antagonist on *Agaricus* induced catalepsy as tested by a new method," *Arc. Int. Pharmacodyn.*, vol. 295, pp. 40-51, 1988.

[6] Ferreira A.L., J.C.M. Santos and M.A. Ross, "Effects of Alcohol Ingestion on Adrenergic Nerve Endings of Rat Atrioventricular valves", *Experientia*, vol. 31, pp.82-83, 1975.

[7] Sukul N.C., "Increase in seretonin and dopamine metabolites in mouse hypothalamus following oral administration of *Agaricus muscarius* 12, a homoeopathic drug", *Sci. Cult*, vol. 56, pp. 134-137, 1990.

[8] Sukul N.C., A. Batuave, V. Sabanov and N. Kurzina, "Neuronal activity in the lateral hypothalamus of the cat and the medial frontal cortex of the rat in response

to homoeopathic drug", *Ind. Biologist*, vol. 23, pp. 17-20, 1991.
[9] Konig J.F.R. and R.A. Klippel, *The Rat Brain. A Stereotoxic Atlas of the Forebrain and Lower Parts of the Brain Stem*, Baltimore, MD, Williams and Willkins, 1963.
[10] Sukul N.C., "Mechanical agitation, the main factor in increasing the efficacy of *Agaricus muscarius*, a homoeopathic drug", *Environ. Ecol.*, vol. 10, pp. 7-10, 1992.

HIGLY DILUTED AGITATED SILVER NITRATE AND WHEAT SEEDLING DEVELOPMENT EFFECT KINETICS OF A PROCESS OF SUCCESSIVE AGITATION PHASES

WALTRAUD PONGRATZ
Institute for Plant Physiology of the University of Graz, Austria and Ludwing Boltzmann Institute for Homoeopathy, Graz

ANDREA NOGRASEK
Institute for Botanics of the University of Graz

P. CHRISTIAN ENDLER
Institute for Holistic Medicine, University of Urbino, Italy

ABSTRACT

Silver nitrate is well-known to be generally biostatic. However, it is also discussed as a biologically vital trace substance. This substance has been prepared in a standardized process of stepwise dilution and agitation. Its influence on the development of wheat seedlings has been studied. When seedlings are exposed to silver nitrate at high molecular concentration, growth is normally *inhibited.* Interestingly, growth was *stimulated* by a dilution log 24 of this probe. In our experiments, addition of one more standardized step of dilution and agitation, or agitation alone, to the dilution log 24, significantly diminished the effect of the test dilution, whereas addition of two steps again enhanced the stimulatory effect ("double switch" kinetics). This was found by different independent researchers. Furthermore, optimized conditions for performance of the experiment are discussed.

KEY WORDS

Biostatic substance, wheat, high dilution, agitation phases, inverted effect, double switch kinetics

1. Introduction

The effect of a substance which, in common biology and medicine, is known to be biostatic, but which is also discussed as a necessary trace substance, namely silver nitrate [1,2], on the development of wheat seedlings was investigated. In a previous study,

seedlings were exposed to silver nitrate at high molecular concentration (10 mg per ml water, dilution log 2); germination and growth were drastically *inhibited* (unpublished material). However, at lower concentrations (e.g. in a dilution log 7) silver nitrate is able to exert a stimulating effect on botanical systems [1]. In two multicenter studies, seedlings were exposed either to an aqueous test dilution log 24 of this molecular solution, specially prepared by stepwise dilution (23 steps 1: 10) and vigorous agitation (24 agitation phases), called 'silver nitrate D24' or to analogously prepared solvent water (or to unprepared solvent water). In these studies, growth was *stimulated* by the test dilution with statistical significance. This was shown by four independent researchers including researchers from the Universities of Graz and of Vienna [3,4].
The study presented in this paper is based on findings reported in the literature, namely that adding one more step of dilution and agitation to D24 (= D25) significantly *diminishes* the effect of the test dilution, whereas adding two more steps (= D26) again *enhances* the stimulating effect [5]. A 'double switch kinetics' - pattern was postulated (D24: significant stimulation with regard to reference - D25: less or no stimulation - D26: again significant stimulation). In this experiment, obviously, both water submitted to 24 agitation phases as well as unprepared water can be used as reference (unpublished data).
In the study presented here, the effect pattern of the three dilutions silver nitrate D24, D25 and D26 was observed in experiments at different times of observation. The aim of this was to learn about the conditions of the 'double switch effect' (see above) with regard to the organisms involved. Furthermore, starting from a dilution silver nitrate D24, the process of stepwise dilution and vigorous agitation was replaced by a process of stepwise agitation without dilution. The aim of this was to learn about the conditions of the 'double switch effect' with regard to the dilution involved.

2. Methods

Plants: Unbroken, unsorted grains of winter wheat (grown without an application of herbicides or pesticides) (Triticum aestivum, sorte: Mephisto) were used.

Observed development: It is reported that biological systems in circumscript developmental transitions show maximal sensitivity [6; 7]. In our study, the initial development of stalks was observed.

Preparation of test solutions: The grains were observed under the influence of specially prepared aqueous solutions of silver nitrate (Merck). The solution that was used as the stock solution was also used in the previous experiments concerning the influence of the molecular substance on wheat seedlings (see introduction). It contained 10 mg silver nitrate per ml distilled (not deionized) water, and it was diluted in distilled water in steps of 1: 10 (parts of volume).
The stock solution and each of the different successive dilutions 1: 10 were vigorously agitated: at every step, a disposable sterile hardglass bottle with a volume of 20 ml, filled with 9 ml of distilled water was used. Each time 1 ml of the silver nitrate dilution from the previous step was added to this bottle, which then was pushed down 50 times at intervals of 1.2 seconds (= 1 minute) (agitation by hand, pushing

down against the left palm) to create mechanical shocks. For the last steps of dilution, larger bottles were used, but the ratio [volume : content] was as above. After each *dilution plus agitation* - period, the solution was given a rest of 1.25 minutes. One test dilution was prepared with 24 steps of agitation (agitation of the stock solution plus agitation of 23 further dilutions 1:10) (called 'D24'), one with 25 steps ('D25') and one with 26 steps ('D26'). *Annotation: due to a redactional error, the respective protocol in reference [4], p. 24, line 38 is incorrect. It should read "aqueous solution 1 : 10^{24} part of the weight of the stock solution of silver nitrate" as above.*

Several analogously prepared sets of ranges of the dilutions were used.

In a further part of the study, starting from a dilution silver nitrate D24, the process of stepwise dilution and agitation was replaced by a process of stepwise vigorous agitation (without dilution). The agitation phases were always separated by resting phases. Three test substances were called silver nitrate D24/1 (which is similar to D24), D24/2 and D24/3.

Untreated solvent (water) was used as reference. All sets were applied blindly, i.e. the substances were all coded.

Laboratories involved: Experiments were performed in the laboratories of W. Pongratz, C. Endler and A. Nograsek. For further details, see the Results Section.

Circadian specificity: With regard to older literature, it was assumed that the experiments should be started at noon. Thus, most of the experiments were started between 8 a.m. and 1 p.m.

Exposition to probes and Data base: The grains with the germination furrow downwards were put on floss paper in sterile glass dishes (diameter 11cm). They were arranged in the form of a circle with all germination pools towards its center. This was done in order to make observation of the stalks easier. However, previous experiments have shown that also a random distribution of the grains can be used. Each dish contained 20 ml of the respective probe. It was covered by a glass beaker with a volume of 1000 ml. Dish and beaker were wrapped in stannium foil and kept at 20°C in a temperature stabilized chamber at. In one part of the study, the stalk lengths were measured after 7 days, in another part after 5 days. In the experiments, three sets of dishes were always used for the treatment with the silver nitrate dilutions plus one set for the water control. 20 - 30 seedlings were put into one dish; the respective number of grains per dish was equal for all groups (for total numbers, see below).

Evaluation of the data: The arithmetic mean was used to describe the stalk length, and the lengths were compared by one way analysis of variance (Fisher-test). 1 S.D. of the mean was calculated. The groups of seedlings treated with silver nitrate D24, D25 and D26 were compared to the water control. In addition to this, with regard to the hypothesis of a double switch kinetics, the groups treated with silver nitrate D24 and D26, respectively, were statistically compared to the D25 - group. Here, only results at the 1%-level were considered for interpretation. The D24/1, D24/2 and D24/3 - groups and water control were compared analogously.

3. Results

Experiments Set I, II and III on step by step diluted agitated substances

3.1 SET I: Observation after 7 days (W. Pongratz)

Experiments with 4 groups of 2140 grains each, treated with silver nitrate D24, D25, D26 and unprepared solvent, respectively, were performed. When the data of all experiments were pooled, the comparison of the mean stalk length showed a characteristic effect pattern (see introduction) of the three dilutions. The values for the silver nitrate - groups were generally above the control - value (see Fig. 1). When the test substances were mutually compared, both the groups treated with the dilution D24 as well as with D26 showed more growth than the group treated with the dilution D25 (+ 4.3% and + 4.3%, respectively). This difference is statistically significant (P < 0.01 and < 0.01; for a comprehensive survey over P - values, see Tab. 5).

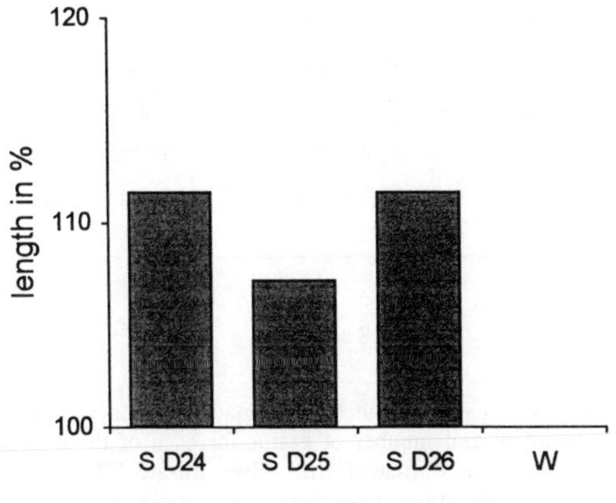

Fig. 1: The influence of the test dilutions silver nitrate D24, D25 and D26 in experiments that lasted 7 days. Ordinate: mean stalk length. 100% refers to the stalk length under the influence of unprepared solvent. Abscisses: S, silver nitrate. For further information, see text and table 5.

However, as can be seen in Tab. 1, the overall effect pattern in this study has in fact been caused by only about half of the single experiments (marked with *x* in Tab. 1).

SET I

Duration of growth 7 days

Experiment nr. & N_{grains}	S D24	S D25	S D26	water
1: (4 x 120)	86.8 ± 27.9	84.2 ± 15.6	71.1 ± 40.5	31.4 ± 27.1
2: (4 x 160)	93.8 ± 33.6	69.3 ± 45.15	87.2 ± 23.2 x	41.3 ± 12.6
3: (4 x 120)	85.3 ± 25.9	68.4 ± 13.4	98.0 ± 15.3 x	54.6 ± 29.5
4: (4 x 160)	67.9 ± 17.2	57.5 ± 31.8	62.1 ± 31.5 x	56.8 ± 19.6
5: (4 x 200)	80.2 ± 27.8	76.3 ± 25.9	83.7 ± 28.2 x	59.8 ± 33.1
6: (4 x 120)	88.4 ± 24.8	86.3 ± 25.3	97.7 ± 13.6 x	67.8 ± 30.7
7: (4 x 120)	80.6 ± 26.8	73.9 ± 26.6	83.9 ± 22.6 x	69.6 ± 36.0
8: (4 x 200)	87.3 ± 28.1	80.6 ± 30.0	82.7 ± 30.1 x	74.4 ± 28.7
9: (4 x 180)	102.1 ± 34.8	98.0 ± 30.9	96.3 ± 41.6	74.7 ± 27.3
10: (4 x 120)	78.2 ± 15.7	87.4 ± 26.2	79.5 ± 18.4	75.8 ± 19.3
11: (4 x 240)	90.8 ± 29.2	94.9 ± 35.7	97.0 ± 30.6 x	92.3 ± 33.5
12: (4 x 200)	91.5 ± 28.5	99.0 ± 33.9	99.1 ± 31.5	94.8 ± 32.5
13: (4 x 200)	99.5 ± 38.0	105.3 ± 38.6	102.1 ± 36.6	96.4 ± 32.0
1 - 13: (4 x 2140)	91.3 ± 31.1	87.8 ± 33.4	91.3 ± 32.3	81.9 ± 35.0

Tab. 1: Stalk lengths in experiments on diluted and agitated substances that lasted 7 days. Left column: Number of experiment and number of grains for each of the 4 groups. Arithmetical mean (mm) + 1 SD for the groups treated with silver nitrate D24, D25, D26 and water, respectively. The experiments are arranged according to the stalk lengths in the control group. For further information, see text.

Interestingly, all experiments that showed the pattern discussed above were characterized by a stalk length of the control group between 40 and 75 mm.

3.2 SET II: Observation after 5 days, first experiments (W. Pongratz)

Experiments with 3 groups of 890 grains each, treated with silver nitrate D24, D25 and D26 and one group of 690 grains, treated with unprepared solvent, respectively, were performed.

Again, the values for the silver nitrate-groups are generally above the control-value (see Fig. 2). When the test substances were mutually compared, both the group treated with the dilution D24 as well as with the dilution D26 showed more growth than the group treated with the dilution D25 (+ 8.7% and +10.6%). This difference is statistically significant ($P < 0.01$ and < 0.01).

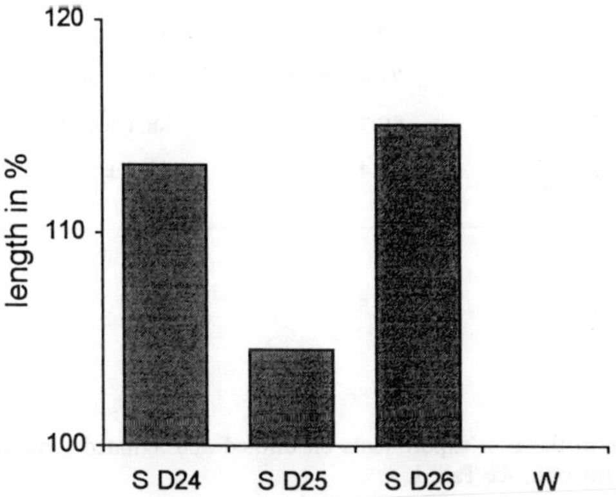

Fig. 2: The influence of the test dilutions in the first experiments that lasted 5 days. For further information, see Fig. 1.

All of the separate experiments of this study are characterized by a stalk length of the control group between 40 and 70 mm (see above) and all but one of them show the effect pattern 'stimulation - less stimulation - stimulation' as is presented in Tab. 2.

SET II

Duration of growth 5 days

Experiment & N_{grains}	S D24	S D25	S D26	water
14: (3 x 240)	09.1 ± 15.2	04.3 ± 08.5	10.0 ± 17.2	-
15: (3 x 60)	no separate protocol available			
16: (4 x 80)	84.6 ± 24.4	81.6 ± 22.3	86.5 ± 27.2	57.4 ± 31.9
17: (4 x 100)	83.8 ± 24.9	80.0 ± 23.5	86.0 ± 26.3	59.8 ± 33.1
18: (4 x 60)	84.4 ± 23.9	86.0 ± 23.0	82.7 ± 29.2	60.4 ± 34.2
19: (4 x 60)	86.8 ± 26.4	79.6 ± 20.6	88.9 ± 29.2	56.8 ± 32.6
20: (4 x 50)	82.1 ± 26.8	73.6 ± 21.4	92.8 ± 20.1	64.7 ± 33.6
21: (4 x 100)	83.2 ± 23.8	80.3 ± 24.0	88.4 ± 19.8	59.4 ± 32.0
22: (4 x 140)	30.6 ± 28.7	20.4 ± 26.3	29.7 ± 29.9	44.0 ± 37.4
14 - 22: (3 x 890, 1 x 690)	60.1	55.5	61.1	53.1

Tab. 2: Stalk lengths in the first experiments on diluted and agitated substances that lasted 5 days. For further information, see Tab. 1.

3.3. SET III: Observation after 5 days, control experiments (W. Pongratz, C. Endler, A. Nograsek)

Control experiments with a total of 4 groups of 510 grains each, treated with silver nitrate D24, D25 or D26 or with unprepared solvent, respectively, were performed in the laboratories of W. Pongratz, C. Endler and A. Nograsek (Tab. 3). When these data were pooled, again, the values for the silver nitrate - groups were generally above the control - value (Fig. 3). When the test substances were mutually compared, both the group treated with the dilution D24 as well as with the dilution D26 showed more growth than the group treated with the dilution D25 (+ 16% and + 10%). This difference is statistically significant ($P < 0.01$ and < 0.01).

SET III

Control experiments

Experiment & N_{grains}	S D24	S D25	S D26	water
Laboratory: W. Pongratz				
23: (4 x 200)	64.2 ± 23.8	55.6 ± 25.2	56.0 ± 27.3	47.8 ± 22.8
Laboratory: C. Endler				
24: (4 x 210)	33.0 ± 22.1	29.1 ± 20.7	31.3 ± 21.8	32.4 ± 21.4
Laboratory: A. Nograsek				
25: (4 x 100)	54.9 ± 20.4	45.5 ± 19.0	59.7 ± 19.6	42.7 ± 21.4

Tab. 3: Stalk lengths in control experiments on diluted and agitated substances that lasted 5 days. For further information, see Tab. 1

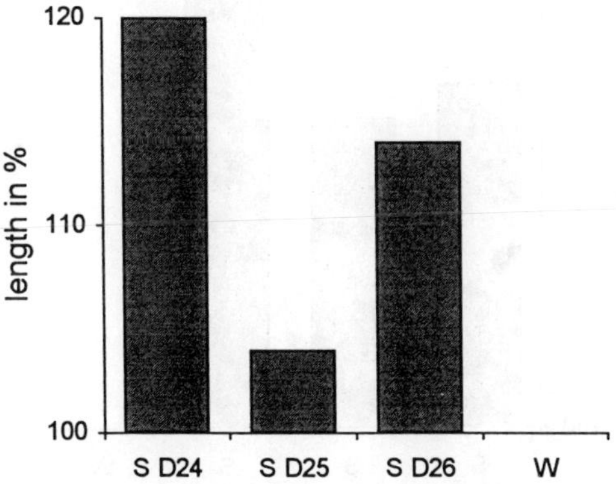

Fig. 3: The influence of the test dilutions in SET III. For further information, see Fig. 1.

In the experiments of Pongratz, the typical "V"-form pattern is found, but only the difference between the groups treated with the dilutions D24 and D25 is statistically significant ($P < 0.01$).

In the experiments of Endler, the typical "V"-form pattern is found, but at the 1%-level, the difference is not statistically significant (D24 vs D25: $P < 0.05$).

In the experiments of Nograsek, the typical "V"-form pattern is found and both the difference between the groups treated with the dilutions D24 and D25 as well as between D25 and D26 is statistically significant ($P < 0.01$ and < 0.01).

3.4 Experiments SET IV on step by step agitated substances

(C. Endler, W. Pongratz)

Furthermore, experiments with 4 groups of 750 grains each, treated with silver nitrate D24/1, D24/2, D24/3 and unprepared solvent, respectively, were performed. The comparison of the mean stalk length again showed a characteristic effect pattern of the three test substances. The values for the silver nitrate - groups were generally above the control - value (see Fig. 4). The difference is statistically significant. When the test substances were mutually compared, both the group treated with the dilution D24/1 as well as with D24/3 showed more growth than the group treated with the dilution D24/2 (+ 2.2% and + 9.6%). The difference between the groups treated with dilutions D24/1 and D24/2 is statistically significant ($P < 0.01$).

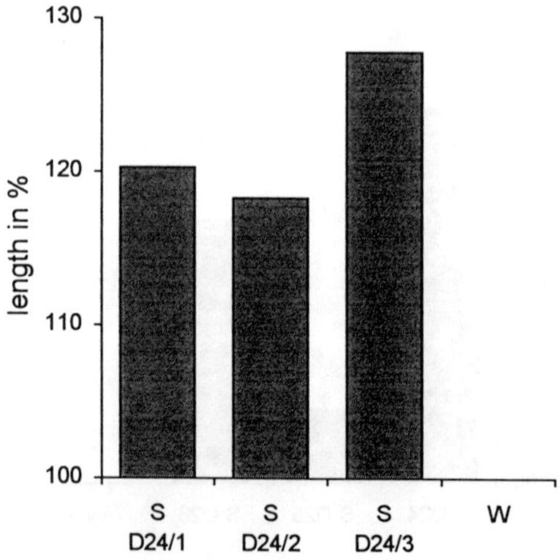

Fig. 4: The influence of the test dilutions silver nitrate D24/1, D24/2 and D24/3. For further information, see Fig. 1 and text.

SET IV

Duration of growth 5 days

Experiment & N_{grains}	S D24/1	S D24/2	S D24/3	water
Laboratory: C. Endler				
A: (4 x 120)	31.4 ± 23.0	26.7 ± 23.2	28.6 ± 23.1	27.5 ± 21.1
Laboratory: W. Pongratz				
B: (4 x 120)	89.0 ± 35.0	86.6 ± 35.7	97.0 ± 45.8	61.4 ± 50.7
C: (4 x 120)	83.5 ±41.6	94.3 ± 34.8	108.6 ± 42.5	68.4 ± 32.3
D: (4 x 200)	78.6 ± 25.1	73.6 ± 31.4	76.3 ± 29.1	72.4 ± 24.4
E: (4 x 160)	69.8 ± 29.9	66.7 ± 29.2	70.0 ± 28.0	61.3 ± 42.3
A - E: (4 x 750)	69.2 ± 37.8	68.0 ± 39.1	73.5 ± 43.0	57.5 ± 38.7

Tab. 4: Stalk lengths in experiments on step by step agitated substances. For further information, see Tab. 1.

	SET I	SET II	SET III				SET IV
	Pongratz	Pongratz	Pongratz	Endler	Nograsek		
24-25	++	++	++	+	++	24/1-24/2	-
24-26	-	-	++	-	-	24/1-24/3	++
24-water	++	++	++	-	++	24/1-water	++
25-26	++	++	-	-	++	24/2-24/3	++
25-water	++	-	++	+	-	24/2-water	++
26-water	++	+	++	-	++	24/3-water	++

Tab. 5: P - values in the experiments SET I - IV. ++, $P < 0.01$; +, $P < 0.05$; -, not significant (Fisher-test). For further information, see text.

4. Discussion

The study presented here is based on two previous findings: (a) that a concentrated solution of silver nitrate (10 mg per ml water), *inhibits* the germination of wheat seedlings (unpublished material) and (b) that this very solution, when diluted 10^{23}-fold in a special step by step process including 24 phases of vigorous agitation (called "D24") *stimulates* stalk growth. The latter was found in two previous multicenter blind studies (3,4). In the blind study presented in this paper, we have focused on the comparison of three different dilutions, namely silver nitrate D24, D25 and D26. In different series of experiments, it was found that adding one more step of dilution and agitation to D24 (= "D25") significantly *diminished* the effect of the test dilution, whereas adding two more steps (= "D26") again *enhanced* the stimulating effect. This was found by three independent researchers in three Austrian laboratories. The finding is also consistent with previous findings reported in literature [5, see also 8].

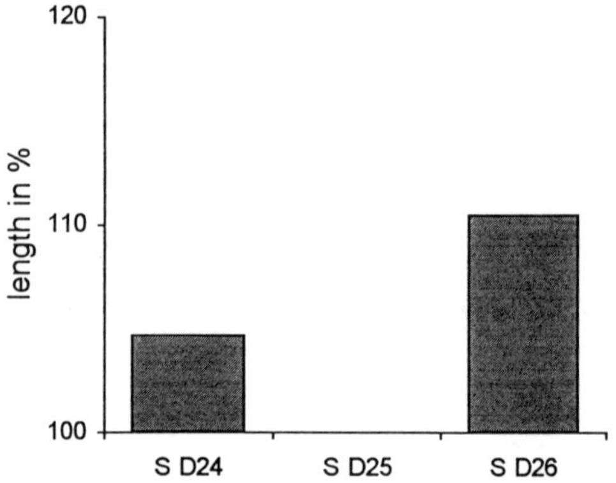

Fig. 5: The influence of the test dilutions D24, D25 and D26 (Kolisko 1926). 100% refers to the stalk length under the influence of silver nitrate D25. For further information, see Fig. 1. (Adopted from [5], modified.)

A "double switch" kinetic was postulated (D24: significant stimulation with regard to reference - D25: less or no stimulation - D26: again significant stimulation). This "V"-form effect pattern was investigated in experiments under different conditions of observation time. The pattern was less marked when growth was given a time of 7 days than when growth was given a time of 5 days. It was postulated that a final stalk length of 40 - 75 mm (water control group) was optimal to observe a clear effect. Further experiments with an observation time of 3 days and final stalk lengths of < 10 mm showed no difference between the groups treated with silver nitrate D24, D25 and D26 (unpublished material).

In the further blind experiments (duration 5 days), starting from a dilution silver nitrate D24, the process of step by step *dilution and agitation* was replaced by a process of only step by step *agitation*, alternating with resting times, without further dilution. In these experiments the pattern 'stimulation - less stimulation - stimulation' was found as above. This points to the conclusion that - in the range of the test substances D24 to D26 - it was not the process of step by step dilution, but of agitation, that caused the difference in the biological effect of the probes.

Further (preliminary) experiments were performed on a possible circadian specificity of the starting time of the experiments. Obviously, it is a conditio sine qua non in order to find the "V"-form effect pattern discussed here, that the experiments are started in the morning or noon, but not in the late afternoon (unpublished material).

At present, we have only very limited insight into the physical properties of such test dilutions [9-11 review in 15] and into their physiological interconnections with the sensitive living system [12-15]. Further physiological and biophysical experiments have to be performed on the topic.

5. Acknowledgements

We would especially like to thank G. Kastberger, R. van Wijk, H. Walach, M. Moser and M. Haidvogl; as well as E. Lauppert who performed the experimental work in the laboratory of C.E.

6. References

[1] Strasburger E. New York: G Fischer Verlag 1990
[2] Haidvogl M, Graz. Personal communication
[3] Pongratz W et al.. In: Stacher A (ed). Wien: Facultas Universitätsverlag 1991, 385
[4] Pongratz W and Endler PC. In [15]
[5] Kolisko L. Dornach: Goetheanum Verlag 1926
[6] Smith CW. In: ZDN (ed). Essen: Verlag für Ganzheitsmedizin 1990; in [15]
[7] Popp FA. In [15]
[8] Lauppert E. Thesis. University of Graz 1995
[9] Demangeat JL et al. J Med Nucl Biophys 1992, 16/2, 135
[10] Kratky K and Milavec T. Besteht ein qualitativer Unterschied zwischen Hochpotenz und reinem Lösungsmittel? In: Bergsmann O (ed). Wien: Facultas 1994.
[11] Weingärtner O. Berlin: Springer 1992
[12] Benveniste J et al. FASEB J 1994, 8/4, 398
[13] Endler PC et al. J Vet & Human Toxicol 1994, 36/1, 56; 1995, 37/3, 259; FASEB J 1994, 8/4, 400
[14] Youbicier - Simo BJ, Boudard F, Mekaouche M et al. Int J Immunotherapy 1993, IX/3, 169
[15] Endler PC and Schulte J (eds). Dordrecht - Boston - London: Kluwer Academic Publishers 1994

DOSE RESPONSE IN A HIGH DILUTION OF *AGARICUS MUSCARIUS* L ON ALBINO MICE

N.C. SUKUL, S. GHOSH and S.P. SINHA BABU
Department of Zoology, Visva-Bharati University,
Santiniketan-731 235, West Bengal, India

ABSTRACT

Agaricus muscarius 30s, a potentized drug prepared by successive dilution with 90% alcohol followed by sonication, when given through oral route, suppressed haloperidol-induced catalepsy in Swiss albino mice significantly. This anticataleptic effect was dose dependent being maximum with the undiluted *Agaricus* 30s and minimum with its dilution 1:20,000. Higher dilutions like 1:40,000 and 1:50,000 did not produce any anticataleptic effect. The effect reappeared when the dilution 1:50,000 was sonicated. An amount of 0.05 ml of the dilution $2x10^{-4}$ of *Agaricus* 30s is quantitatively equivalent to 0.075 ml of the dilution $3x10^{-4}$ and to 0.16ml of the dilution $4x10^{-4}$. Three groups of mice were pretreated with the three dilutions of *Agaricus* 30s in such a way that each group received an equivalent volume dose of the drug. Of the three dilutions, only the lowest one $2x10^{-4}$ was effective in reducing haloperidol has an active physical entity which can be attenuated by dilution and multiplied by mechanical agitation or sonication. It is also suggested that potentized *Agaricus* acts through tongue receptors.

1. Introduction

Agaricus 30, produced by successive dilution and mechanical agitation serially for the 30th time, is reported to produce mild catalepsy in normal rats and mice [1-5]. We have observed anticataleptic effect of potentized *Agaricus* at a dilutin of 10^{-60} in mice against haloperidol induced catalepsy [6]. Since the potentized *Agaricus* shows biological effect at a dilution beyond the Avogadro number it is worthwhile to explore the physical basis of this high dilution. If this dilution has any physical entity, it should show dose response, a fundamental principle of pharmacology. We have, therfore, tried to see if a potency of *Agaricus* could suppress haloperidol-induced catalepsy in a dose- dependent fashion.

2. Materials and Methods

The mother tincture of *Agaricus muscarius*, a fungus was obtained from Boericke and Tafel Co., Philadelphia, U.S.A. Its 30th potency was prepared in the laboratory by the conventional method [7]. The method involves diluting the mother tincture with 90% ethanol in the proportion of 1:100 and giving the mixture 10 downward strokes to prepare the 1st potency called *Agaricus* 1. But here each dilutin of *Agaricus*, instead of being subjected to 10 manual strokes, was sonicated at 20 Hz for 30 sec in an ultrasonic homogenizer (Labsonic 2000, Braun, Germany). It was observed earlier that sonication produced a more effective potency than manual agitation [8]. This potency of *Agaricus* produced by sonication is designated a *Agaricus* 30s.

This potency was diluted with 90% ethanol in the proportions 1:1000, 1:10,000. 1:20,000, 1:30,000, 1:40,000 and 1:50,000 without any further sonication or machanical agitation. An aliquot of the last dilution 1:50,000 was sonicated as

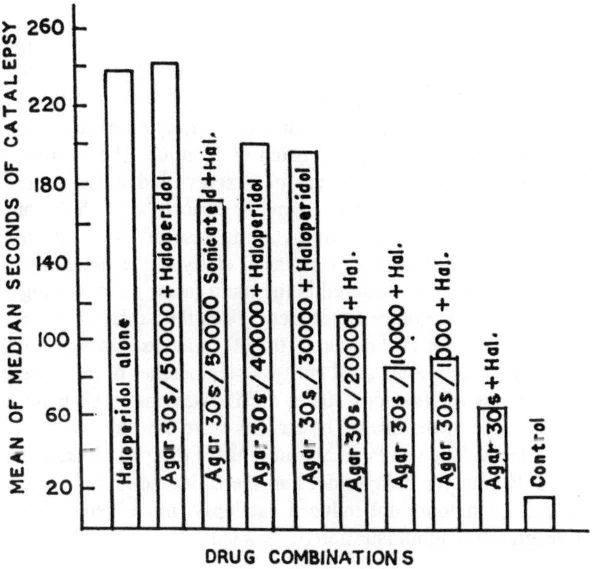

Figure 1 - Anticataleptic effect of *Agaricus* 30s (oral) at dilutions upto 1:20,000 against haloperidol catalepsy. No effect with higher dilutions of *Agaricus*. *Agaricus* 30s at 1:50,000 (sonicated) reduces haloperidol catalepsy.

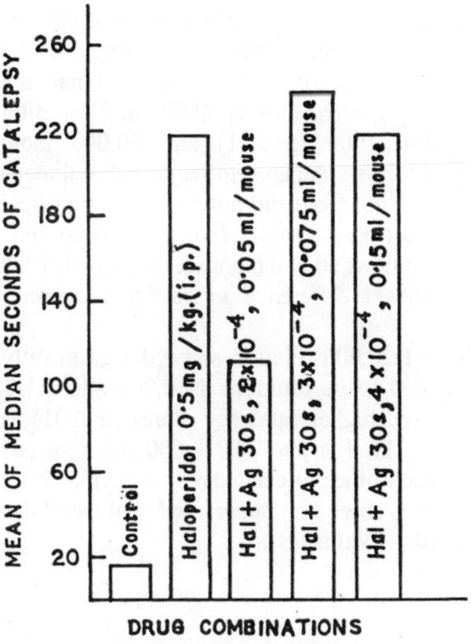

Figure 2 - Reduction of haloperidol catalepsy by *Agaricus* 30 (oral) at a dilution 1:20,000 only. No effect was observed with higher dilutions although equivalent amount of the three dilutions were administered on albino mice.

above.

Catalepsy tests were performed on Swiss albino mice (25-30g) kept in groups of 5 in conventional cages in an animal house with natural light and temperature of 29±2°C. *Agaricus* 30s and its 7 dilutions were mixed with distilled water in the proportion of 0.05 ml/ml of water and administered orally on mice at the rate of 0.05 ml/mouse. Each dilution was given to a batch of 10 mice. The control consisted of 10 mice which were given blank 90% ethanol mixed with distilled water in the same proportion. Haloperidol, obtained from Sigma Chemical Co., U.S.A., and the saline control were given by i.p. at 5 mg/kg in a volume dose of 10 ml/kg. In order to see whether the concentration of the active physical entity in the potentized drug or the total amount of that entity administered is the main factor in producing a biological effect, we have performed another experiment. Thus 0.05 ml, 0.075 ml and 0.16 ml of the dilutions 2×10^{-4}, 3×10^{-4} and 4×10^{-4} of *Agaricus* 30s, respectively contain the same amount of the active physical entity. Three batches of mice were pretreated with the three dilutions of *Agaricus* 30s in such a way that each batch received an equivalent volume dose of the drug. The batches were then treated with haloperidol and tested for catalepsy. The test latency from time of drug administration was 4.5 hr for *Agaricus* and 30 min for haloperidol and the saline control. In order to see the effect of *Agaricus* on haloperidol induced catalepsy, mice were injected i.p. with haloperidol 4 hr after oral administration of *Agaricus*.

3. Results

Means of 2 median numbers of seconds of catalepsy for each treatment and the control are shown in histograms in Fig. 1 and Fig. 2. Haloperidol alone and haloperidol with such dilutions of *Agaricus* 30s as 50,000 (unsonicated), 40,000 and 30,000 produced significant cataleptic effect as compared to the control ($p<0.01$, Kruskal-Wallis test). Posthoc t-test for drug pairs showed that catalepsy scores were significantly reduced with *Agaricus* 30s ($p<0.01$) and its dilutions like 20,000 ($p<0.01$), 10,000 ($p<0.01$), 1000 ($p<0.01$) and 50,000 (sonicated $p<0.1$) in combination with haloperidol as compared to haloperidol alone (Fig. 1). However, catalepsy scores with haloperidol alone did not show any significant difference from those with *Agar* 30s/50,000 (unsonicated), *Agar* 30s/40,000 and *Agar* 30s/30,000. But *Agaricus* 30s and its dilutions 50,000 (sonicated), 20,000, 10,000 and 1,000 with haloperidol showed significant difference in catalepsy scores from the control ($p<0.01$).

Again *Agaricus* 30s at 1:20,000 dilution showed significantly higher catalepsy scores ($p<0.01$) as compared to its dilutions 1:10,000 and 1:1,000. The dilutions 1:10,000 and 1:1,000 showed higher catalepsy scores ($p<0.01$) as compared to the undiluted *Agaricus* 30s (Fig. 1), *Agaricus* 30s/10,000 and *Agaricus* 30s/1000 did not differ significantly from each other in catalepsy scores. Fig. 2 shows that only the dilution 2×10^{-4} of *Agaricus* 30s suppressed haloperidol-induced catalepsy significantly ($p<0.01$, Kruskal-Wallis test).

4. Discussion

The results of the present experiment very clearly indicate that potentized *Agaricus* is effective up to a dilution of 1:20,000 and no more. Thus potentized *Agaricus* has some active entity which is dilutable and which obeys the fundamental

principles of pharmacology in showing a dose response. The results further show that it is the concentration of the active entity and not its total amount that is responsible for the anticataleptic effect. Now, the question is what could be the
active physical entity in *Agaricus* 30s , the dilution of which is 10^{-60} and is evidently beyond the Avogadro number.

Solvent molecules may form clathrates around drug molecules in the starting solution or the mother tincture [9]. Due to mechanical agitation, the molecules in the centres of clathrates may be dislodged and new clathrates around the vacant ones appear [10, 11]. It is these clathrates that may carry the message of drug molecules when the latter disappear in course of successive dilution. We have already demonstrated that potentized *Agaricus* act only through the oral route and not through parenteral ones [4, 6]. The clathrates present in the solvent may act on the receptors of lingual epithelium wherefrom the impulse is carried to the brain. These clathrates during successive dilution without any mechanical agitation are rapidly depleted, and for this, the dilutions of *Agaricus* 30s above 1:20,000 had no anticataleptic effect. Again, when such a dilution was agitated as with the sonication of 1:50,000 dilution, the effect reappeared. Thus mechanical agitation is a source for the generation or multiplication of the physical entity in the potentized *Agaricus*. We have observed earlier that mechanical agitation increased the efficacy of potentized *Agaricus* [8].

The involvement of tongue receptors is further evidenced by the fact that the concentration of the potentized drug is important and not the amount. It is intersting to note that more than 60, probably 100 G-protein coupled receptors occur in the lingual epithelia [12, 13]. G-protein involved here is alphagustducin [14]. The large number of receptors may respond to various types of potentized drugs. As potentized *Agaricus* is put on the tongue, the structure of the specific clathrate is converted into a pattern of action potentials of receptor neurons, which is processed by the brain in its complex neural net-work and finally a physiological effect is produced.

References

[1] Sukul N.C., B. Bhattacharya and S.K. Bala, "Influence of homoeopathic drugs on neuro-transmiter systems", *Sci. Cult.* vol. 51, pp. 161-163, 1985.

[2] Sukul N.C., S.K. Bala and B. Bhattacharya, "Prolonged cataleptogenic effects of potentized homoeopathic drugs" *Psychopharmacology*, vol. 89, pp. 338-339, 1986a.

[3] Sukul N.C., B. Bhattaharya and S.K. Bala, "Differentiation of homoeopathic potencies by experimental catalepsy", *Environ. Ecol.* vol. 4, pp. 585-588, 1986b.

[4] Sukul N.C., S.K. Bala and B. Bhattacharya, "Cataleptogenic effect of a homoeopathic drug mediated through the alimentary tract", *Environ. Ecol.*, vol. 5, pp. 411-415, 1987.

[5] Sukul N.C. and W.R. Klemm, "Influence of dopamine agonists and an opiate antagonist on Agaricus-induced catalepsy, as tested by a new method", *Arch. int. pharmacodyn.*, vol. 295, pp. 40-51, 1988.

[6] Sukul N.C., A. Paul, S.P. Sinha Babu and S. Ghosh, "Haloperidol-induced catalepsy in mice and rats suppressed by orally pre-administered *Agaricus*", *Br. Hom. J.*, vol. 84, pp. 6-11, 1995.

[7] Anonymous, *The American Homoeopathic pharmacopoeia*, 9th edn. Philadelphia, Boericke and Tafel, 1920.

[8] Sukul N.C., "Mechanical agitation, the main factor in increasing the efficacy of

Agaricus muscarius, a homoeopathic drug", *Environ. Ecol.*, vol. 10, pp.7-10, 1992.

[9] Wei S., Z. Shi and A.W. Castleman, "Mixed cluster ions as a structure probe: Experimental evidence for clathrate structure of $(H_2O)_{20}H^+$," *J. Chem. Phys.*, vol. 94, pp. 3268-3273, 1991.

[10] Anagnostatos G.S., G. Vithoulkos, P. Garzonis and C. Tavovxoglou, "A working hypothesis for homoeopathic microdiluted remedies". *Berlin J. Res. Hom.*, vol. 1, pp.141-147, 1991.

[11] Anagnostatos G.S., P. Pissis and K. Viras, "Possible water cluster formation by dilution and succussions", in: *Atomic and Nuclear Clusters*, proc. Second International conference at Santorini, Greece, 1993, pp. 215-217.

[12] Keiko, Abe, Yuko, Kusakabe, Tanemura Kentaro, Yasufumi Emori and Soichi Arai, "Multiple genes for G protein-coupled receptors and their expression in lingual epithelia", *FEBS Lett.*, vol. 316, pp. 253-256, 1993.

[13] Watson S.P., "Factors to consider in the naming of a G-protien coupled receptor subtype", *J. Recep. Sig. Transduc. Res.*, vol. 15, pp. 5-17, 1995.

[14] McLaughlin K., P.J. McKinnon, and R.F. Margolskee, "Gustducin is a taste -cell-specific G-protein closely related to the transducins", *Nature*, vol. 357, pp. 563-569, 1992.

EFFECTS OF HIGHLY DILUTED DRUGS ON THE PREVENTION AND CONTROL OF MICE TERATOGENICITY

CLOE TADDEI-FERRETTI and ANTONIO COTUGNO
Istituto di Cibernetica, CNR, 80072 Arco Felice (NA), Italy

ABSTRACT

Various substances, among which some purine derivatives, notably caffeine and adenine, exert teratogenic effect on mice. If administered during pregnancy, the litters present anatomical, biochemical and behavioural consequences; at higher doses, the death and resorption of foetuses occur. We investigated the possibility to prevent and/or abolish the teratogenic effect of caffeine and adenine in mice by means of high dilutions of substances chosen according to the isopathic or homeopathic criterion (either the same teratogenic agent, or other substances): the effect has been either prevented, or counteracted by highly diluted substances administered either before or after the intoxication by the teratogenic agent. The amount of the effect depends on the type of substance used and on the degree of its dilution.

1. Introduction

It is known that various substances exert teratogenic effects on mice [*e.g.*, 21-22, 24, 30, 39, 55]. In particular, caffeine [1, 6, 14, 17-19, 25, 28, 31-34, 36, 46, 52] and adenine [15], which are purine derivatives, are teratogenic for mice: if they are administered to female mice during pregnancy, the litters will present anomalies at the biochemical, anatomical (terata with cleft lip, cleft plate, micrognathia, agnathia, open eyelids, subcutaneous hematoma, digital defects, axial skeletal defects, gross malformations) and behavioural levels; at higher doses, death and resorption of foetuses occur.

Caffeine has teratogenic effects also on other organisms, from rats to humans [2, 5, 7, 10, 12, 35, 37, 40-41, 43, 54], up to the second generation [46], and postnatal effects on rats [53]. Caffeine [4, 16, 26-27] and other substances [23] interact with the teratogenic effects exerted by different agents.

Pretreatment with some substances reduces the teratogenic effects of caffeine in mice [18]. A chronic administration of caffeine at a 0,05% diluition to female mice during some months before pregnancy reduces the teratogenic effects of caffeine administered later during pregnancy; the mitigation of the teratogenic effects in this case is probably due to an increased metabolic rate of caffeine degradation [52]. Repeated administration of caffeine induces caffeine tolerance [11, 13, 38]. Also the teratogenic effects obtained by fasting [50] or physical activity [51] of pregnant mice is reduced by treatment of females before pregnancy respectively with fasting or physical activity.

The *placebo* groups of female mice, which in some experiments of Terada and Nishimura did not receive the injection of 150 or 250 mg/kg of caffeine, presented a low frequency of open eyelids; the occurrence of this natural type of teratogenicity was lowered by the chronic administration of 0,05% caffeine to the female mice before pregnancy [52].

The last data are related to the control of natural teratogenicity in mice by chronic administration of low doses of the same substance that at high doses induces the same type of teratogenicity. This observation leads to consider the parallel observation - on which the homoeopathic medicine is based [20] - that a natural disease in humans can be treated by the administration of very low doses of the same substance that, if administered at high doses to healthy persons, produces a complex of symptoms very similar to the symptoms presented during the desease itself. Thus all the therapeutic drugs used as remedies in the homoeopathic medicine are choosen after experimental observations of their general and particular effects at psychic and physical levels on healty persons. In the particular case of the isopathic practice, the identical substance that has caused a desease is used to control the desease itself.

At this point it seemed interesting to investigate the possiblity to preventing and/or abolishing the teratogenic effect of caffeine and adenine in mice by the administration of substances chosen according to the isopathic or homoeopathic criterion and administered at the doses routinely used in homoeopathic medicine.

2. Materials and Methods

Albino mices (purchased from Morini, S. Polo d'Enza, RE, Italy, and breeded in our laboratory) were used. Three months old virgin females (mean weight 30 g) were mated overnight with one male each and checked for vaginal plug, indicating the occurrence of copulation [9, 16, 24, 30, 39, 51, 52], in the following morning. Upon finding of the plug (day 1 of gestation) each female, always singularly caged, was assigned to one of the three groups of each experiment, having all the same number of females.

Group 1 females of each experiment received each one a subcutaneous injection either of caffeine (K&K, Plainvew, NY, USA; 250 mg/kg in aqueous solution, in a volume of 25 ml/kg) on day 13 of gestation, or of adenine chloridrate (Biagini, Pisa, Italy; 265 mg/kg in 0,7% saline, in a volume of 65 ml/kg, corresponding to 200 mg/kg adenine, *i.e.* 1.15 the molar dose of caffeine) on day 12 of gestation. The day for the caffeine, or adenine, injection was chosen as the more sensible day, in which the teratogenic agents exert their maximum effects [15, 36].

Group 2 females received, in addition to the caffeine or adenine injection, a highly diluted drug, either purchased, or diluted by us according to the homoeopathic pharmacopraxy. The drug was administered in single oral doses (by means of a pipette) on successive days, either before the teratogenic injection day, or at the injection day and after it, or before and after such day. The vehicle of the stored higly diluted drugs was either glucose previously imbibed by 95° ethylic alcohol solution of a higly diluted drug, or 95° ethylic alcohol. The vehicle of such drug when they were administered orally to mice was either glucose 0.1% in distilled water or 95° ethylic alcohol 1% in distilled water. Prevous experiment had shown that the vehicle of the administered drugs did not have effect upon the parameters under examination.

Group 3 control females received only a subcutaneous injection of the wehicle of the caffeine or adenine injection at the same volume and same gestation day of groups 1 and 2 of the same experiment.

According to the homoeopathic pharmacopraxy, the total dilution of a homoeopathic drug is achieved by a number (hereafter) indicated as "N" of successive dilutions in 95° ethylic alcohol, each one followed by repeated strong shaking. Both the number of successive dilutions and the type of dilution (either 1/10, or 1/100, or 1/50,000) used in that succession may vary for each total dilution [20, 42, 44].

We used different numbers of successive dilutions for the total dilution of each utilized drug, while the type of dilution was always 1/100 (except 1/50,000 in 1 case).

The utilized drugs are as follows. In the case of caffeine injection: either Coffea cruda (which is a therapeutic homoeopathic substance prepared from the total extract of the seeds of *Coffea arabica* L.), or nitrate of strychnine, or Sulfur (which is a therapeutic homoeopathic substance prepared from triturated sulphur), or caffeine (which is the teratogenic agent itself). In the case of adenine injection: only adenine (which is the teratogenic agent itself).

The sources of the homoeopatic drugs are as follows. Coffea cruda was purchased diluted from Unda, Harzé, Belgique, except that diluted with 1/50,000 steps, which was purchased diluted from Omit, Cisterna di Latina, Italy. The nitrate of strychnine was purchased from BDH, Poole, UK, and successively diluted either by us in the case of the first experiment reported in Table 1, or by Omit in the case of the two following ones. Sulfur was purchased from Unda and successively diluted by courtesy of Dr Alma Rodriguez, CeMON, Napoli, Italy. The caffeine was purchased from K&K, Plainvew, NY, USA, and successively diluted either by us in the case of the first two experiments of Table 1, or by Omit in the case of the three following ones. The adenine was purchased from Fluka, Buchs, Switzerland, and successively diluted by us.

We recorded for each group of each experiment: the number of alive newborn mice and the number of dead or cannibalized newborn mice (within 12 hours after partuition); the number of the dead females to take into account the low lethal effect on the mothers imputable to the teratogenic agents at the utilized doses and the number of litters with respect to the number of observed plugs, *i.e.* with respect to the number of used females with sign of gestation, irrespective to the fact that they gave litters or not.

We estimated indirectly the relative numbers of embryonic and foetal resorptions of the three groups without checking it *in situ.* They were calculated from the value of the percent of total litters over females with plug of groups 3, and of the numbers of alive and of dead or cannibalized newborn mice of groups 1, 2 and 3 [48]. The estimation was made on the basis of the assumption that the number of the foetus implantation sites should have been the same in all three groups and of the arbitrary hypothesis that the number of resorptions of groups 3 were zero; if, on the contrary, such number was actually more than zero, the only diference should have been the addition of this same number to those of the resoptions of all three groups.

3. Results

3.1. General observations

The value of the mean percent ratio of pregnant females over females with plug in groups 3 was consistent with the values obtained by other researchers [9].

A low lethal effect on the females of groups 1 was observed after the injection of caffeine or adenine and a lowering of such effect was obtained in groups 2 by the higly diluited drugs (Df of Tab. 1).

The number of dams was greatly lowered in groups 1 with respect to control groups 3 and was shifted towards normal value in groups 2 after the use of highly diluited drugs (L and L% of Tab. 1).

The number of newborn mice alive within 12 hours after partuition was much lower in groups 1 than in groups 3, and it had an intermediate value in groups 2 (A of Tab. 1).

The number of newborn mice dead or cannibalized within 12 hours after partuition was higher in groups 1 and lower in groups 3 than in groups 2 (DC of Tab. 1)

The number of all newborn mice alive, dead or cannibalized within 12 hours after partuition was low in groups 1, medium in groups 2 and high in groups 3 (ADC of Tab. 1).

Table 1
Results of the experiments.

T	H	N	G	P	Df			L			L%			A			DC			ADC		
					+	x	°	+	x	°	+	x	°	+	x	°	+	x	°	+	x	°
caffeine	Coffea cruda	5	B A	27	4	1	0	7	13	17	26	48	63	45	66	124	14	21	15	59	87	139
		30	B A	5	0	1	0	2	3	4	40	60	80	15	17	33	4	0	1	19	17	34
		200-1000	B	7	1	2	0	3	2	2	43	29	29	9	15	16	10	1	0	19	16	16
		30 (1/50,000)	B	8	2	0	0	5	7	7	63	88	88	16	27	62	17	8	0	33	35	62
	strychnine nitrate	9	B A	11	0	0	0	5	8	10	45	73	91	27	71	91	4	2	1	31	73	92
		4-5-7-9-11-13 15	B A	8	0	0	0	2	5	4	25	63	50	18	22	33	0	11	0	18	33	33
		5-7-9-11-13 15-30	B A	15	0	0	0	7	6	10	47	40	67	33	31	72	16	5	1	49	36	73
	Sulfur	5	B A	5	0	0	0	2	4	4	40	80	80	15	27	33	4	2	1	19	29	34
	caffeine	5	B A	6	6	5	0	0	1	5	0	17	83	0	5	32	0	0	0	0	5	32
		9	B A	5	0	0	0	1	2	5	20	40	100	0	13	50	2	2	0	2	15	50
		5-7-9-11-13-15	A	15	3	0	0	8	10	9	53	67	60	25	36	78	27	16	0	52	52	78
		4-5-7-9-11-13 15	B A	8	0	0	0	2	6	4	25	75	50	18	36	33	0	7	0	18	43	33
		5-7-9-11-13 15-30	B A	15	0	0	0	7	9	10	47	60	67	33	59	72	16	9	1	49	68	73
adenine	adenine	4-9-30	A	9	6	1	0	2	3	4	22	33	44	16	15	33	1	10	1	17	25	34
		4	A	6	2	0	0	3	0	5	50	0	83	14	0	47	0	0	0	14	0	47
		30	A	6	2	2	0	3	3	5	50	50	83	14	26	47	0	0	0	14	6	47
TOTAL				156	26	12	0	59	82	105	38	53	67	298	466	856	115	94	21	413	560	877

Legend of the symbols used in Table 1

T	teratogenic substance
H	hyghly diluted drug
N	number of 1/100 dilution steps accomplished to obtain the H total dilution utilized (see Materials and Methods). When many Ns are indicated, the H was administered in different successive days at different increasing total dilutions. The notation 1/50,000 indicates the type of dilution steps which, only in that case, was used
G	period of gestation at which the H was administered: B = before the day of injection of the T; A = at that day and after it
P	number of females the plug of which was recognized and which were thus utilized for each one of the three groups of an experiment
Df	number of females dead within 24 hours after the injection of the T or of its vehicle
L	number of females that gave litters (alive, dead or cannibalized within 12 hours after partuition)
L%	percent of litters over females with plug (the values of the line corresponding to Total were calculated from the Totals of P and of L)
A	number of alive newborn mice
DC	number of dead or cannibalized newborn mice
ADC	A + DC
+	group 1 (teratogenic substance)
x	group 2 (teratogenic substance and highly diluted drug)
°	group 3 (vehicle of the teratogenic substance)

When considering the means of alive newborn mice, of dead or cannibalized newborn mice, of embryonic or foetal resorptions, and of foetal implantation sites of each dam or potential dam, one can observe that groups 2 are characterized by a lower number of embryonic or foetal resorptions, and of dead or cannibalized newborn mice and a higher number of alive newborn mice if compared with groups 1; the reverse situation holds when comparing groups 2 with groups 3 (Table 2).

Table 2

	+	x	°
A	2.83	4.43	8.15
DC	1.1	0.9	0.2
R	4.42	3.02	0
ADCR	8.35	8.35	8.35

Mean number of alive newborn mice, A, of dead and cannibalized newborn mice, DC, of embryonic or foetal resorptions, R, and of foetus implantation sites, ADCR, for each dam of group 1, +, group 2, x, and group 3, °.

3.2. Particular observations

Few exceptions were found to the trends reported in the General observations. The most significant negative case is that obtained with adenine when N of dilution step accomplished to obtain the total dilution of the higly diluted adenine was 4: a teratogenic effect was observed in group 1 with respect to group 3 for what it concerns the number of females that gave litters, the percent of litters over females with plug and the number of newborn mice, but no mitigation of the teratogenic effect was observed and 1; no dams were obtained in group 2, that received the higly diluted adenine in addition to the adenine injection, with respect to group.

When considering the number of alive newborn mice (A of Tab. 1), the highest mitigation of the teratogenic effect was obtained with higly diluted Coffea cruda (N = 200 and 1,000), Sulfur (N = 5), caffeine (N from 5 to 30 and from 4 to 15) and strychnine nitrate (N = 9).

4. Discussion

The first result reported gives the assurance of a normal fertility of the used mice strains.

The other results, which confirm and extend preliminary ones [47, 49], indicate a general tendency of the used higly diluted drugs to counteract the effect of the teratogenic agents, either if given only before, or only after, or before and after the teratogenic injection, thus acting both with a mechanism of prevention and of treatment.

This tendency can be observed for what it concerns the number of dead females, of litters with respect to females with plug, of alive newborn mice, of dead or cannibalized newborn mice (Tab. 1), as well as of embryonic and foetal resorptions (which is a parameter as important as the number of alive newborn animals and that of dead newborn animals in order to judge the embryolethal effect of a substance [16, 22, 24, 30, 39, 51, 52] and foetus implantations sites (Tab. 2).

Many of the exceptions to this general tendency are of little significance, since the the differences between the values of groups 1 and 2 are low. Many of the exceptional data relative to the dead or cannibalized mice could be considered normal if one takes into account the total number of alive, dead, or cannibalized mice: in fact, the temporal threshold used to assigne newborn mice to the class of alive or not alive ones was arbitrary and not precise (a period within 12 hours after the time of partuition, which was often unknown).

However, one could hypothesize that when adenine (N = 4) was used in group 2 the total dilution of adenine, was not sufficiently high, if related to the molecular weight (*i.e.*, to the dilution expressed as molarity value and not simply as mg/ml), to obtain the reversal of the pathologic effect which is expected at high dilutions. A confirmation of this hypothesis is given by the facts that: i) a lower unsuccess was obtained with adenine (N = 4, 9, 30), when the dilution N = 4 was administered the day of the teratogenic injection (when the teratogenic agent effect was maximum) and the day after, and the higher dilution N = 9 was administered two days after the teratogenic injection, and later the N =30 dilution was administered; ii) a high mitigation of the teratogenic effect was obtained when only the dilution N = 30 of adenine was used in group 2; iii) an absence of unsuccess of the higly diluted drugs was observed both with strychnine nitrate (N from 4 to 15) and with caffeine (N from 4 to 15), although also the dilution N = 4 was used, but such dilution was used before the day of the teratogenic injection, thus exerting a prevention effect [52], and the ratio between the molarity of highly diluted agent and the molarity of the teratogenic agent was 5 times lower in the case of strychnine nitrate (N = 4) with respect to adenine (N = 4) and 2.5 times lower in the case of caffeine (N = 4) with respect to adenine (N = 4), in accordance with the dose dependence of the effect sign observed by Bastide *et al.* [3], Burlakova *et al.* [8] and Youbicier-Simo *et al.* [57].

As a conclusion, the reported experiments prove that the highly diluted drugs had an effect (which certainly was not due to subjective suggestion) both in the prevention and control of the teratogenicity of caffeine and adenine in mice. However, further experiments performed with the appropriate methodologies applied at the molecular or lower level, are reqired to investigate how the effect of higly diluted drugs is obtained.

One has to stress also that the highly diluted drugs were applied taking into account only one characteristic. In the case of higly diluted caffeine or adenine, the teratogenic agent itself was used to test the reversal of the effects of substances at low doses, according only to the isopathic criterion. Coffea cruda was chosen because it is a therapeutic homoeopathic substance prepared from the total extract of the seeds of *Coffea arabica* L, of which caffeine, used here as teratogenic agent, is an alchaloid. The nitrate of strychnine was utilized because strychnine is an alchaloid of *Strychnos nux-vomica* L, from the total extract of which the therapeutic homoepathic substance Nux vomica is prepared, which is utilized also in severe intoxications, and successfully tested to counteract the caffeine effects [56]. Sulphur was utilized because it is used as therapeutic homoepathic substance also for its property to "bringing outside" internal deseases.

However, for a proper use of all these drugs according to the "*simillimum* law" [20], one should have taken into account many other characteristics. According to this law, the drug more suitable to control at high dilutions the global set of symptoms of an organism which is in a pathological situation is the drug that, when tested at higher doses on healthy individuals of the same type of organism, produces the same set of symptoms. In all cases, the large set of symptoms, of hierarchical values, presented by an organism at each phase of its existence, is the overt expression of the state of the global physiological interactions exerted in such organism at such phase [29].

The above type of methodology, in order to the applied to the above reported experiments on mice, should have required both the existence of many large sets of symptoms produced by many substances previously tested at high doses on healty mice, and the consideration of the complete large set of symptoms (not only the teratogenic symptoms) presented by each single tested female mice. Obviously, such type of methodology could not be utilized in the case of mice. We should thus hypothesize that this is the reason for which, apart the case of adenine (N=4), the above results were positive, but not totally positive.

Acknowledgements

We wish to thank Prof. A. Negro (Roma, Italy), who suggested the present research and supported it with discussions and criticism, Dr A. Rodriguez (Napoli, Italy) for the supply of Sulfur dilutions and for other facilities, and Dr M. Briggs-Smith (Napoli, Italy) for the correction of the English manuscript.

References

[1] Ajarem, J.S. & M. Ahmad, "Behavioral and biochemical conseqqence of perinatal exposure of mice to instant coffee: a correlative evaluation", *Pharmacology Biochemistry and Behavior* 40(4), 1991, pp. 847-852.

[2] Al-Hachim, G.M., "Teratogenicity of caffeine: A review", *Eur. J. Obsbet. Gynecol. Reprod. Biol.* 31(3), 1989, pp. 237-247.

[3] Bastide, M., V. Daurat, M. Doucet-Jaboeuf, A. Pelegrin & P. Dorfman, "Immunomodulatory activity of very low doses of thymulin in mice", *Int. J. Immunotherapy* 3, 1987, pp. 191-200.

[4] Beck, S.L. & C.M. Urbano, "Potentiating effect of caffeine on the teratogenicity of acetazolamide in C57BL/6J mice", *Teratology* 44(3), 1991, pp. 241-250.
[5] Bhatnagar, M. & S. Bhatnagar, "Maternal caffeine and postnatal modulation of brain and behaviour in animal models", *Indian J. Physiol. Allied Sci.* 42(4), 1988, pp. 130-135.
[6] Bertrand, N., J. Girod & M.F. Rigaud, "Ectrodactylie provoquée par la caféine chez les rongeure", *C. R. Soc. Biol.* 164, 1970, pp. 1488-1489.
[7] Beyer, A., "Effects of caffeine consumption on pregnancy outcome", *J. Reprod. Med.* 33(12), 1988, pp. 945-956.
[8] Burlakova, E.B., Konradov A.A., & Khudyakov I.V., "Effect of chemical agents in ultralow doses on biological objects", *Journal of nonlinear Biology* 1(1), 1990, pp. 77-91.
[9] Chernoff, N., E. Rogers, B. Carver, R.Kavlock & E. Gray, "The fetotoxic potential of municipal drinking water in the mouse", *Teratology* 19(2), 1979, pp. 165-170.
[10] Collins, T.F.X., J.J. Welsh, T.N. Black & D.I. Ruggles, "The teratogenic potential of caffeine ingested in drinking-water", *Food Chem. Toxicol.* 21(6), 1983, pp. 763-778.
[11] Colton, T., R.E. Gosselin & R.P. Smith, "The tolerance of coffee drinkers to caffeine", *Clin. Pharmacol. Ther.* 9, 1968, pp. 31-39.
[12] Driscoll, P.G., F. Joseph Jnr & T. Nakamoto, "Prenatal effects of maternal caffeine intake and dietary high development in fetal rats", *Br. J. Nutr.* 63(2), 1990, pp. 285-292.
[13] Eddy, N.B. & A.W. Downs, "Tolerance and cross-tolerance in the human subjects to the diuretic effect of caffeine, theobromine and theophylline", *J. Pharmacol. Exp. Ther.* 33, 1928, pp. 167-174.
[14] Elmazar, M.M.A., P.R. McElhatton & F.M. Sullivan, "Teratogenic effects of different oral preparation of caffeine in mice", *Toxicology* 23(1), 1982, pp. 57-72.
[15] Fujii, T., "Comparison of the teratogenic action of adenine and methylxanthine compounds in mice", *J. Kansai Med. Univ.* 28, 1976, pp. 901-918.
[16] Fujii, T. & T. Nakatsuka, "Potentiating effects of caffeine on teratogenicity of alkylating agents in mice", *Teratology* 28, 1983, pp. 29-33.
[17] Fujii, T. & H. Nishimura, "Teratogenic action of some methylated xanthines in mice", *Okajiamas Folia Anat. Jap.* 46, 1969, pp. 167-175.
[18] Fujii, T. & H. Nishimura, "Reduction of frequency of fetopahtic effect of caffeine in mice by pretreatement with propranolol", *Teratology* 10, 1974, pp. 149-152.
[19] Fujii, T., H. Sasaki & H. Nishimura, "Teratogenicity of caffeine in mice related to its mode of administration", *Jap. J. Pharmacol.* 19, 1969, pp. 134-138.
[20] Hahnemann, S., *Organon der Heilkunst*, 6th posthumous edition 19211, Stuttgart: Hohstetter, 1978.
[21] Hamburgh, M., "Malformations in mouse embryos induced by Trypan Blue", *Nature* 169, 1952, p. 27.
[22] Hamburgh, M., M. Erlich, G. Nathanson & I. Pesetsky, "Some additional observations relating to the mechanism of Trypan Blue teratogenesis", *J. Exp. Zool.* 192, 1975, pp. 1-12.
[23] Harbison, R.D., B. Mantilla-Plata & D.J. Lubin, "Alteration of Δ9-tetrahydrocannabinol-induced teratogenicity by stimulation and inibition of its metabolism", *J. Pharmacol. Exp. Therap.* 202(2), 1977, pp. 455-565.
[24] Kaufman, M.H. & K.S. O'Shea, "Induction of monozygotic twinning in the mouse", *Nature* 276, 1978, pp. 707-708.

[25] Knoche, C. & J König, "Zur pränatalen Toxizität von Diphenilpyralin-8-chlortheophillinat unter Berucksichtigung von Erfahrungen mit Thalidomid und Caffein", *Arzneimittelforschung* 14, 1964, pp. 415-424.

[26] Kurishita A, "Histological study of cell death in digital malformations induced by 5-azacytidine; Suppressive effect of caffeine", *Teratology* 39(2), 1989, pp. 163-272.

[27] Kusama, T., N. Sugiura, M. Kai & Y. Yoshizawa, "Combined effects of radiation and caffeine on embryonic development in mice", *Radiat. Res.* 117(2), 1989, pp. 273-281.

[28] Kwasigroch, T. E. & R.G. Skalko, "Estimation of creatine kinase isoenzymes: The effects of caffeine and retinoic acid on creatine kinase levels in fetal mouse limbs", *Toxicol. Lett.* (AMST), 21(3) 1984, pp. 339-348.

[29] Marotta, P. & C. Taddei-Ferretti, "I principi della medicina omeopatica", in F. Allocati, C. Musio & C. Taddei-Ferretti eds, *Cibernetica Biologica*, Torino: Rosemberg & Sellier, 1994, pp. 332-336.

[30] Miller, R.P. & B.A. Becker, "Teratogenicity of oral Diazepam and diphenylhdantoin in mice", *Toxic. Appl. Pharmac.* 32, 1975, pp. 53-61.

[31] Mulvihill, J.J., "Caffeine as teratogen and mutagen", *Teratology* 8, 1973, pp. 69-72.

[32] Nakai, K., "Congenital anomalies in mouse embryos induced by administration of a large dose of caffeine during pregnancy", *Acta Anat. Nippon.* 36, 1961, pp. 452-464.

[33] Nakai, K., "Relation between occurrance of hematomas and malformations induced in mouse embryos by administration of large dosage of caffeine during pregnancy", *Acta Anat. Nippon.* 36, 1961, pp. 521-528.

[34] Nanjo, H., "Maldevelopement of the fetuses caused by maternal administration of caffeine during pregnancy in relation to maternal age", *Cong. Anom.* 4, 1964, pp. 23-26.

[35] Nash, J.E. & T.V.N. Persaud, "Influence of nicotine and caffeine on rat embryonic development", *Histol. Histopathol.* 3(4), 1988, pp. 377-388.

[36] Nishimura, H. & K. Nakai, "Congenital malformations in offspring of mice treated with caffeine", *Proc. Soc. Exp. Biol. Med.* 104, 1960, pp. 140-142.

[37] Oei, S.G., R.P.L. Vosters & N.L.J. van der Hagen, "Fetal arrhythmia caused by excessive intake of caffeine by pregnant woman", *Br. Med. J.* 298(6673), 1989, p. 568.

[38] Pieper, K.H., "Weakening of the action of caffeine after repeated administration", *Arzneimittelforschung* 13, 1963, pp. 585-588.

[39] Randall C.L. & W.J. Taylor, "Prenatal ethanol exposure in mice: teratogenic effects", *Teratology* 19(3), 1979, pp. 305-312.

[40] Salvador, H.S. & B.Y. Koos, "Effects of regular and decaffeined coffee on fetal breathing and heart rate", *Am. J. Obstet. Ginecol.* 160(5), 1989, pp 1043-1047.

[41] Sasahara, H., H. Yamano & T. Nakamoto, "Effect of maternal caffeine with zinc intake during gestation and location on bone development in newborn rats", *Arch. Oral Biol.* 35(6), 1990, pp. 425-430.

[42] Schmidt, P., "The hidden treasures of the last *Organon*", *British Homoeopathic Journal*, July, 1954, pp. 134-156.

[43] Schneider, P.E., H.I. Miller & T. Nakamoto, "Effects of caffeine intake during gestation and lactation on bones of young growing rats", *Res. Exp. Med.* 190(2), 1990, pp. 131-136.

[44] Schwabe, W., *Pharmacopea homoeopathica polyglotta*, Leipzig: Schwabe, 1879.

[45] Sinton, C.M., "Preliminary indications that functional effects of fetal caffeine exposure can be expressed in a second generation", *Neurotoxical Teratol.* 11(4), 1989, pp. 357-362.

[46] Snigorska, B. & H. Bartel, "Studies on the teratogenic influence of caffeine on the fetuses of white mice", *Folia Morph.* 29, 1970, pp. 353-363.

[47] Taddei Ferretti, C. & A. Cotugno, "Treatment of teratogenicity induced by purine derivatives", *International Congress of Homeopathic Medicine*, Roma, Italy, 1981.

[48] Taddei Ferretti, C. & A. Cotugno, "Treatment of the teratogenicity induced in mice by caffeine or adenine", *Progr. Biochem. Biotechnol.* 3(1), 1995, in press.

[49] Taddei Ferretti, C. & A. Cotugno, "Control of embriolethality experimentally induced in mice", *Deutsches Journal für Homeopathie*, 1996, in press.

[50] Terada, M., "Effect of intermittent fasting before pregnancy upon maternal fasting as a teratogen in mice", *J. Nutrition* 100(7), 1970, pp. 767-772.

[51] Terada, M., "Effect of physical activity before pregnancy on fetuses of mice exercised forcibly during pregnancy", *Teratology* 10(2), 1974, pp. 141-144.

[52] Terada, M. & H. Nishimura, "Mitigation of caffeine-induced teratogenicity in mice by prior chronic caffeine ingestion", *Teratology* 12, 1975, pp. 79-82.

[53] Vornanen, M., "Effects of caffeine on the mechanical properties of developing rat heart ventricles", *Comp. Biochem. Physiol.* 78C(2), 1984, pp. 329-334.

[54] Wilcox, A.J., "Caffeinated beverages and decreased fertility", *Lancet* 1(8642), 1989, p. 840.

[55] Waddington, C.H. & T.C. Carter, "Malformations in mouse embryos induced by Trypan Blue", *Nature* 169, 1952, pp. 27-28.

[56] Wurmser, L., "Un essai de pharmacologie homoeopathique: Antagonisme entre *Nux vomica* et la Caféine", *Homoeopathie française* 6, 1948, pp. 17-24.

[57] Youbicier-Simo, B.J., F. Boudard, M. Mekaouche, M. Bastide & J.D. Baylé, "Effects of embryonic bursectomy and *in ovo* administration of highly diluted bursin on adrenocorticotropic and immune response of chickens", *Int. J. Immunotherapy* 9, 1993, pp. 169-180.

ASSESSMENT OF ANTI-INFLAMMATORY ACTIVITY OF TWO DILUTED DRUGS ON SKIN: *URTICA URENS* AND *APIS MELLIFICA*

J. HANDSCHUH, A. DEMISCAUT and M.C. POELMAN
Laboratoire de Dermopharmacie et de Biophysique cutanée
Faculté des Sciences Pharmaceutiques et Biologique
4, avenue de l'Observatoire, Paris

The aim of this study was to investigate the anti-inflammatory activity of two homeopathic diluted drugs : Urtica urens and Apis mellifica, by a non invasive and non subjective cutaneous technique. This method consisted to induce a skin inflammation, provided by an application of an aqueous methyl nicotinate solution (o.5%). The vasodilation induced was quantified in terms of cutaneous blood flow by a Laser Dopler Velocimeter (L.D.V.). This vasodilation could be inhibited, in part, by a skin pretreatment with homeopathic dilutions. The efficacy of these dilutions was expreessed by the value of the inhibition percentage, which was calculated by comparing responses supplied by the L.D.V., for the methyl nicotinate score, to the homeopathic dilutions scores.Ten adult volonteers of either sex, aged from 20 - 50 years, took part.Under this protocol, homeopathic dilutions, placebo and positive control were assessed. The different dilutions gave significant results but with a different amplitude in the response : Urtica urens 4 CH exhibited the strongest anti-inflammatory effect reaching an inhibition of 37 % ; Urtica 7 CH reduced blood flow by up to 30,8 % and Apis 4 CH to 21,2 %. No significant difference was noted for Apis 7 CH and between the dilutions.

The aim of this study was to investigate the anti-inflammatory activity of two homeopathic diluted drugs : Urtica urens and Apis mellifica, by a non invasive and non subjective cutaneous technique.

These two drugs were chosen for their homeopathic activity : they are considered, in fact, as active against "redness", "heat" , "pain" and "oedema". These four reactions are caracteristic of an inflammatory type reaction. It is well known that, Apis presents anti-inflammatory and anti-histaminic properties (1, 2) ; Urtica urens is classically used in homeopathy against local inflammatory symptoms (5). Dilutions 4 and 7 CH, were chosen according to homeopathic use, which recommands low and mean dilutions for a local and a general action (4). The selected method was described and validated by Poelman and al (3).

The principle of this method consists to induce a skin inflammation, provided by an application of an aqueous methyl nicotinate solution (o.5%). The inflammatory response (vasodilation) is quantified in terms of cutaneous blood flow by a Laser Dopler Velocimeter (L.D.V.). This vasodilation could be inhibited, in part, by a skin pretreatment with homeopathic dilutions.The efficacy of these dilutions fits with the value of the inhibition percentage, which is calculated by comparing responses supplied by the L.D.V., for the methyl nicotinate score, to the homeopathic dilutions scores, on the pretreated and untreated skin sites.

Materials and methods :

Subjects :

Ten adult volonteers of either sex, aged from 20 - 50 years, took part. One week before the study, they were asked to avoid any injestion or skin application of anti-inflammatory drugs.

Preparations :

Urtica urens and Apis mellifica 4 and 7 CH were prepared from Mother Tinctures by successive dilutions in hydro-alcoholic solvent and succussions according to french Pharmacopea protocol.Then, the last dilutions and the placebo were gelified by carbomer 974P. Formulations are summarized on Table I.

Components	gel 1	gel 2	gel 3	gel 4	gel 5
Urtica urens 3CH	1%	0	0	0	0
Urtica urens 6CH	0	1%	0	0	0
Apis mellifica 3CH	0	0	1%	0	0
Apis mellifica 6CH	0	0	0	1%	0
Carbomer 974P	0.50 %				
Triethanolamine	QS pH 5.8				
Ethanol (30%V/V)	QS 100 %				

Legend : Gel 1 = Urtica urens 4 CH ; Gel 2 = Urtica urens 7 CH ; Gel 3 = Apis mellifica 4 CH ; Gel 4 = Apis mellifica 7 CH ; Gel 5 = Placebo gel

Table 1 :Composition of different gels :

Protocol :

The four formulations were tested versus a control placebo gel and a control positive gel containing 2.5 % niflumic acid. After the evaluation of the physiological baseline on untreated skin for the ten subjects, these preparations were applied on the ventral surface skin of the two forearms (2 mg/cm^2). The sites were then, occluded by a plastic film for 3 hours. After depatching, each zone was let for one hour rest. Following that, methyl nicotinate solution was applied on the sites and the vasodilation induced , was measured. The vascular responses to methyl nicotinate solution on pretreated sites, were evaluated by laser Dopler velocimetry using a Periflux PF 2B laser Doppler fluxmeter. The recordings of all the areas were carried out over 30 minutes.

Results :

Under this protocol, homeopathic dilutions, vehicule (placebo) and niflumic acid gel (positive control), were assessed.The areas under the response-curve obtained for each drug were calculated and compared with those of the vehicule. Results were expressed as the inhibition percentage of the inflammatory reaction induced by methyl nicotinate solution, calculed from the following equation :

$$\text{inhibition (\%)} = \frac{\text{AUC (V)} - \text{AUC (D)}}{\text{AUC (V)}} \times 100$$

AUC (V) = area under the curve of the vehicule treated site
AUC (D) = area under the curve of the drug treated site

The recording, given as an example in the following diagram (Fig 1), shows the increase of the vasodilation during the first 5 minutes after the application of methyl nicotinate solution and the progressive decrease (curve I). It shows too, the recording obtained with anti-inflammatory drugs treated areas (curve II) and the physiological baseline (curve III).

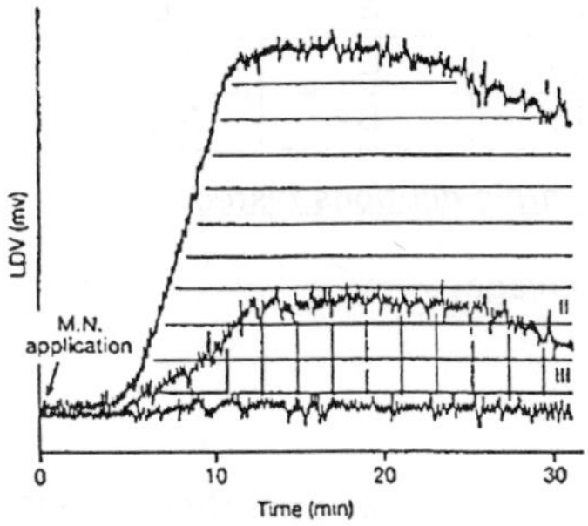

legend : I : M.N. recording
II : A.I. drug treated areas recording
III : Physiological baseline

Fig I : Example of Laser Dopler Velocimeter recording.

These four dilutions gave positive results, with a different amplitude in the response. The variation of the area under the curve, for each dilution, placebo gel and niflumic acid gel, versus placebo, is shown in fig 2 .

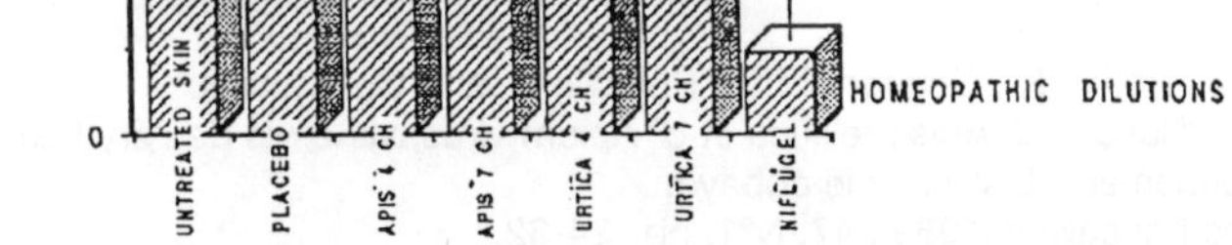

Table 2 : Comparison of aeras responses under the curves of different dilutions

Statistical analysis was realized by paired Student 's t Test.

No significant difference was observed between untreated skin and placebo. Niflugel* (niflumic acid gel) exhibits a significant decrease of skin blood flow. Apis 7 CH was no effective under our conditions. Urtica 4 CH reduced blood flow by up to 37. 1 %, Urtica 7 CH by up to 30. 8 % and the lowest inhibitory effect (21.2 %) was obtained with Apis 4 CH (Table 3), but no statistical result could be noted between these formulations.

Drugs	m	AUC	t	P
Apis 4 CH	21.2 %	175.06	S	0.014
Apis 7CH	20.6%	176.22	NS	0.104
Urtica 4CH	37.1%	139.75	S	0.02
Urtica 7CH	30.8%	153.65	S	0. 007
Niflugel	75. 0%	47.75	S	0. 001

Table 3 : Inhibition percentage of homeopathic dilutions tested

Legend : m = mean of ten results of inibition percentage
A.U.C. = Area Under the Curve

Discussion :

Consequently to this work, two points can be discussed : first, we can note the significant results obtained with Urtica urens 4 and 7 CH, by a non subjective and non invasive cutaneous method. Second, we can remark that, under our conditions, the inhibition percentage of Apis mellifica was not so hight, and that of Apis 7 CH was not significant. However, Bildet in 1989 (1) investigating Apis dilutions on guinea pigs, put in obviousness the anti-inflammatory and anti-histaminic activity of this drug. These general results raise the problem of the choice of the experimental model.

Conclusion :

For the first time we observed the release of homeopathic drugs from an aqueous gel and Urtica urens and Apis mellifica dilutions tested according to this proceeding exhibited an anti-inflammatory activity. The significant results obtained in the course of this work, let foresee a new way of research in homeopathy, even if topical application is rarely used in homeopathic therapy. Nevertheless this kind of research allows to study homeopathic drugs directly in man, into their pharmaceutical forms.

References

1 - Bildet J., Guyot M., Bonini F. , Grignon M.C., Poitevin B. and Quilichini R.
Mise en évidence des dilutions d' Apis mellifica et d' Apium virus, vis-à-vis de l'érythème provoqué par un rayonnement U.V. chez le cobaye.
Ann. pharmaceutiques françaises ,1989 , 47, N°1, pp. 24-32.
2- Poitevin B., Davenas E., Benveniste J.
In vitro immunological degranulation of human basophils is modulated by Lug histamin and Apis mellifica.
Br. J. Clin. Pharmacol., 1988 , 25, 439-444.
3- Poelman M.C., Piot B., Guyon F., Deroni M. and Levêque J.L.
Assessment of topical non-steroidal anti-inflammatory drugs
J. Pharm. Pharmacol., 1989, 41, pp.720-722
4- Matière Médicale Homeopathique
Guermomprez., Pinkas., Forck.
Doin, Paris, 1985.
5- Larue F..
Contribution à l'étude expérimentale de dilutions infinitésimales d'Urtica urens et d'histaminum.
Thèse, Pharmacie, Bordeaux 2, 1985.

SUPERSENSITIVITY OF THE RAT LENS IN A CRITICAL PERIOD OF THE DEVELOPMENT

KONSTANTIN MURANOV
Institute of Chemical Physics, Kosygin St. 4, 117977 Moscow, Russia

1. Introduction

It is known, that the rat lens formation is finished on day 20 post-partum, i.e., to a time of the opening an eye. In a period with 10 till 16 days postnatal the lens is extremely vulnerable. Application this time of the diquat, potassium selenite, or buthionine sulfoximine causes the fast cataract formation. These compounds cause the DNA strain-breaks formation by the free radical mechanism. As a free radical oxidation influences the many cell processes, the main cause of the cataract appearance in a critical period is not understood.

The aim of the investigation was to study the influence of the ASN2-type methylating agent, methyl methanesulfonate (MMS) on the lens development.

2. Materials and Methods

Thirteen days post partum white rat pups received a single subcutaneous injection of MMS. The lens cataract was estimated on the eighteen days postnatal. The cataract lens injury was scored by follow method: Lenses were photographed by slit-lamp biomicroscopy after mydriasis (1% atropine sulphate). Then the photographs were estimated in a five-step scale (0 - clear, 4 - total cataract). The photography negatives were subjected to densitometric image analysis.

Methyl methanesulfonate (Merck), and 0.9% NaCI sterile for injection were used.

Nonparametric pairs Mann-Whitney and Spearman rank correlation tests were used.

3. Results

1) MMS causes the cataract injury of the rat lens during critical period of the postnatal development. The cataract injury was depended from the MMS dose (Fig. 1). The Spearman Rank Correlation Coefficient was significant for the grade estimation of the cataract (r=0.8) in the 0.08-0.8 mmol/kg dose limit. MMS dose 0.08 mmol/kg caused the smallest lens injury (less than grade number 1).

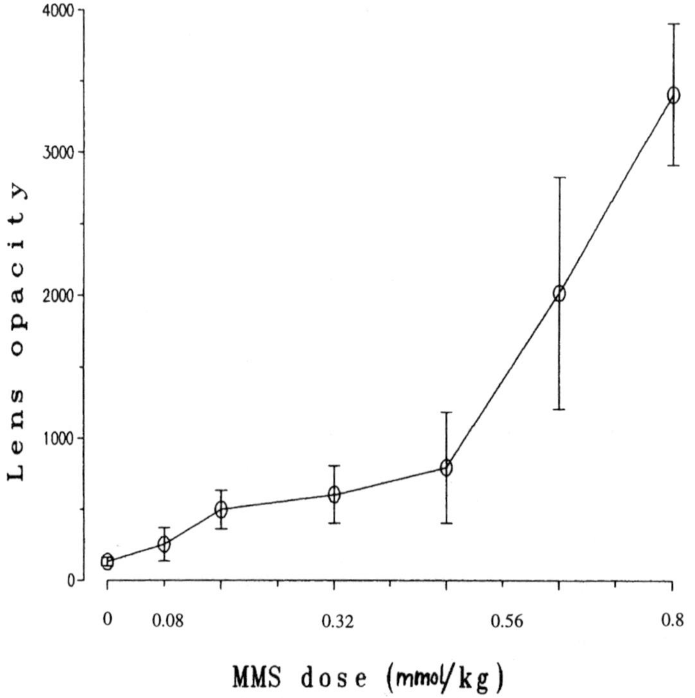

Figure 1 - MMS influence on the rat pups cataractogenesis.

The shadowy dust-like opacities, occasionally with some very small, dense particles were detected. Sometimes the lens opacities were not marked. The frequency of the cataract injury was 15%.

2) The single subcutaneous injection of MMS (0.72 mmol/kg) to one half the litter (the experimental group) caused the cataract formation both in the experimental and control group (Table 1). Nonparametric pairs (Mann-Whitney) test has shown the equality of the cataract injury values of the experimental and control group (two tailed probability of equalling or exceeding Z=0.3485).

	Experimental group (n=18) MMS (0.72mmol/kg)	Control group (n=15) 0.9% NaCl
Grade estimation (Median)	**2**	**2**
Densitrometric estimation (Average ± S.D.)	**1262±220**	**1363±350**
Two tailed probability of equalling or exceeding of experimental and control group lens injury		**Z=0.3485**

Table 1 - Cataract injury of the lens of experimental and control group.

4. Discussion.

The detected phenomenon could be explained by the two manners: (a) coprophagy phenomenon, i.e., the pups of the control group get the MMS via the mother milk; (b) fantastic - direct influence of the MMS damaged animal on the control animal lens development.

Let's consider the first supposition. There are four ways describing the MMS fate in-vivo: (i) Taking part in the methylaion of the different substrates - 90%; (ii) Liver metabolism of the MMS -6%; (iii) Hydrolyse with an inactive products form - 3%; iv) MMS excretion via urine, excrement, and breathed out air - 1%. MMS is detected in the rat blood during the 1.5 hours after injection. [1].

This way the MMS dose received by the untreated rat pups via milk could not be more than 72 pmol/kg. As this dose is more less than the dose causing cataract in a dosedepended experiment (Fig. 1).

It has been proposed: 1) during in-vivo application the MMS is metabolised to any super-active derivative; 2) MMS can act in a low diluted dose.

Bibliography

[1] Swann, P.F. "Rate of breakdown of methyl methanesulfonate and N-methyl-N-nitrosourea in the rat", *Biochem. J.*, (1968), 110(1), 49-52

[2] Barnsley, E.A. "Metabolism of methyl methanesulfonate in the rat", *Biochem. J.*, (1968), 106(2), 18P-19P

[3] Chu B.C.F., "Lawley P.D. Increased urinary excretion of pyrimidine and nicotinamide derivatives in rats treated with methyl methanesulfonate", *Chem. Biol. Interactions*, (1974), 8(2), 65-73

VISUO-VESTIBULAR REFLEXES ADJUSTMENT BY GABA ANTAGONISTS

ANNIE REBER, MARIE-HELENE LEROY
Laboratoire de Neurosciences et Environnement
Université de Rouen, Faculté des Sciences, Mont-Saint-Aignan, France

and

BERNARD POITEVIN
Institut Boiron, 20 rue de la Liberation, Sainte-Foy-Les-Lyon, France

ABSTRACT

We determined if high dilutions of antiGABAergic drugs modulate the visuo-vestibular reflexes responsible for gaze stabilization, and examined a possible target for their biological activity. The horizontal optokinetic and vestibulo-ocular reflexes were first depressed by concentrated antiGABAergic drugs (10^{-3} M purified picrotoxin in 15/15 rats or 10^{-6} M picrotoxin in unpurified mother tincture of Cocculus in 44/59 animals). The reflexes returned normal to compensate visual and vestibular stimulations after a subsequent injection (lml) of Cocculus 4CH (10^{-13}M picrotoxin) in 11/12 rats, but not of the solvent alone (5/7 rats). An amelioration was also observed after an injection of higher dilutions (6/6 rats). These results demonstrate a high sensitivity of the oculomotor system to antiGABAergic drugs with a dose related effect, and a modulation of the toxic effect by diluted agonists. The deficits of the reflexes were correlated with the low level of GABA which was found at the intoxicated state, but the amelioration was not correlated with any change of GABA level. The target of action of high dilutions appears different from that of concentrated drugs. Therefore it appears paradoxal. A possible explanation for physical mechanisms of action of high dilutions could result from biophysical modifications of the solvent as observed in NMR (nuclear magnetic resonnance).

Introduction

Two reflexes are particularly responsible for gaze stability, the optokinetic reflex in which the eye mouvements copy the rotation of the visual environment and the vestibulo-ocular reflex in which the eyes compensate for head displacement. Picrotoxin, a GABA blocker depresses these two reflexes [1]. It is contained in *Cocculus,* the mother tincture (TM) of the vegetal *Anamirta Cocculus,* which is toxic. For its depressive effect on the vestibular responses, picrotoxin has been considered as an antimotion sickness drug [2]. However, more than a depression, an adjustement of the oculomotor response to stimulus condition seems more required to minimize gaze intability detected in different conditions such as in motion sickness.

Interestingly, in-vitro, a small amount of picrotoxin modulates the GABAergic activity in appropriate conditions [3]. Moreover, clinical observations suggest that high dilutions of *Cocculus* are efficient against motion sickness or vertigo. Taken together these data indicate that in-vivo, high dilutions Nof this GABA blocker could be responsible for an adjustement of the visuo-vestibular reflexes, and for further gaze stabilization.

The aim of this study was to verify this hypothesis by an experimental study, and to detect a possible target fo' the substances.

Materials and Methods

The horizontal optokinetic and vestibulo-ocular reflexes (H-OKN and H-VOR) were tested in alert brown rats before (control state) and after intoxication (intoxicated state) by a daily intraperitonal injection of either *Cocculus-TM (0.*15-0.40ml containing 6.6 10^{-5} M picrotoxin of 10 % extract weight in 85% alcoholic solution) or picrotoxin (0.25ml of 3.3 10^{-3}

M - 6.6 10^{-3} M picrotoxin - Sigma - in saline solution). The rats which presented depressed reflexes within 24 to 48 hours, were randomly selected to receive 1ml of a dilution of *Cocculus* TM (either 4 CH, 7CH or 15CH dilution, i.e.centesimal Hahnemannian according to the French Pharmacopea), or the solvent dynamized. A subsequent amelioration of the reflexes was searched 30 minutes later (treated state). For reference, the effect of an injection of 1ml of *Cocculus* 4CH or of the solvent was investigated in rats which received no intoxicating substances.

The eye mouvements (position and velocity) were recorded using a search coil technique [4] during rotation of the surrounding environment (from 1 to 40°/s) or head oscillations (0.05-0.2 hz;+/-30° - +/-60°). Maximal gain (eye velocity/ Stimulus velocity) served to evaluate the responses performances.

Statistical analysis of gain data was developped using a classical distribution free-test (Mann-Whitney statistic) in rats without missing data. A comparison of all the data in each rat at the different state served to detect individual effect of the successive treatments. In the population of rats which display significantly depressed responses, the effect of *Cocculus*-4CH was compared to that of the dynamized solvent.

Results

Fifteen among twenty rats survived after the injection of picrotoxin. Their optokinetic responses were mainly deteriorated 2 hours after the injection and remained deficient for few hours (< 24 hours). The H-OKN was irregular, with velocity of the slow phases smaller than that of the stimulus. As a consequence, the gain fluctuated in a range of low values.

fortynine among fiftyeight rats survived after the injection of *Cocculus*-TM. Despite individual variations, the deficits became progressively aggraved within the first 48 hours. They became large, subtotal and small in 24, 20 and 5 rats respectively. Large, they consisted in a delayed H-OKN with irregular and curvilinear slow phases of small amplitude and low velocity when comparing with the normal responses. Subtotal, they turned to no or residual responses. The gain was almost reduced by a factor of 0.5, and close to zero for these two levels of deficits. No study was continued on rats with small deficits.

No effect of *Cocculus*-4CH nor of solvent in non-intoxicated rats (11 and 6 rats respectively) was detected since the reponses remained identical as in normal rats, being almost regular and of high performances up to 10°/s of visual stimulation.

The effect of *Cocculus*-4CH was tested on large deficits of 3 rats intoxicated by picrotoxin and of 8 rats intoxicated by *Cocculus*-TM up to large (6 rats) or subtotal (2 rats) deficits. 30 to 60 minutes later, the responses of the 3 first rats returned close to normal. The H-OKN returned normal in 2 latter which exhibited subtotal deficits, and was improved but still irregular in the 6 others which displayed large deficits. Quantitatively, the mean gain values returned close to normal in the rats intoxicated by picrotoxin and normal in those intoxicated by *Cocculus*-TM.

In a comparative approach, the effect of the solvent was tested. The deficits remained as large as those of the intoxicated state in the 5 rats intoxicated by picrotoxin and in 5/7 rats intoxicated by *Cocculus*-TM (4 with large deficits, 1 with subtotal ones). An H-OKN either reappeared strongly irregular in a remaining rat and became more regular in a second. Despite these disparities, the mean gain remained small and almost similar to that of the intoxicated state.

Twenty minutes after the injection of *Cocculus* 7CH (3 rats) and of *Cocculus* 15CH (3 rats), the responses were improved in 4 rats and completely restored in the 2 which displayed no more responses after intoxication

The vestibulo-ocular reflex was tested at each state in all the animals. After intoxication, the vestibular nystagmus was abolished or strongly depressed for all the stimulation parameters, leading to a state of deficits identical as for optokinetic responses. After injection of the high dilutions of *Cocculus*, the H-VOR was improved or completely reestablished although after solvent injection, no recovery was observed except in the two cases already described. For a same rat, the amelioration was parallel to that of the optokinetic responses.

Statistically, the decrease of H-OKN gain values was significant for the whole range of visual stimulus velocity tested in 5/5 and 11/12 rats intoxicated by picrotoxin and *Cocculus*-TM respectively (unilateral test; $\alpha = 0.05$; $p < 0.0039$). *Cocculus*-4CH improved significantly the deteriorated responses in 1/2 rats and 5/5 rats intoxicated by picrotoxin and *Cocculus*-TM respectively (unilateral test; $\alpha = 0.05$; $p = 0.010$). The gain returned statistically even not different from control in 2 of them for the whole range of the velocities of visual stimulations. After solvant injection in 7 rats, the responses did not exhibit a significant direction of change (5 and 2 rats respectively). In the population of rats with no missing gain values for the three states, and with significantly depressed gain in intoxicated state, the responses obtained after an injection of *Cocculus*-4CH (n=20 and n=37 by picrotoxin or *Cocculus*-TM) were significantly enhanced in comparison of those obtained after solvant injection (n=19 and n= 30). This was made objective by the very low p value (p=0.0001 and p=0.0000) obtained by the unilateral test conducted for $\alpha=0.05$. Taken together all these results demonstrate a strong ameliorating effect of the high dilutions of *Cocculus* in comparison with a solvent effect.

Discussion

The oculomotor deficits induced by *Cocculus*-TM are similar to those induced by picrotoxin and therefore suggest that the effect of *Cocculus*-TM is due to picrotoxin, which is known to impair the reflex systems [5]. Our results indicate that qualitative and quantitative parameters are pertubated, with gain drop and time-constant increase. However, the variable intensity of the deficits may be attributed to a variable sensitivity to picrotoxin of the vestibular nuclei neurons mediating both optokinetic and vestibular inputs [1,6]. Such a variable intensity of the symptoms could be paralled to the variable lower firing in vestibular nuclei during visuo-vestibular impairement of motion sickness [7]. The symptoms that we observed have probably a functional and not a morphological origin since they are reversed by high dilutions of *Cocculus*.

The modulation of the reflexes by high dilutions of GABA blockers could be related to the unequal sensitivity of the different types of GABAergic receptors in respect to ionic concentration [8], and to modulation of GABAergic activity by low concentration of picrotoxin in different chemical or electrophysiological experimental conditions as observed in vitro [9].

However in a parallel study, we showed that the level of GABA droped in the intoxicated state but was very variable after injection of *Cocculus*-4CH or of solvent even in rats with well restored reflexes. In addition the depression of the reflexes was obtained in one or two days althought the recovery appeared in about 20-30 minutes. These results suggest that the appropriate adjustment of the reflexes to stimulations could involve not only GABAergic systems but other integrated systems.

Concerning the biophysical action of high dilutions used in this protocol, there are numerous hypotheses published on the physicochemical process in high dilutions, but there is few experimental facts [10,11]. Recently, a detailed physical study has been carried out into high dilutions of silicea using nuclear magnetic resonance (NMR). The study revealed a statistically-significant increase in the proton's T1 relaxation time, for potencies equal to or above 9CH [12]. Among the different hypotheses emitted to explain these findings, the existence of a modification of the solvent in the sense of a specific de-structuralization of free water has been discussed.

Whatever biophysical mechanisms of action of high dilutions, our results demonstrate a modulation of toxic effect by high dilutions of *Cocculus* in rats intoxicated by high concentration of the agonist.

Acknowledgements

This work was supported by a grant of the Conseil Regional de Haute Normandie and the Institut Boiron.

References

[1] Felix D, Ehrenberger K. The action of GABA and acetylcholine in the labyrinth of the cat. in Portmann M and Aran JM eds. Inner ear Biology. INSERM: Paris, 1977: 68 147-154.1977.

[2] Ehrenberger K, Benkoe E, Felix D. Suppressive action of picrotoxin, a GABA antagonist, on labyrinthine spontaneous nystagmus and vertigo in man. Acta Otolaryngol. 93,269-273. 1982.

[3] Belhage B, Damgaard I, Saederup E, Squires RF, Schousboe A. High and low affinity GABA receptors in cultured cerebellar granule cells regulate transmitter release by different mechanisms. Neurochem. Int. 19,4,475-482. 1991.

[4] Sarrau JM, Reber A, Mazari B, Carnet J. Capteur angulaire de mouvements oculaires. Paris Conf. internat. capteurs. 54-62. 1989.

[5] Cazin L, Precht W, Lannou J. Pathways mediating optokinetic reflex. Pflugers Arch. 384, 19-29. 1980.

[6] Precht W, Schwindt PC, Baker R. Removal of vestibular commissural inhibition by antagonists of GABA and Glycine. Brain Res. 62,222-226.1973.

[7] Horn KM, Miller SW, Neilson HC. Visual modulation of neuronal activity within the rat vestibular nuclei. Exp. Brain Res. 52, 311-313..1983.

[8] Dutia M.B, Johnston A.R, McQueen D.S. Tonic activity of rat medial vestibular nucleus neurones in vitro and its inhibition by GABA. Exp. Brain Res. 88, 466-472 .1992.

[9] Chen JCT, Chesler M. Extracellular alkalinization evoked by GABA and its relationship to activity dependant pH shifts in turtle cerebellum. J. Physiol. 442,431-446. 1991.

[10] Poitevin B. Mechanism of action of homeopathic medicines. Recent findings and hypothesis I. Physico-chemical mechanisms. British Homeopathic J. 84,1,32-39. 1995.

[11] Smith CW, Endler PC. Resonance phenomena of an UHD, In "Ultra High dilution, Physiology and Physics, Eds Endler and Shulte, pp203-207. 1995.

[12] Demangeat JL, Demangeat C. Gries P., Poitevin B., Constantinesco A. Modifications du temps de relaxation RMN à 4 MHz des protons du solvant dans les très hautes dilutions salines de silice/lactose. J: med. nucl. Biophys, 16,2, 135-145. 1995.

Illustration

Mean gain of optokinetic reflex in function of step of visual stimulation.

Dotted lines: control state. Open symbols: intoxicated state. full squares: treated state by *Cocculus* 4CH. Open squares: tretead state by solvent.

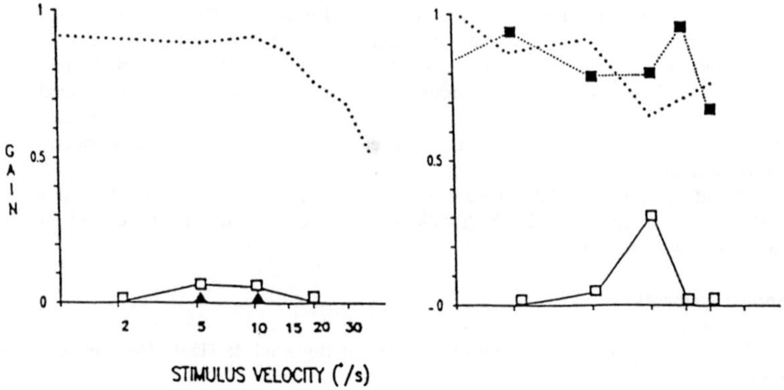

HYPOTHALAMIC NEURONAL RESPONSES OF ALBINO RATS ON SALTY DIET TO HIGH DILUTIONS OF SODIUM CHLORIDE AND PHOSPHORUS

S.P. SINHA BABU, A. PAUL, N.C. SUKUL
Department of Zoology, Visva-Bharati University, Santiniketan-731235, West Bengal, India

and

N. KURZINA, A.S. BATUEV
Department of Higher Nervous Activity, St.Petersburg University St. Petersburg, Russia

ABSTRACT

An 1% solution of sodium chloride was prepared in 90% ethanol. This solution was agitated manually by ten downward strokes. This solution was serially diluted and agitated in a similar way in thirty steps to produce a dilution 10^{-60} of sodium chloride.Phosphorus was also similarly prepared to produce a dilution 10^{-400} . Sodium chloride at a dilution of 10^{-60} , applied on the tongue of albino rats kept on 4% salty food and 6% salt solution as drinks for 7 days, inhibited the spontaneous neuronal activity in lateral hypothalamic area(LHA) of those rats. The rats were on normal food and drink for one day before the test. Distilled water dropped on the tongue increased the rate of discharge of the LHA and this increase was more pronounced in rats which were on 8% salty diet continuously without a break before treatment. Sodium chloride at the dilution 10^{-60} and phosphorus at the dilution 10^{-400} reduced firing rate of single neurons of the LHA of the rats which were given the same dilution of sodium chloride, one dose daily for 18 days.The rats were under anaesthesia during the recording of the neuronal activity. The results indicate that high dilution of sodium chloride, prepared with agitation could elicit very consistent and often prolonged response from the hypothalamic neurons of albino rats either on salty diet or on the same dilution of sodium chloride. Thus diet with excessive salt and sodium chloride at the dilution 10^{-60} caused central and peripheral reorganization in the taste sensor system of the rats which became responsive to the high dilution of sodium chloride.

1. Introduction

There is clinical evidence that potentized medicines act on human patients (1-5).The effect of potentized medicines could also be observed in an animal model, particularly rats and mice (6-8).The action of the medicines appears to be mediated through the receptors of the gastrointestinal tract (9) and consequently through the autonomic nervous system which is centrally regulated in the hypothalamus.Potentized *Agaricus muscarius* extract increases serotonin and dopamine metabolises in the mouse hypothalamus (10).*Arnica 30* and *Hypericum 200*,applied on the tongue of anaesthetized cats, reduce the firing rate of lateral hypothalamic neurons while *Arsenic 30* increases it (11).

Potentized *Natrum mur*, prepared from common salt, NaCl, is effective on patients who crave for salty food (12-13).The number of 'taste' neurons responding to salt solution in the lateral hypothalamic area (LHA) of rats subjected to prolonged salt diet increases as compared to those in rats on normal diet (14). Moreover, neurons in the former group respond to a much wider range of sodium chloride solution as compared to the control. The purpose of the present experiment is to see whether neurons in the LHA of rats on salty diet purpose of the present experiment is to see whether neurons in the LHA of rats on salty diet respond to the potentized *Natrum* mur. Neuronal responses of rats administered with *Natrum mur 30* regularly and its antidote *Phosphorus* (13) have been studied.

2. Materials and Methods

Charles Foster rats of either sex (175 - 210 g) were kept on Bengal gram soaked with 4% NaCl solution, and 6% NaCl solution as drink in cages for 7 days. Another batch of rats were kept on 8 % salty food and drink for 7 days. Rats were on normal food and drink for one day before electrophysiological recording except 4 rats which were on 8% salty food and drink. A third batch of rats was given orally *Natrum mur 30*, one dose daily, for 18 days and single unit discharges in the LHA of these rats were recorded. Each batch consisted of 10 rats. The rats were anaesthetized with paraldehyde (I.M.,1.3 ml/kg) and fixed on a stereotaxic table. The skull was exposed and trephined by means of an electric drilling machine at the point of lateral hypothalamus. The point was determined with the help of stereotaxic coordinates (A,4.6 - 0.5,L,1.5 - 0.5,H,2.7 - 0.7) according to the atlas of Konig and Klippel (15). A glass- coated silver electrode (tip 0.5 mm,resistance 0.002 Ω) was inserted into the LHA and sealed on the surface of the skull by a dental acrylic.Unit activityNin the LHA was also recorded with a glasscoated silver micro-electrode (tip 3.5 um in diam.,resistance 1.5 - 2.5 MΩ). The animal was placed in a Faraday cage and its electrode was connected through a head stage (AK -100 1st stage pre-amplifier: Digitimer), a preamplifier and a filter (Digitimer) to an oscilloscope (L&T Gould). The neuronal activity of LHA was observed on the oscilloscope screen and recorded by means of a two-channel pen recorder (L&T Gould).After recording the spontaneous activity for a fixed period,a potentized drug was applied on the tongue of the rat by a glass pipette.Drugs,obtained in the form of sucrose globules soaked with liquid drug (90% alcohol) from M. Bhattacharya & Co.,Calcutta, were dissolved in distilled water at the rate of 1 globule (average wt.6.18 mg)/0.5 ml and applied to the tongue with 0.1 ml each time.The control was prepared by dissolving sucrose globules, soaked with 90% alcohol, (es vehicle) in distilled water in the same proportion. While gross activity could be recorded in 8 rats, unit activity was recorded in 7 rats. After each experiment , the animal was perfused with saline and 10% formalin through the heart, and the recording site was examined histologically.

3. Results

In measuring gross activity,a spike of a fixed amplitude was selected and its frequency per 40s was counted. In case of unit activity impulses per 40s were counted.Gross activity in the LHA of three rats is given in Figure l (A,B,C,).*Natrum mur 30* produced reduction in the frequency of spikes (-51.72 + 12%, p<O.05) and the effect lasted from 5.16 to 57.67 min. A second application of *Natrum mur 30* produced further reduction in the spike frequency in all the 6 cases tested in which the drug was applied for the second time. Application of the control did not show any marked change in the spike frequency.

Application of distilled water on tongue increased spike frequency (+160.52 + 22.6 % p<0.01) in 4 rats on 8 % salt diet. These rats did not have normal diet before test. Application of *Natrum mur 30* reduced the spike frequency to near pretreatment level(Fig.lA).Application of water on the tongue produced excitatory effect on the hypothalamic neurons(Fig.lA). *Natrum mur 30* could inhibit the effect.

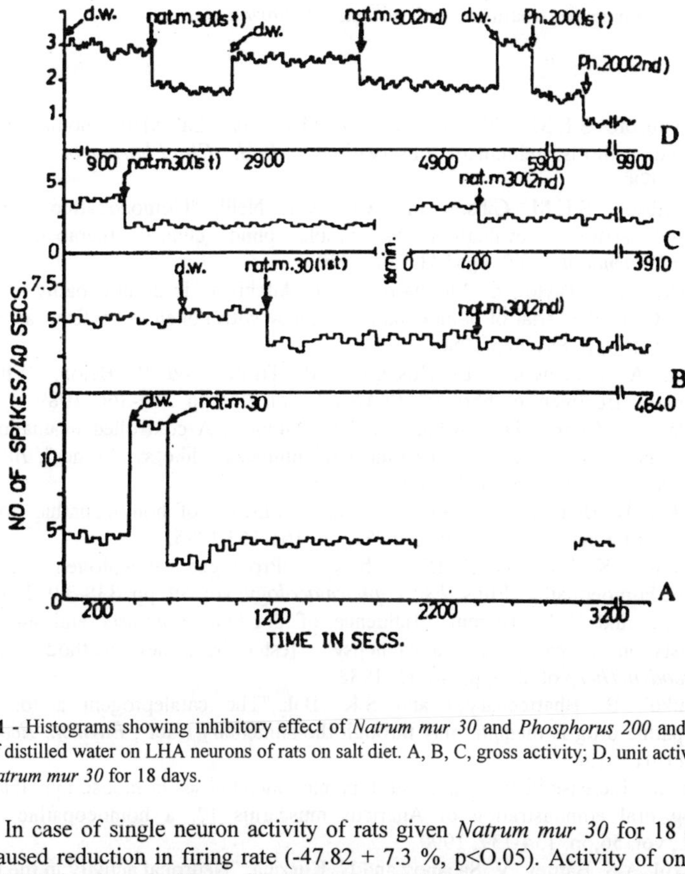

Figure 1 - Histograms showing inhibitory effect of *Natrum mur 30* and *Phosphorus 200* and excitatory effect of distilled water on LHA neurons of rats on salt diet. A, B, C, gross activity; D, unit activity in a rat given *Natrum mur 30* for 18 days.

In case of single neuron activity of rats given *Natrum mur 30* for 18 days, the drug caused reduction in firing rate (-47.82 + 7.3 %, p<O.05). Activity of one neuron in~the LHA is given in Figure lD. Here also distilled water increased the firing rate (Fig.lD). Phosphorus 200, which is an antidote to *Natrum mur* also reduced the firing rate (-54.68 + 8.1 %, p<O.05) of this neuron. A second application of *Phosphorus* further reduced the firing rate (Fig.lD). It appears that the rats, which were given Natrum mur 30, behaved as if they were on salt diet. Their hypothalamic neurons responded to *Natrum mur 30* in a similar way as those of rats on salt diet.

4. Discussion

It is evident that *Natrum mur 30* produced an inhibitory effect on the neurons of LHA of rats on salt diet. Intensity of thirst and cosequently water intake increase with the increase in plasma osmolarity which is perceived by the osmoreceptors of the hypothalamus (16). Since salty food and drink increased plasma osmolarity,the rats became very thirsty for fresh water. This was also evident from their behaviour. The results indicate that potentized *Natrum mur* could elicit very consistent and often prolonged response from the hypothalamic neurons of rats either on salty diet or on *Natrum mur 30* . Thus salty diet and *Natrum mur 30* caused central and peripheral reorganization in the taste sensor system of test animals which became suitable subjects for treatment with potentized *Natrum mur*.

References

[1] R.E.G. Gibson, S.L.M. Gibson and A.D. Mac Neill,"Salicylates homaeopathy in rheumatoid arthritis:preliminary observations", *Br.J. Clin.Pharmacol.* vol.6, pp. 391-395,1978.

[2] R.E.G. Gibson, S.L.M. Gibson and A.D. Mac Neill, "Homoeopathic therapy in rheumatoid arthritis: evaluation by double blind clinical therapeutic trial", *Br.J.Clin.Pharmacol.*,vol.9, pp. 543-559, 1980.

[3] D.T. Reilly, M.A. Taylor, C. McSharry and T. Aitchison,"Is homoeopathy a placebo response? Controlled trial of homoeopathic potency with pollen in hayfever as model", *Lancet*, vol. 18, pp. 881-885,1986.

[4] P. Fisher, A. Greenwood, E. Huskisson, P. Turner and P. Belon, "Effect of homoeopathic treatment on fibrosis", *Br.Med.J.*, vol. 299 ,pp. 365-366, 1989.

[5] J.P. Ferley, D. Zmirou, D. D'Adhmar and P. Balducci,"A controlled evaluation of a homoeopathic preparation in the treatment of influenza - like syndromes", *Br.J.Clin. Pharmacol.*vol. 27, pp. 329-335,1989

[6] N.C. Sukul, B. Bhattacharyya and S.K. Bala,"Influence of homoeopathic drugs on neurotransmitter systems", *Sci.Cult.*,vol.'51, pp. 161-163,1985.

[7] N.C. Sukul, S.K. Bala and B. Bhattacharyya, "Prolonged cataleptogenic effects of potentized homoeopathic drugs" *Psychopharmacology*, vol. 89, pp. 338-339,1986.

[8] N.C. Sukul and W.R. Klemm, "Influence of dopamine agonists and an opiate antagonists on *Agaricus* induced catalepsy a tested by a new method", *Arc.Int. Pharmacodyn.Ther*,vol. 295, pp.40-51, 1988.

[9] N.C. Sukul, B. Bhattacharyya and S.K. Bala,"The cataleptogeni action of a homoeopathic drug is mediate only through the alimentary tract", *Environ. Ecol.*,vol. 5, pp. 411-415, 1987.

[10] N.C. Sukul, "Increase in serotonin and dopamine metabolises in mouse hypothalamus following oral administration of Agaricus muscarius 12, a homoeopathic drug", *Sci.Cult.*, vol. 56,pp. 134-137, 1990.

[11] N.C. Sukul, A.S. Batuev, V. Sabanov and N. Kurzina, "Neuronal activity in the lateral hypothlamus of the cat and the medial frontal cortex of the rat in response to homoeopathic drugs", *Indian Biologist* vol. 23, pp.1720, 1991.

[12] J.T. Kent, *Lectures on Homoeopathic Materia Medica*, Indian ed.(1962), Sett Dey & Co., Calcutta, 1911.

[13] W. Boericke, *Pocket Manual of Homoeopathic Materia Medica*, Indian ed.(1976), Sett Dey & Co.,Calcutta,1927.

[14] A.S. Batuev and Y.S. Lee, "Neuronal mechanisms of hypothalamus in disturbance of the water salt balance" *Schnov Physiol.J.USSR*, vol. 73, pp. 317-322,1987.

[15] J.R.F. Konig and R.A. Klippel, *The Rat Brain. A Stereotaxic Atlas of the Forebrain and Lower parts of the Brain Stem*, Baltimore, Williams and Wilkins, 1963.

[16] W.F. Ganong, *Review of Medical Physiology*, Appelton & Lange, Prentic-Hall International Inc. New Delhi, 1989.

USE OF HIGH DILUTED DRUGS FOR DREAM RECALL IN PSYCHOTHERAPY

SERGIO SPORTIELLO
Servizio Speciale di Psicoterapia, Facoltà di Medicina
Università "Federico II", Napoli, Italy

1. Introduction

The prevailing aspect of modern medicine is to concentrate the attention on the study of disease when trying to find the right diagnosis and therapy. This situation produces a high level of specialization that leads to a more and more organistic vision, which may loose the relationship between physician and patient and make it difficult to understand the global pathology and the morbid individuality. On the contrary, in the psychotherapeutic approach the focus is always on the patient and his/her sufferings and not on the diseases of his/her single organs. This type of approach creates a relationship - unique in the medical field - between the observer, who is trying to understand, and the patient, who is trying to relax, which determines the interobjective penetration. This guide line animates also the homeopathic medicine, which, starting from a condition of ill-being, works to restore the well-being condition in a sick person, taking into account the global set of his/her symptoms in order to choose the appropriate therapy.

The author of this work, a psychotherapeutist-physician, in his clinical practice utilizes the psychodramatization of dreams. Psychodrama is a psychotherapeutic technique [19-20], set up by the psychologist J.L. Moreno [13, 14 vol. III pp. 1670-1692, 15] since 1921, which consists in a dramatic improvisation of a theme performed by the patients, who behave as actors under the participant observation of the therapeutist, who strives to obtain a reduction of unconscious conflicts [6, 12]. In the process of dramatization, the author uses the patients' dreams as a method that makes it possible to move, elaborate and utilize the deepest emotions, which otherwise could remain psychologically encapsulated. However, many persons do not remember their dreams, thus preventing the utilization of this kind of approach. In order to realized his therapeutic work also with this type of patients and to stimulate their memory, the author utilizes high diluted drugs of the official homeopathic pharmacopeia as medical support to the psychodramatization, also when the patients are sure that they do not dream at all [2, 17].

2. Materials, methods and subjects

The drugs utilized are ten. Six come from the mineral kingdom: Borax ($Na_2B_4O_7.10H_2O$) [4 vol. I pp. 294-298], Calcarea Carbonica ($CaCO_3$) [4 vol. I pp. 337-351], Natrum Muriaticum (NaCl) [4 vol. II pp. 549-562], Phosphorus (P) [4 vol. II pp. 772-793], Silicea (SiO_2) [4 vol. II pp. 1175-1189], Sulphur (S) [4 vol. II pp. 1299-1324]. Two drugs come from the animal kingdom: Lachesis (*Trigonocephalus lachesis*, *Ophidia*) [4 vol. II pp. 210-227], Sepia (*Sepia officinalis*, *Cephalopoda*) [4 vol. II pp. 1158-1174]. Two drugs come from the vegetable kingdom: Lycopodium (*Lycopodium clavatum*, *Lycopodiaceae*) [4 vol. II pp. 329-348], Thuja (*Thuja occidentalis*, *Coniferae*) [4 vol. II pp. 1419-1437]. The drug dilution was 10^{-400} g (200 CH of the homeopathic pharmacopeia); globules of lactose where then imbibed by the diluted solution for the administration.

This type of treatment was carried out since 1988 and up to now. The time of observation of each patient covered a total of forty hours during an average of one solar year. The choice of the drugs to be used followed the classic homeopathic praxis [1, 3, 5, 8-9, 11, 16, 18]; this means that each drug was chosen not only according to the symptom "unrembered dreams" [10 p. 1234] (which for some of the drugs used does not appear in the specialized literature), but according to all the symptoms of each treated patient, as prescribed by the "*simillimum* law" [7 p. 47]. Each drug was assumed in a single sublingual administration, two hours after dinner. The mineral drugs were given only once, at the beginning of the psychotherapeutic treatment, while the vegetable and animal drug administration was repeated every two months.

16 subjects were treated: 11 female, 5 male; age range 18-50. They had not used psychomedecines and they had no organic diseases. As an example, the case of a woman treated with Silicea is reported here. She is a 36 year old married woman, who looks older than she is; she has two children and has been working in an office for 19 years where she is very much appreciated for her professionalism; she asked to be treated for a headache, from which she has been suffering for 15 years, almost every week-end, with nausea and vomit; very scrupulous and anxious about her own health, she underwent many laboratory tests which did not point out any organic pathology, while from cervical radiographs emerged only a moderate arthritis; the headache, generally nucal and occipital, irradiates on the forehead and the right upper eye-socket and is so terrible as to force her to stay immobile in the dark and in complete silence for the whole day; the first symptoms are nausea, vomit and eye troubles, which make her very nervous and intolerant to all types of environmental stimulation, so that she has to be left alone even by her own family; her husband reports that she is normally a hypersensitive, touchy person, not very effusive, who does not accept being contradicted; however, at the same time she cares a lot about the needs of others, showing herself available and sweet, asking for company; her headache becomes worse because of weather instability; physically the patient is short, lean, with globose abdomen, fibromatous uterus, a yellowish complexion and flat warts on her toes; her armpits and feet sweat acridly; she is usually a constipated person, except when she eats meat or drinks milk; in her early medical history a perforating otitis at the age of ten and many caries were found, while she has a bilateral perceptive ipoacusie; she uses to talk in her sleep, sitting on her bed, and she wakes up quite tired in the morning, without remembering her dreams. For such a patient the prescription was Silicea 200 CH [7 pp. 150-153] in a single administration.

3. Results

In 2 cases, both males, treated one with Lycopodium and the other with Thuja, no effect was achieved, even after repeated administration every two months: the patients did not remember their dreams. In 3 cases, all females, treated one with Lycopodium, another with Sepia and the third with Lachesis, some mild positive effects were obtained: 1 or 2 dreams were remembered each week. In 5 cases, 2 males, one treated with Calcarea Carbonica and the second with Sulphur, and 3 females, one treated with Natrum Muriaticum, another with Phosphorus and one with Silicea, some medium positive effects were obtained: 3 or 4 dreams were remembered each week. In 6 cases, 1 male, treated with Sulphur, and 5 females, treated one with Borax, another with Natrum Muriaticum, the third with Phosphorus and two with Silicea, some excellent effects were obtained: 5 to 7 dreams were remembered each week. The drugs from the mineral kingdom had a more lasting efficacy, which made their further administration unnecessary.

In the case reported in the above section, during the week following the Silicea 200 CH administration, the patient reported that both general symptoms and headache were reduced, even if she still did not remember her dreams, while she was thinking that she could not remember them because they were "all bad dreams". She was encouraged to write her dreams and the feelings related to them as soon as she awaked. One month later she reported many and different dreams, on the basis of which the psychodramatization could begin.

4. Discussion

This experimental work demonstrates that high dilution drugs permitted the patients to remove the inhibition of the memory recall mechanism and to remember the dream contents. Such drugs, which acted at the unconscious level by removing the obstacles to the free play of the memory storage, were chosen according to the *similia*-principle.

Occasional observations of dream recall after treatment with highly diluted drugs without any contemporary psychotherapeutic treatment should confirm the above results on the effects of highly diluted drugs; it should be worthwile to observe systematically a treated group without any psychotherapeutic aid.

Acknowledgements

I would like to thank the late Prof. Antonio D'Errico for the opportunity and the trust he gave me by allowing me to gain my experience at the University where I work, Prof. Antonio Negro and the late Dr Thomas P. Paschero, my teachers in homeopathy. I would thank also UNDA, Grumo Nevano (Naples), and Dr Anna Scurti, Laboratoire Boiron, Rome, for the literature they provided for me.

References

[1] Allen, T.F. *Handbook of Materia Medica*, Jain: New Delhi, 1990.
[2] Bertini, M. & C. Violani. *Cervello e sogno*, Feltrinelli: Milano, 1982.
[3] Candegabe, E.F. *Materia Medica Comparada*, Albatros: Buenos Aires, 1983.
[4] Clarke, J.H. *A dictionary of practical Materia Medica*, Health Science Press: Rustington, Sussex, 1962.
[5] Gallavardin, J.P. *Psychisme et homoéopathie*, Ternet-Martin Vienne: Isère, 1960.
[6] Guidano V.F. & M.A. Reda, *Cognitivismo e psicoterapia*, Angeli: Milano, 1985.
[7] Hahnemann, S. *Organon dell'arte del guarire*, Ital. transl. of the VI German edition, C.E.L.A.S.: Belluno, 1941.
[8] Hahnemann, S. *Traité des maladies chroniques*, French transl., Maisonneuve: Paris, 1969.
[9] Hering, C. *The guiding symptoms of our Materia Medica*, Jain: New Delhi, 1991.
[10] Kent, J.T. *Repertory of the homoeopathic Materia Medica*, Hahnemann publishing Co.: Calcutta, 1973.
[11] Lathoud, J.A. *Études de matière médicale homéopathique*, Franche-Comtè: Levier, 1981.

[12] Lemoine, G. & P. Lemoine, *Le psychodrame*, Laffont: Paris, 1972 (Ital. transl.: *Lo psicodramma*, Feltrinelli: Milano, 1982).
[13] Moreno, J.L. *Fondament de la sociemétrie*, P.U.F.: Paris, 1954.
[14] Moreno, J.L. "Lo psicodramma", in: S. Arieti, ed., *Manuale di psichiatria*, Vol. III, pp. 1670-1692, Ital. transl., Boringhieri: Torino, 1969.
[15] Moreno, J.L. *Manuale di psicodramma*, Voll. I and II, Ital. transl., Astrolabio: Roma, 1969.
[16] Paschero, T.P. *Homeopatia*, Talleres Graficos Gral San Martin: Buenos Aires, 1973.
[17] Pearls, S.F. *Gestalt therapy verbatim*, Real People Press, 1969 (Ital. transl.: *La terapia gestaltica*, Astrolabio-Ubaldini: Roma, 1980).
[18] Roberts, H.A. *The principles and art of cure of homoépathy*, Jain: New Delhi, 1976.
[19] Schutzemberger, A.A. *Introduzione allo psicodramma e al gioco di ruolo*, Ital. transl., Astrolabio: Roma, 1977.
[20] Yablonsky, L. *Psicodramma*, Ital. transl., Astrolabio: Roma, 1978.

STRUCTURE AND COMMUNICATION

PRESSURE-INDUCED ELECTRICAL SKIN CONDUCTIVITY AS A SENSITIVE MEAN FOR THE DETECTION BY HUMAN-HUMAN INTERACTION OF SUBTLE BODILY CHANGES

R. VAN WIJK
Department of Molecular Cell Biology, Utrecht University
P.O Box 80.056, 3508 TB Utrecht, The Netherlands

ABSTRACT

Several reports on animals, plants and cells suggest significant biological responses following exposure to agonists so highly diluted that very few molecules of the original substance, if any, could have remained. However, there is a large gap in our kwowledge on the information transfer mechanism, although a few theoretical studies have explained the effect in terms of a biophysical model. In our studies, we proposed that the effect of ultra-high dilutions, if any, is not dependent on molecular interaction, e.g., mixing, with the organism. We focussed our attention on a human test model. Using the control of subtle muscular movements as a parameter we studied human interactions and the effect of the proximity of specific ultra-high dilutions in sealed glass vials on this interaction.

1. Introduction

More and more, as physiology has advanced, we have learned how almost infinitely numerous and subtle the bodily movements are which accompany awareness and emotion. These bodily accompaniments are much more far-reaching and complicated than we ordinarily suppose. Not only the heart, but the entire circulatory system forms a sort of sounding-board, which every change of our awareness may make reverberate. Hardly a sensation comes to us without sending waves of alternate constriction and dilatation down the arteries of our arms. The blood-vessels of the abdomen act reciprocally with those of the more outward parts. The bladder and bowels, the glands of the mouth, throat and skin and the liver are known to be affected at least transiently in emotions. And what is really equally prominent but less likely to be admitted until special attention is drawn to the fact, is the continuous co-operation of the voluntary muscles in our emotional states. Even when no change of outward attitudes is produced their inward tension is felt as a difference of tone or of strain.

In general human creatures will react in some way or another to other human creatures. The question can be asked whether it is possible to apply these subtle human muscular movements to detect changes in the state of another human's disorder. A particular experimental approach consisted of the treatment of a human being with a low dose of a substance that is toxic at higher doses. A small degree of disorder or upsetting of the organism is an important perception of any potential damage in the future. The question is whether another person (indicated as tester) is aware of this (low degree of) upsetting of the person (indicated as testperson) and possibly registrates it by a change in muscle tension. This article is an introduction to our research on this subject.

2. Subtle differences in pressure application are detectable through a pressure-resistance regulatory system of the skin as demonstrated by a loss of electrical conductivity

The basic principle of the detection method is the measurement of pressure during the voluntary and conscious movement of the tester's arm when approaching closely a test-person and carrying out pressure on the skin of the test-person. In case the test-person has been upset the tester might be aware of it and show it in subtle differences in arm movement and the application of pressure. For its detection the pressure is transduced in a change of electrical skin resistance by applying the pressure with an electrode that was part of an electric circuit. The awareness of the upsetting of the tested person can then be recognized by registrating a change in the pressure-induced resistance pattern during the measurement of skin resistance of the test-person. This method of testing is explained first.

3. DC measuring equipment and measuring procedure

A device called the ACU-1, designed by Drs. P.J. Romeyn of Electronische Instrumenten voor de Geneeskunde Company, based in Zeist, The Netherlands, was used for the measurements with direct current. This device has been used in previous studies (La Rivière, 1988; Van Os-Bossagh, 1989; Van Wijk en Wiegant, 1989; Van Wijk 1992). The ACU-1 is connected to a two-channel Linseis recorder. The pressure on the active electrode is recorded by means of a measuring electrode with a pressure reader. The pressure reader is connected to the recorder's second channel. In this way current and pressure are recorded at the same time. The measurement set-up has previously been described in more detail (Van Wijk and Wiegant, 1989; Van Wijk, 1992).

The measurements were taken at the so-called Ting points on the hands. The skin under the active electrode is moistened with tap water. The usual method involves the measuring of the pressure-induced loss of electrical conduction at those skin points. The analysis of conduction and conductivity loss is illustrated in *Figure 1*. After a sharp increase in conductivity with the application of pressure on the moistened point, a maximum value was reached. This value was called the initial conductivity. Subsequently a more or less sharp loss in conductivity can occur. If no loss occurs, conductivity remains at the maximum level and shows an angle of 90 degrees with the sudden increase. The loss was expressed in μA per time unit.

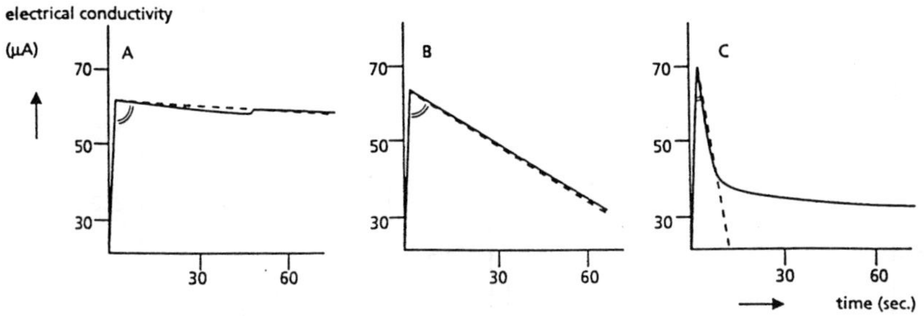

Figure 1 - Analysis of conductivity loss: three examples. Two types of calculation were applied. 1) In terms of degrees. The indicated angle is 90° or less. The angle is no absolute measure since the speed of recording determines the slope. 2) In terms of μA per time unit. The slope of the loss in conductivity is oμA per time unit or more. In example A the conductivity loss is less, the decrease in terms of μA/sec is low and the angle in degrees is large. Example C shows the reverse situation, while example B represents an intermediate condition.

4. Factors determining loss of conductivity

Several experiments have been performed to study the influence of pressure on changes in conductivity using fine-mechanical devices in which the pressure-registering electrode could be moved vertically with the help of a micrometer screw and a rotating switch. The finger of the subjects was positioned in such a way that in a downward movement, the pressure electrode experienced resistance in exactly the opposite direction (the tangential pressure is thus at a minimum). A more detailed description of the experimental set-up has been published elsewhere (Van Wijk, 1992).

From the obtained data (described extensively in Van Wijk, 1992; chapter 7), the following conclusions may be drawn:

(a) the sudden application of pressure results in an increase in conductivity, with the amount of pressure applied being directly correlated to conductivity;

(b) after pressure has been applied and maintained at a constant level, conductivity does not remain constant but falls;

(c) the application of greater amounts of pressure does not produce further rises in conductivity, irrespective of the pressure applied (up to 280 g), conductivity rises to its level at 120 g;

(d) conductivity loss is reduced as existing pressure is increased.

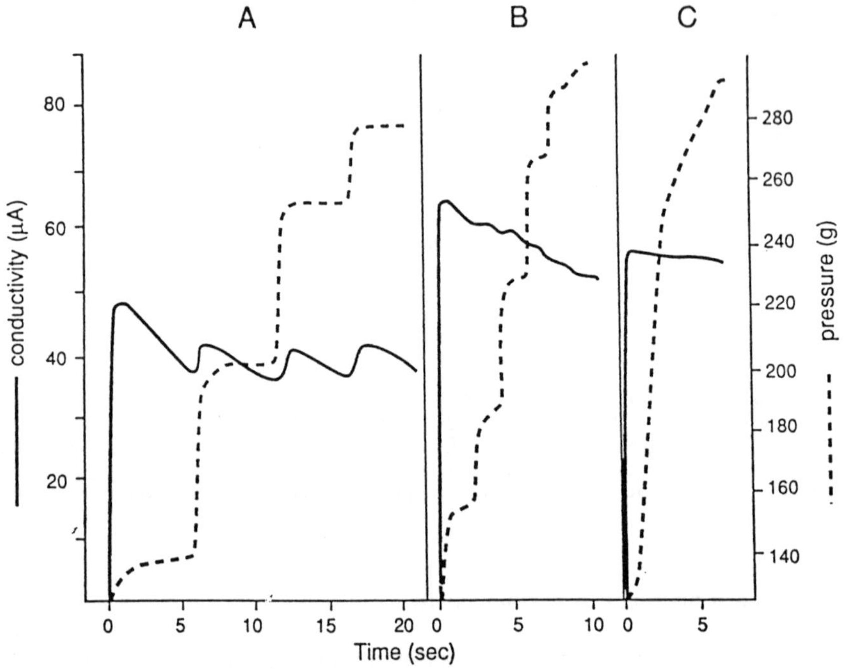

Figure 2 - Example of application of pressure from one subject in one experiment. A) Increase of pressure with a 5 seconds interval. B) Increase of pressure with a 2 seconds interval. C) Gradual increase of pressure.

A few representative examples from one subject in one experiment are provided here to illustrate this phenomenon. As shown in Figure 2, in this experiment pressure was increased in stages and the stable periods between the stages were varied.

At first the pressure was quickly increased after the initial application. After the second pressure had been applied for 5 seconds, it was increased again and then increased for a fourth time. The results show that conductivity changed in jumps (Figure 2A). At the first increase, conductivity fell quickly, but at every subsequent increase conductivity rose. Each time the rise was temporary, however, and conductivity started falling again when the pressure applied was kept constant.

In the second approach (Figure 2B), pressure was increased in stages as before. But in this instance the period between increases was shortened to 2 seconds. Under these conditions conductivity loss occurred more gradually.

In the third approach (Figure 2C), pressure was gradually increased in such a way that conductivity remained constant.

5. Attunement by tester

The phenomenon of pressure tolerance can be of great importance for the study of changes in conductivity during the study of subtle muscular tension changes. In general it is not easy (or possible) to suppress any conductivity loss completely. The tester has to search for the appropriate pressure application. In order to fully suppress the conductivity the subject had to be 'normalized' or influenced. An existing conductivity loss was suppressed on a trial and error basis by removing cosmetics, synthetic materials and garments, an if possible bracelets and rings, electrical equipment, or creating a magnetic field. The tester continues his trials until an optimal suppression of conductivity loss has been achieved. 'Normalization'is that after one or several procedures no loss of conductivity was observed during the measurement of the test-person. With this procedure the tester optimally attuned himself to the state of the test-person. When the suppression of conductivity loss was achieved, the measured conductivity of the patient became very sensitive to any contact of the tested person with toxic substances. These can in turn bring about a conductivity loss.

It is evident that the skill and experience of the tester is important. There are several courses specifically aimed at this test procedure, in particular with respect to its application in electroacupuncture (Kramer, 1971; Voll, 1971,1980). Obviously it cannot be assumed that all electroacupuncturists possess the same skills. Experience, practice, the circumstances of the task and individual constitution all play a significant role. An example: it has been found that cutting out visual contact between tester and test substances easily leads to irritations of the tester. Less experienced testers often found it difficult to keep to the researcher's instructions. Because an experiment series usually takes between three to four hours and requires heightened concentration from the tester, tendencies towards tiredness were inevitable. Our subjective impression was that beginners are less sure of their measurements and take longer to look for active preparations than testers with many years of experience. Only members of the latter group, which may be described as experts, were considered to take part in our studies.

6. An experimental test system

An experimental model was devised in which an indicator drop was artificially induced in a subject who in the preliminary treatment showed no more symptoms of conductivity loss.

To illustrate this, diphenyl has been used in these studies. Diphenyl or phenyl benzene has fungicidal properties and is effective against a limited number of moulds. The toxic effects of diphenyl on workers who were exposed to it have been described. Workers who had come into contact with diphenyl developed symptoms of nausea, vomiting, irritation of the eyes and nose, headaches, diffuse abdominal pains, aching of the limbs and general exhaustion.

An experimental model of this kind offers many advantages for the study of the antagonistic effect of an ultra-low dose of a substance with similar effects if applied at high dose. For it will be known in advance which potentized remedies are capable of suppressing toxin-induced conductivity loss. It thus becomes possible to prepare valid control samples with which to conduct blind testings. The development of such an artificial testing system for inducing and then suppressing conductivity loss, has been described previously (Van Wijk and Van der Molen, 1990; Van Wijk 1992). In short: the pre-treated indicator drop-free subject was given diphenyl orally. Indicator drop induced in this way was then modulated with the help of sulphur potencies.

The choice for an ultra-low dose of sulphur came from the remedy picture of sulphur. The overall picture of the remedy indicates which physical reactions it can produce. A remedy picture is the 'portrait' of a 'special human being', which shows all typical effects of the remedy in question.
Many and detailed observations of this kind have been made about sulphur. Among the typical physiological changes it produces are lightness in the head, stabbing pains in the forehead, inflammations in the nose, loss of appetite or rejection of food, nausea and sometimes vomiting, upset stomach, pains in the liver region and increased bile secretion, diarrhea, skin disorders with itching, hot flushes, increased perspiration and profuse sweating.

Many symptoms caused by diphenyl can be traced back to the typical medication profile of sulphur. This correspondence has important consequences for the homeopathic similia approach. According to the 'like cures like' principle, it should be possible to suppress diphenyl-induced changes with sulphur potencies. In practical work the view that the application of sulphur, in particular sulphur D12, reduces complains has also gained currency.

Placing sealed phials sulphur D12 in the medicine holder produced a suppression of conductivity loss. The specificity of potentized sulphur D12 was the subject of a previous investigation (Van Wijk and Van der Molen, 1990). The conductivity loss in question was that which occurred after subjects who showed no further loss through the choice of remedies by the tester were given 0.1 μg of diphenyl orally. The suppressive effect of sulphur D12 was observed when the phial was placed in the holder and disappeared with the removal of the phial from the holder. It was shown that the phial could be placed in and removed from the holder many times without affecting the results. This opened the way to setting up blind testings. The central question was to what extent the tester was able to identiy phials of sulphur D12 on the basis of observed conductivity loss.

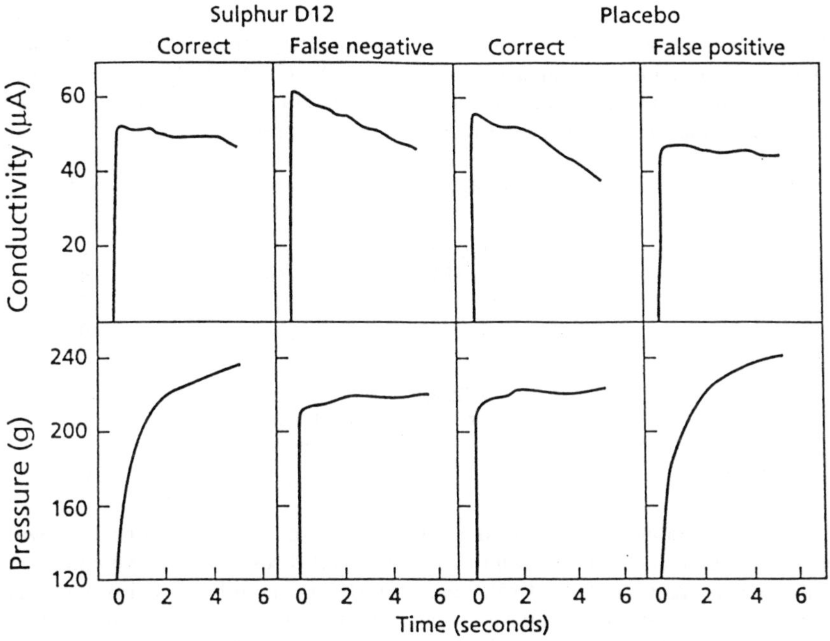

Figure 3 - Simultaneous recording of pressure and conductivity in blind testings (8 experiments).

7. Blind testing of the effect of sulphur D12 (in closed phial) by means of diphenyl-induced conductivity loss

In some blind tests the pressure and conductivity were recorded simultaneously. The curves this produced are shown in *Figure 3*. The conductivity loss was slow with a secondary increase in pressure. When pressure was applied quickly a sudden loss occurred. The correlation was independent of the correct/false scores in the blind tests. The false scores with sulphur D12 showed no secondary increase in pressure, while the false scores with placebo did.

These experiments show that the results of the test are directly linked to the way in which pressure increases.

Five experimental sessions were held on the basis of the above procedure. For analysis blind verum values were compared with blind placebo values. Open verum values and open placebo values were also compared. The data were compared on the basis of the average of the four groups.

Table 1 shows the average values of initial conductivity (I.V.) and of conductivity loss (Z.A.) of all sessions. A significant deviation between placebo and verum values in open measurements was observed. The blind values ranged between the two extremes. Thus the difference between the averages of the blind verum and placebo values was smaller than that of the open measurements.

Table 1 *Total set of data*

	Open test		
	Verum (n=137)	**Placebo (n=92)**	**p.**
I.V.	54.01 ± 9.9	59.07 ± 10.2	0.000
Z.A. (Rate)	0.97 ± 1.1	4.06 ± 2.1	0.000
	Blind test		
	Verum (n=379)	**Placebo (n=371)**	**p.**
I.V.	55.88 ± 10.1	57.40 ± 10.0	0.038
Z.A. (Rate)	1.95 ± 2.1	2.77 ± 2.1	0.000

The cross-tabulation of the data of all open and blind experiments with respect to verum/placebo are shown in *Table 2, A and B* respectively. We compared the false and correct values and calculated the deviation with the chi-square test. Since overlap between the two groups was found a limit of 1.5 μA/min was used in order to applicate the chi-square test. As shown in *Table 2, A and B*, the results are statistically highly significant.

8. Conclusion

In our studies we proposed that the effect of ultra-high dilutions is not dependent on molecular interaction, e.g., mixing, with the testperson. Using the control of subtle muscular movements as a parameter we studied human interactions and the effect of proximity of specific ultra-high dilution in sealed glass vials on the interaction. We performed our studies with a method based on pressure-induced skin resistance loss. Studies on the pressure-induced resistance shows that this physiological response is explained by the existence of a regulation mechanism in the human skin which causes increased skin reistance upon pressure; the change of the skin resistance was dependent on the kinetics of the pressure application.

The control of muscular activity is influenced by human interaction. We propose that the tester, measuring skin resistance changes at the skin of a tested person, is influenced by this second person. Furthermore, we propose that this is a learned capacity that requires a 'fall of the threshold' of normal awareness. With this technique a tester is able to discriminate very subtle changes in the test-person.

Table 2 Cross tabulation of total data set of open and blind trials limit value 1.5 μA/min

(a) Open trial

	Placebo	Verum	Row	Total (%)
No effect	87	29	116	50.7
Effect	5	108	113	49.3
Column	92	137	229	
Total (%)	40.2	59.8		100.0

Chi-Square	D.F.	Significance
118.62275	1	0.0000

(b) Blind trial

	Placebo	Verum	Row	Total (%)
No effect	278	172	450	60.0
Effect	93	207	300	40.0
Column	371	379	750	
Total (%)	49.5	50.5		100.0

Chi-Square	D.F.	Significance
68.21130	1	0.0000

In order to repeat these studies it is essential to know that a tester with learned capacity to shift its awareness during the applied technique cannot immediately function in blind experiments in which conditions differ from the learned normal context. For this reason it is necessary to search for experimental conditions which (a) did not disturb this state of attention, and (b) make it possible to blind the experiments. This required additional time for learning to function under the experimental conditions.

Acknowledgement
The financial support of the HomInt organisation is gratefully acknowledged.

Literature

Kramer, F. (1971) Beitrag zur Beweisführung der Medikamententestung. In: Electroakupunktur - Anderthalb Jahrzehnte Forschung und Erfahrung in Diagnostik und Therapie. Med. Lit. Verlagsgesch., Uelzen, Germany, pp. 241-251.

La Rivière, B.M. (1988) Weerstandsmeting aan de huid in relatie tot electroacupunctuur.

Verslag van onderzoek bij de Vakgroep Biomedische Natuurkunde en Technologie, Erasmus Universiteit, Rotterdam.
Van Oss-Bossagh, P. (1989) Electroacupunctuur. Verslag van onderzoek bij de Vakgroep Biomedische Natuurkunde en Technologie, Erasmus Universiteit, Rotterdam.
Van Wijk, R. & Wiegant, F.A.C. (1989) Homeopathic remedies and pressure-induced changes in the galvanic resistance of the skin. VSM Geneesmiddelen bv, Alkmaar, The Netherlands.
Van Wijk, R. & Van der Molen C. (1990). Biological effects and physiological characteristics of potentized high dilutions of Sulphur; development of a model for fundamental studies in homeopathy. In: Homeopathy in Focus (Schlebusch K.P. (ed.)), VGM Verlag, Essen, pp. 122-147.
Van Wijk, R. (1992) Homeopathic medicines in closed phials tested by changes in the conductivity of the skin: a critical evaluation; blind testing and partial elucidation of the mechanisms. VSM Geneesmiddelen bv, Alkmaar, The Netherlands.
Voll, R. (1971) Medikamententestung, Beweise für die Richtigkeit und Probleme des Grundlagenforschungs der Medikamententestung. In: Electroakupunktur - Anderthalb Jahrzehnte forschung und Erfahrung in Diagnostik und Therapie, pp. 200-215. Med. Lit. Verlagsgesch., Uelzen, Germany.
Voll, R. (1980) The phenomenon of medicine testing in electroacupuncture according to Voll. Am. J. Acupuncture; 8 (2): 97-104.

NON-LINEAR EFFECTS AT LOW LEVELS OF SIGNAL INTENSITIES AND BIOCOMMUNICATION

F.A. POPP
Institute of Biophysics, The Technology Center, Sauerwiesen 6
Kaiserslautern, Germany

1. Introduction

About twenty years ago the scientific community was not as aware as it is now that nonlinearity plays an essential role in understanding the physics of life, or, more generally, of cooperative phenomena. Indeed, with the discovery of the laser, we started to learn that the unusual optical effects of laser light have to be traced back essentially to the non-linearity between the electromagnetic field-amplitudes and the charge-density of the interacting matter. However, with increased understanding of fashionable new investigations such as Prigogine's dissipative structures, non-linear Schrödinger equations, catastrophe theory, chaos, and the singularities in elementary physics, we have become more and more familiar with the possibilities of non-linear effects at low and even the lowest levels of signal intensities. Actually, "non-linearities" are not confined to high amplitudes, as some people think. The necessary and sufficient condition for non-linearities is an "autocatalytic" feed-back loop between an energy source and an entropy source, where the one amplifies or weakens the "signals" of the other such that there are regions where an infinitesmally small deviation of the coupling can change the dynamic behavior (= time dependence of the measurable events) of the system in practically unpredictable ways.

Non-linearities at the lowest level may occur, for instance, simply by deviations from thermal equilibrium: Consider the case of a system that emits photons in the UV-range even at physiological temperatures of about 300 K. The probability of emitting a UV photon at thermal equilibrium is in the range of the Boltzmann factor $\exp\left(-\frac{h\nu}{kT}\right)$, where $h\nu$ is the energy of the photon and kT the mean thermal energy, rather low, since $\frac{h\nu}{kT}$ is then of the order 100. Since the signal intensity $\dot{s}$ has always to be compared with the noise level N (which is proportional to $\exp\left(-\frac{h\nu}{kT}\right)$), in order to get the decisive quantity "signal/noise-ratio" $\frac{\dot{s}}{N}$, we may get extremely high values in this optical range even at rather low absolute intensities $\dot{s}$.

Thus, it is well known that living biological systems constantly emit ultraviolet (and visible) light ("biophotons") of an intensity $\dot{s}$ of a few up to some hundred

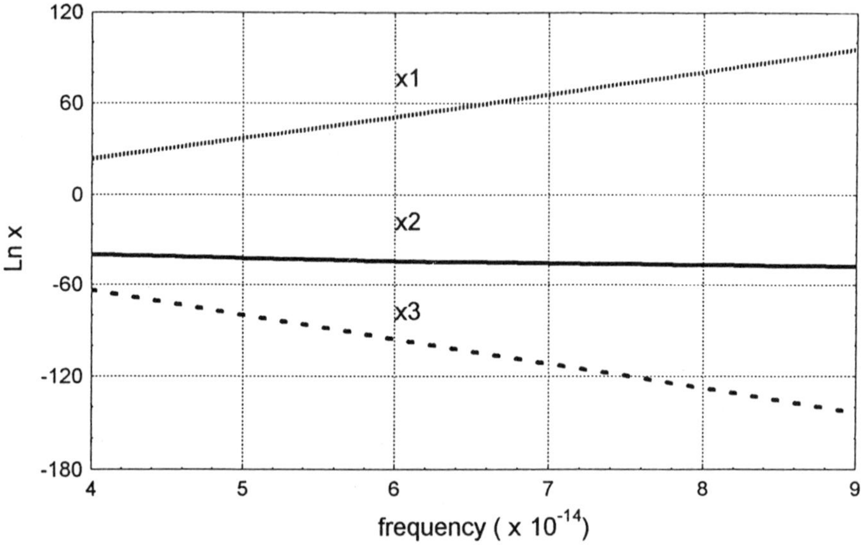

Figure 1: The three curves represent different terms which are decisive for the signal/noise-ratio. x3 is the logarithm of the Boltzmann factor (- $\frac{h\nu}{kT}$) at physiological temperature in the optical frequency range. x2 is the logarithm of the corresponding factor which is calculated from the spectral intensities of biophoton emission. It shows that biological systems occupy their phase-space in the average independent of the frequency. x1=x2-x3 is the logarithm of the signal/noise-ratio in case that biophoton emission is referred to a heatbath-noise.

photons/(s cm^2) [1]. By calculating the signal/noise-ratio of the biophoton emission of cucumber seedlings from the measured values for the different wavelengths, one gets the surprising results of Fig. 1.

These rather high experimental values of $\frac{s}{N}$ show that "low level luminescence" (biophoton emission) is an important, if not the most important factor for communication within biological systems, since it is the only possible candidate for the triggering of biochemical reactions in cells. Actually, these facts show evidence that the delivery of activation energy for the transition state complexes (including those of enzymatic reactions) is $\frac{s}{N}$ - times higher in living biological systems than under thermal equilibrium conditions. It means that, despite their low intensity, biophotons may work effectively for the regulation of all reactions in living tissues. Take, for instance, 10^5 reactions/s of a cell and assume that a single photon triggers the first reaction (taking

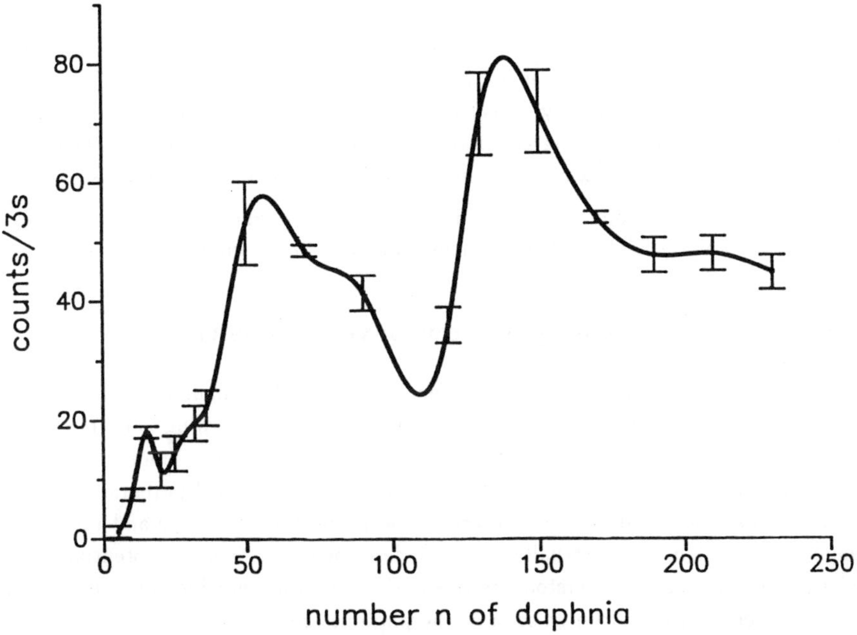

Figure 2: The photon intensity of daphnia displays an interference-like dependence on the number n of daphnia. It turns out that this "pattern" can be described in terms of integer ratios of the distance between the animals.

about 10^{-9} s), then after having been borrowed without thermalization triggers the second reaction, and so on. This means then that even 10^9 reactions/s can be triggered by a single photon, provided it delivers at the right time the right amount of energy without getting thermalized. Actually, Cilento has shown that those reactions do take place in a cell [2], and it cannot be ruled out that all reactions of a cell are governed by this mechanism. We see from these examples that the assumption of extremely high signal/noise-ratios is not unrealistic for living systems even at very low signal intensities. However, in order to understand the "non-linearity" and the basic significance of these effects, we will now show in addition an example where with the change of an extensive parameter x another extensive parameter y will increase or decrease but not in a linear way. Take, for instance, the biophoton emission of Daphnia magna in its dependence on the number of daphnia in the measurement cuvette [3] (Fig. 2).

It displays clearly non-linear behavior in the intensity of photon emission of the system under investigation, where maxima and minima of photon intensity appear at definite numbers of animals, corresponding to definite mean distances between them. This example and others, as, for instance the hyperbolic relaxation of "delayed luminescence" from biological systems (indicating a non-linear ependence of intensity and the number of rescattering centers) [4] as well as the extinction coefficient of biophotons [5] are further examples of non-linearities at very low intensities.

Let us discuss now some theoretical models for understanding the physical background as well as the biological consequences of this kind of interaction.

2. Basis of Nonlinearity at Low-Level Intensities

H.Fröhlich [6] introduced a chemical potential into the Boltzmann factor in order to describe just the deviations from thermal equilibrium which according to his postulate of long-range coherence and energy storage in biological systems should take place in the microwave region. Actually, the introduction of a chemical potential μ of the order $\mu = h\nu$ leads to dramatic changes in the properties of matter and radiation. Biophoton emission, which actually has to be described by a chemical potential of the order $\mu = h\nu$ [7], can be understood on this basis as a radiation which may sensitively change the entropy of matter as soon as the photons are absorbed.

In terms of the maximum entropy principle, we showed that a possible model that accounts for the spectral distribution of biophotons represents an "ideal open system" [1]. Actually, the maximum entropy under the constraint of energy conservation leads exactly to the Boltzmann distribution. To give up the constraint of energy conservation means that the system has by use of an external or internal energy pump always enough energy available in order to achieve the maximum entropy. This leads obviously to $f(\nu) =$ constant which has actually been measured (see Fig. 1, x2). By means of mode-coupling, the system can then exhibit the puzzling property of displaying as well the absolute entropy value $S = 0$ at the same time as the relative maximum! In this way, the system gets the theoretically highest sensitivity.

A further consistent model, based on the non-linear coupled equations of (1) energy balance between an exciplex system and photons and of (2) Einstein's balance equation between matter and radiation, can be made dynamic and leads to the following equations of "deterministic chaos" [8].

$$f_{i+1} = f_i + \frac{\beta}{2+y_i}(1 + f_i(1 - y_i)) \tag{1}$$

$$y_{i+1} = y_i + \frac{y_i + 2}{(\kappa - 2)} \frac{\frac{f_{i+1}}{f_i} - 1}{1 + f_i} \tag{2}$$

where β is a constant, and

f_i: i-th value of the occupation of photons of the phase space, where f is proportional to the photon density.

y_i: i-th value of the ratio $\frac{N_1}{N_2}$, where N_1 and N_2 are the numbers of exciplexes in the ground state and the excited state, respectively.

$$\kappa = \frac{\mu}{kT} ,$$

where μ is the effective chemical potential and kT the mean thermal energy.

The equations (1) and (2) describe the "dialog" between a two-level exciplex system which absorbs and reemits photons and the photon field at the constant energy of the whole system, which means that the loss is always balanced by the photon influx. In the case where the chemical potential μ becomes 0, thermal equilibrium is obtained. For $\mu \neq 0$, the system can reach different attractors:

$$\begin{aligned} \mu > 2kT &: \ y = 1 + \frac{1}{f} \\ \mu < 2kT &: \ f \to 0; y \to \infty \\ &\quad\ f \to \infty; y \to 0 \end{aligned}$$

At $\mu \approx 2kT$ the system arrives at a strongly non-linear behavior, displaying maximum sensitivity.

This thermodynamical approach is already able to explain non-linearity at low intensities in terms of certain deviations from thermal equilibrium. Although this gives us some insight into possible mechanisms, it does not tell us, for instance, the reason for the organizational force originating from those non-linearities as it is expressed, for instance, in the experiment (daphnia Fig.2)

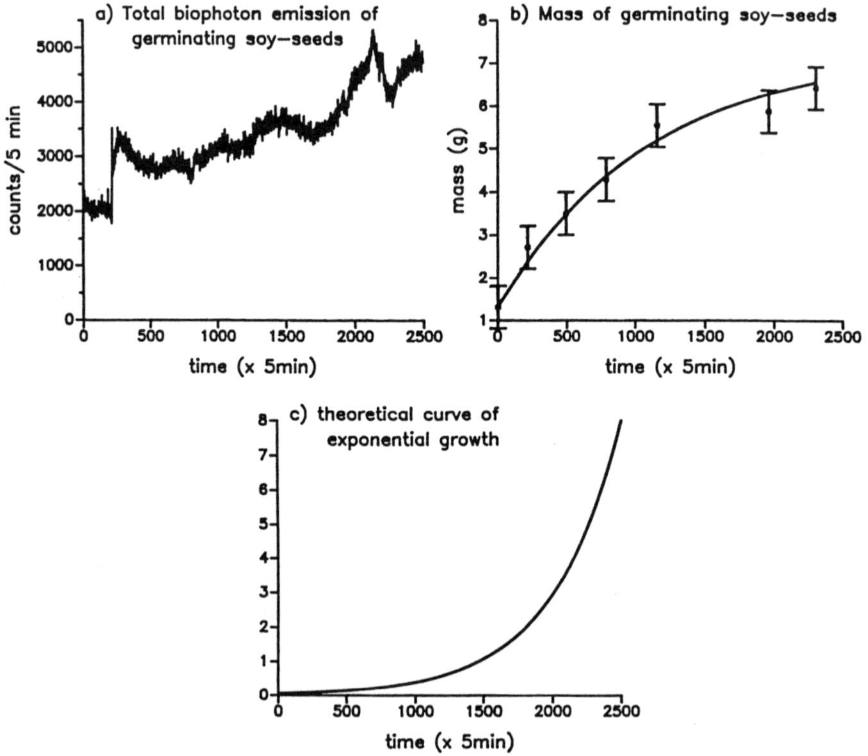

Figure 3: a) The biophoton intensity from 9 soy-seeds increases during germination in darkness. b) The mass m during germination increases according to the curve, where the measured values (points with error bars) habe been approximated by the growth curve $m = \frac{A}{B+(1-B)\exp(-Ct)}$, with $A = 1.3g$, corresponding to the mass of 9 soy-seeds, $B = 0.178, C = 2.6 * 10^{-4}$ (min^{-1}, t representing the time. c) If exponential growth took place, the mass would follow this time dependence instead of the nonlinear increase according to figure 3b.

3. Cell Growth as a Result of Nonlinearity at Low Intensities

Fig. 3 displays the biophoton emission of germinating soy-seeds (left side), where on the right side of this figure the simultaneous development of mass is plotted. It shows that there is a clear correlation between biophoton emission and growth. It even seems that biophotons indicate the biological process of growth rate more accurately than the change of mass. Nevertheless, the important result is the fact that both biophoton emission and increase of cell mass cannot be described in terms of linear behavior (like the growth curve of Fig. 3c) but have to take account of the non-linear differential equation whose result (Eq. 3) describes the growth curve Fig. 3b quantitatively.

$$n(t) = \frac{1}{B + (1 - B)\exp\left(-\frac{t}{\tau}\right)}, \tag{3}$$

where $n(t)$ is the number of cells on time t, B and τ are constant values depending.

A further striking example is the difference between the "delayed luminescence" of normal and tumor cells. While with increasing cell density normal cells display decreasing delayed luminescence intensity as well as higher agreement with hyperbolic relaxation behavior, just the opposite holds for tumor cells (Fig. 4).

Consequently, we ask for a mechanism which explains both the non-linear behavior of daphnia and/or cells in a population and the growth of cells.

The most reasonable answer is the Dicke theory [9]. According to R.H.Dicke, matter starts to reemit coherent photons as soon as typical diameters of the absorbing units get small compared with the wavelength of radiation. Take, for instance, typical distances of molecules in biopolymers (which are of the order of some Angstroem units) and compare them to the wavelength of visible light, and the Dicke condition is fairly well satisfied. In this case the spontaneous luminescence bifurcates into the branches superradiance, where flash-like coherent light is reemitted, and subradiance, where delayed luminescence with a high degree of coherence takes place. Both, superradiance and subradiance, are capable of interference, however on different time scales. For long-term measurements of biophotons, subradiance plays the dominant role. Consequently, lower or higher intensities of biophoton emission in the extracellular space reflect under the constraint of destructive interference (where the waves cancel out in view of opposite phase relations) a higher or lower degree of coherence, respectively. As a result, the minimum of biophoton emission of about 120 daphnia has to be assigned to the highest capacity of destructive interference in the extra-

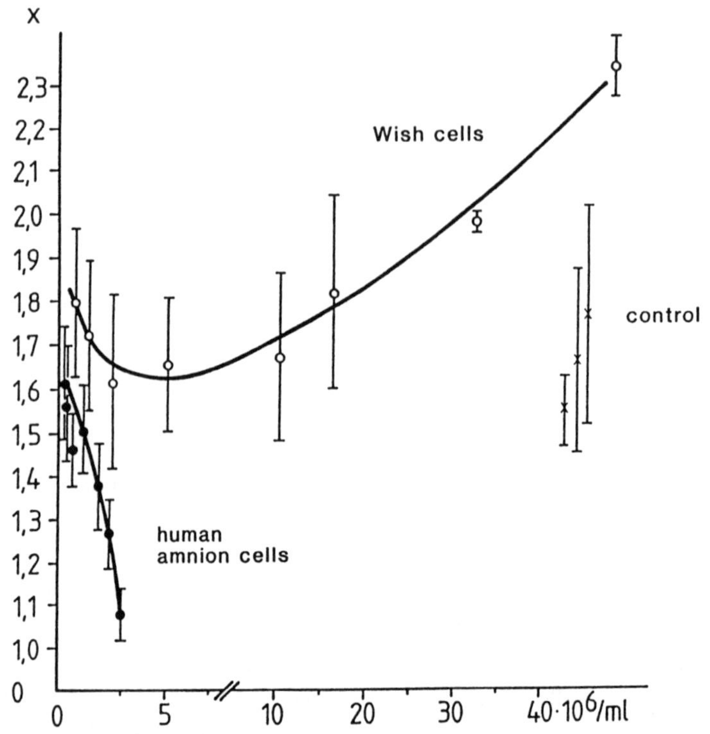

Figure 4: Tumor cells (upper curve) display higher and higher deviations from exponential relaxation with increasing cell density. The opposite is true for normal cells (lower curve).

cellular coherence volume (Fig. 5), while the maxima correspond to the strongest disturbance of this capacity to display destructive interference.

Actually, the daphnia start to build up superorganisms (swarming) as soon as they are collecting in populations with an average distance that corresponds to just this number at the minimum of biophoton emission. One may conclude that under normal conditions also cells show this behavior which makes it understandable that (1) with increasing cell mass the "delayed luminescence" can even decrease and (2) the agreement with hyperbolic relaxation improves in view of the increasing degree of coherence. On the other hand, it indicates the decreasing capacity of tumor cells for destructive interference of biophotons and consequently the incapacity of coherence and communication. At the same time the non-linear growth curve of Eq. 3 becomes completely understandable while, on the other hand, an exponential growth of tumor cells is a logical consequence.

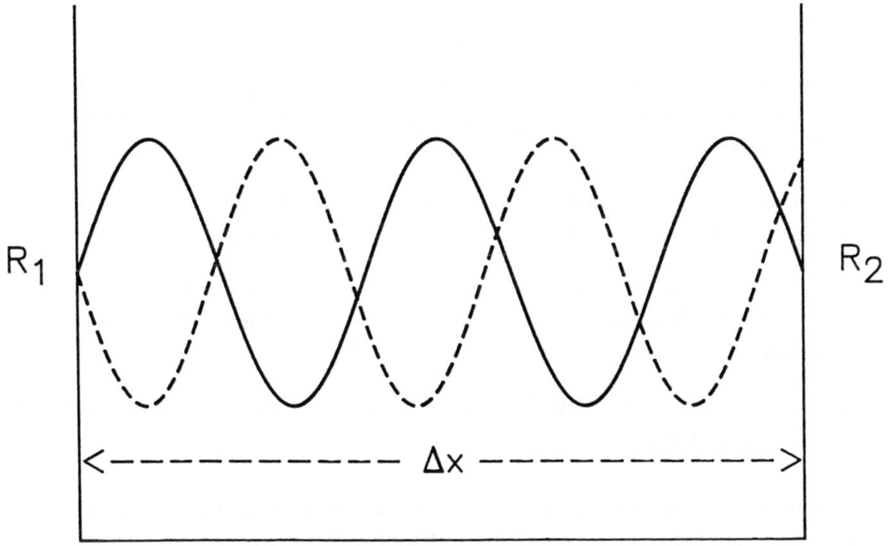

Figure 5: Destructive (and constructive) interference between radiating sources of biological systems (biomolecules, organelles, cells, organs, organisms, populations) may play the most important role of biological organization. This principle combines uncertainty relation and energy conservation.

4. Coherence and Communication

As soon as we accept the model of destructive interference as a basis of cellular organization, it is not difficult to extend this theory to cell communication. Provided that destructive interference of biophotons (which are not confined to the optical region but can range into all the other parts of spectrum) takes place, it explains at the same time

- the capacity of cells and organisms, to identify each other simply by means of the specific interference pattern which is built up by the superposition of the specific waves,
- the capacity for ideal communication, since in case of ideal destructive interference the signal/noise-ratio can approach an infinite value,
- the capacity for establishing communicative populations by means of the radiation potential which arises between zones of destructive and constructive interference,
- the capacity for extending the range of communication by means of extending the coherence volume.

We were successful in showing evidence of optical communication between living systems in case of dinoflagellates [10] and blood cells (neutrophils undergoing respiratory burst [11]). The results can be traced back again to Dicke's theory, where living systems are able to use the quantum effects of non-classical light in terms of destructive (and/or constructive) interference for regulation, organization and communication. It has been shown that even the establishment and evolution of consciousness can be understood on the basis of the development of a kind of "morphogenetic field", based on the coherence of electromagnetic waves in and between biological systems [12]. On this basis, an explanation of homeopathy has been proposed. Take, for instance, the case that by succusion of the "high dilution," coherent states are created which do not interact with the phonons of the heat bath (as it is valid, for example, in case of solitons [13], then rather stable coherent states modulated with the information of the substance may become created. The amplitude may be so low that only high-sensitivity biological systems, but not technical instruments, can respond to those low-level signals. This is not unrealistic if one takes account of the electroreception which shows clearly that biological systems are actually many orders of magnitudes more sensitive to electromagnetic fields than technical devices. Suppose, on the other hand, that within the body the same things may happen, i.e. the storage of coherent waves due to the uptake of a poison which starts to disregulate the "healthy" pattern of coherent waves within the body. Consequently, the same or a similar pattern of coherent waves will lead to a resonance-like overlap of the waves coming from the body and those coming from the homeopathic. The resulting effect will be an adiabatic excretion of the disturbing coherent waves from the body to those of the remedy which is then partially eliminated by the body. The effect works analogical to two identical pendula or tuning forks, where one is oscillating while the other which originally does not oscillate is hung near the oscillating one. The result is an exchange of the kinetic energy to such a degree that on average just half of the energy is transferred from the oscillating to the non-oscillating pendulum or tuning fork. The effect satisfies Hahnemann's rules of homeopathy, i.e. the simila principle (in view of the resonance condition) and the potency rule (due to the fact that the exchange of energy is the more effective the less oscillation energy is stored in the second pendulum, which must be hung near the oscillating one). For more details please refer to the references [1,14].

References

[1] F.A. Popp, K.H. Li, Q. Gu, *Recent Advances in Biophoton Research and its Applications*, World Scientific, Singapore, 1992.

[2] G. Cilento, "Photobiochemistry without light", *Experientia*, vol. 41, pp. 572 585, 1988.

[3] M. Galle, "Untersuchungen zum dichte- und zeitabhängigen Verhalten der ultraschwachen Photonenemission von parthenogenetischen Weibchen des Wasserflohs Daphnia magna", *Dissertation*, Universität Saarbrücken, 1993.

[4] F.A. Popp, K.H. Li, "Hyperbolic relaxation as a sufficient condition of a fully coherent ergodic field", *Int. J. Theor. Phys.*, vol. 32, pp. 1573-1583, 1993.

[5] F.A. Popp, W. Nagl, K.H. Li, W. Scholz, O. Weingärtner, R. Wolf, "Biophoton Emission - New Evidence for coherence and DNA as source", *Cell Biophysics*, vol. 6, pp. 33-51, 1984.

[6] H. Fröhlich, "Long range coherence and energy storage in biological systems", *Int. J. Quantum Chemistry*, vol. 2, pp. 641-649, 1968.

[7] F.A. Popp, G. Becker, H.L. König, W. Peschka, "Coherent photon storage of biological systems", in *Electromagnetic Bio-Information*, Urban & Schwarzenberg, München 1979.

[8] F.A. Popp and J. Deny, "Biophotonen-Information und Chaostheorie", in *Ganzheitsmedizin.* Zweiter Wiener Dialog, Facultas Universitätsverlag, 1991, pp. 53-66.

[9] R.H. Dicke, "Coherence in spontaneous radiation processes", *Phys. Rev.*, vol. 93, pp. 99-110, 1954.

[10] F.A. Popp, J.J. Chang, Q. Gu, M.W. Ho, "Nonsubstantial biocommunication in terms of Dicke's theory", in *Bioelectrodynamics and Biocommunication*, World Scientific, Singapore, 1994, pp. 293-317.

[11] X. Shen, W. Mei, X. Xu, "Activation of neutrophils by a chemically separated but optically coupled neutrophil population undergoing respiratory burst", *Experientia*, vol. 50, pp. 963-968, 1994.

[12] F.A. Popp, "Evolution as the expansion of coherent states", in *The interrelationsship between mind and matter*, Center for Frontier Sciences at Temple University, Philadelphia, 1992.

[13] A.S. Davydov, "Energy and electron transport in biological systems", in *Bioelectrodynamics and Biocommunication*, World Scientific, Singapore, 1994, pp. 411-425.

[14] F.A. Popp, "Some elements of homeopathy", *Br. J. Homeopathy*, vol. 79, pp. 161-166, 1990.

WATER AND BIO-COMMUNICATION

CYRIL W. SMITH
Department of Electronic and Electrical Engineering, University of Salford
Salford, M5 4WT, England

ABSTRACT

A wide range of biological cells show general oscillatory phenomena and a wide range of biomolecular systems are sensitive to weak environmental electromagnetic fields particularly in the low frequency region or to radiofrequency fields modulated by low frequencies. Human circadian rhythms and temporal morphology imply a high degree of coherence. Frequency information can be 'imprinted' into water and metals and on to a light beam but, deuterium oxide would not 'imprint'. This implies that both the protons and electrons in water must be co-operatively coherent. The frequency information in 'imprinted' water can propagate through aqueous and metallic systems which thus become channels for bio-communication.

1. Water as a Memory

Frequency information can be 'imprinted' into water and metals and on to a light beam [1-3]. It does not imprint into deuterium oxide which means that both the protons and the electrons in water must be co-operatively coherent. The frequency information in 'imprinted' water will propagate through aqueous and metallic systems. 'Imprinted' water can carry long-range electromagnetic bio-information, including such chemical information as may be present at a large distance from a molecule. It cannot mimic the short-range chemical bonding interactions that characterise a chemical reaction. It is only at short-range that a molecule can distinguish whether it is receiving information from 'imprinted' water, electromagnetic radiation or another molecule.

2. Water and the Coherence of Biological Frequencies

The cell division cycle of the typical adult human cell takes 18-24 hours, the higher plants have a 10-30 hour cycle, embryonic cells and bacteria may have cycles between 25-70 minutes. There are techniques for synchronizing cell division to facitilate experimentation. The triggering events that initiate DNA synthesis are not known, they may be electrical [4]. What is clear is that there are "windows" [5] during the biological cycle in which the system is particularly sensitive to its electrical environment and especially to any frequencies present during certain cell activities.

Alternating magnetic fields can affect the lac operon system of *E. coli* [6]. This involves going to the DNA to get the instructions for the synthesis of beta-galactosidase. The core of DNA is high in protons, so any local disturbance such as the deposition of energy selectively into DNA protons might be expected to lead to a measurable cell division rate change.

Oscillatory chemical reactions can occur if the reaction rate is limited by the concentration and the diffusion of a reaction component. External magnetic fields can affect concentrations through singlet-to-triplet transitions [7], electromagnetic radiation might be similarly effective through cooperative interactions with strongly excited polar molecular states [8].

Any oscillator must have a process involving positive feedback, i.e. a stimulatory signal from the output reaching the input. In the absence of external stimulation. Positive feedback will give amplification which may result in oscillations triggered by the noise inherent in any system [9]. Weak external signals can synchronise an oscillator, whereas stronger signals will saturate the system. In general, if a coherent oscillation is involved, then a sine wave is the appropriate synchronizing stimulus, if magnetic flux quanta are involved, then a square or pulse-train waveform is appropriate.

A wide range of biological cells show general oscillatory phenomena [10]. The first (1983) direct detection of radiofrequency emissions from yeast cells [11] has now been mostly replicated [12]. Ross Adey [5] has presented evidence of the sensitivity of a wide range of biomolecular systems to weak environmental electromagnetic fields either in the low frequency region or, with radiofrequency fields modulated by low frequencies. Breithaupt [4] has considered the bio-information transfer properties of biological rhythms and their function in temporal morphology. Wever [13] carried out a 25-year program of research into human circadian rhythms which imply a high degree of coherence. This involved persons living for prolonged periods in underground dwellings shielded from environmental fields. Unexpectedly, the deep body temperature and the sleep-wake cycles had a remarkable sensitivity to weak 10 Hz fields which were able to entrain free-running body rhythms and even force a 23-hour daily cycle.

It was known to the ancient Chinese that the acupuncture meridians in the human body have regular periods of activity (Yin), and passivity (Yang) over the 24-hour period. The effects of the solar and lunar tides and environmental low-frequency fields on circadian rhythms have been summarized by Smith and Best [1]. Plants respond electrically to the variations in the geomagnetic field which preceed day-break at ground level, or to an artificial field having the same rate of change [1]. Dubrov [14] has described a very wide range of effects of the geomagnetic fields on biological rhythms and circadian rhythms. Semm, Schneide and Vollrath [15] broadened their investigations of the pineal involvement in circadian rhythms and its reactions to light and darkness, to consider the effects of magnetic field changes comparable to the strength of the geomagnetic field (0.5 G, 50 μT). Reiter [16] has written extensively on the pineal, and on melatonin as the chemical expression of darkness. Power supply fields (which are highly coherent) have also been found to affect the melatonin rhythm [17] and the writer has detected resonances at both 50 Hz and 60 Hz in a sample of pineal extract, which in turn suggests that the worst possible frequencies were chosen for the world's power supply systems since, melatonin is also an anti-cancer agent.

3. Electromagnetic Fields in Biological Systems

In general, routine clinical equipment for measuring the electrical signals from the body (ECG, EEG, EMG), is restricted in its frequency coverage and analysis capabilities. The heart-beat is a prime example of a highly coherent oscillation in a living system and small variations in the periodicity have been used diagnostically by Rein and co-workers [18, 19]. Living systems have cyclical patterns of frequency variation over a wide range of frequencies and these can be altered by external electromagnetic fields at levels typical of those in the environment.

When biologists get round to using the magnetic vector potenetial from a toroidal coil instead of the magnetic field of a Helmholtz coil to investigate a living system they may find the results surprising. Dr. Mae-Wan Ho writes, "The results with the toroidal coil are quite tantalizing. Despite the fact that the magnetic field is negligible, significant increases in abnormalities (in *Drosophila* embryos) are found over matched controls, and both when the embryos are in place before or after the power supply is switched on" [22].

4. Frequencies in Biological Systems

4.1. A Coherence Frequency Measurement Technique

The frequencies of resonances in water, potencies and biological systems can be measured using the the experimenter's autonomic nervous system as a highly sensitive detector. The sample needs to be excited with an alternating field (magnetic or magnetic vector potential) from a solenoid or toroid. The sample, which might be a tube containing potentised water, and the toroid need to be aligned on a North-South axis with an accuracy of ±45° at fields well above the threshold of detection, although a tolerance of ±5° is needed close to the threshold for detection. This orientation may be required to avoid interference from the vector potential of the geo-magnetic field which is in the East-West direction and which seems to be the reference vector for biological coherence effects.

For the writer to be able to detect a 1 kHz resonance in water, a magnetic vector potential **A** >17 nWb/m (calculated rms value) is required at the sample. The detection threshold increases linearly with distance from the toroid between 0.3 m and at least 4 m [2].

The hands and arms of the experimenter are also on the North-South axis, and should be one on each side of the sample to be measured. This leaves one hand free to tune the oscillator feeding the toroid or solenoid. The other hand can hold a pendulum which gives a clear indication when a resonance is reached by amplifying the increased natural muscle tremor. The resonant frequency of the pendulum should matches the natural (2 Hz) resonance in the autonomic nervous system. The fields in a resonant specimen produce an left-to-right unbalance in the experimenter [2]. The (left-handed) writer gets a counterclockwise rotation of the pendulum at resonance when facing West. A water resonance giving this rotation corresponds clinically to a therapeutic frequency while, that giving the opposite rotation produces stressful reactions.

If the experimenter changes from facing West to facing East, the sense of all the responses is reversed. Most substances and potencies show alternate counter-clockwise and clockwise rotations at successive resonances, beginning at the lowest resonance frequency and ending at the highest with a counter-clockwise rotations.

The objective detection of resonances in water has been achieved and repeated on rare occasions using a technique whereby a pair of gold wire electrodes is dipped into a cuvette containing the sample of potentised water, these are connected to a suitable low-noise, high-gain, narrow-band amplifier. Similar results have been obtained at Wekroma AG, Switzerland, using a Fourier Transform Spectrometer with signal averaging over a large number of measurements [2]. The frequencies measured in this way were the same as those originally imprinted into the water and the same as measured on the sample by the above subjective method.

The problem of repeatability may be that of finding out how to couple an electrode to self-trapped radiation without destroying it.

4.2. Results of Frequency Measurements

Each successive dilution and succussion adds one more stimulatory and one more depressive resonance to those found in the previous potency [23]. For all the materials and biological systems so far measured the harmonics were not the usual octave apart but, were spaced according to a high-order power-law for which the only reasonable model is a fractal equation.

There are some exceptions to the above:

(i) Water held and potentised by a healer contained a sequence of only therapeutic frequencies, starting at the heart frequency and becoming a continuum at high frequencies.
(ii) Dilutions of hydrogen peroxide contained only therapeutic frequencies which, if successed became a continuum. With hydrogen peroxide, there was no potentisation if diluted with air-free water.
(iii) Dilutions of formaldehyde contained only stressful frequencies which, if successed became a continuum.
(iv) Deuterium oxide would not potentise when successed in an alternating magnetic field or magnetic vector potential until diluted with water > 5%, <10% (by volume).
(v) Electrotherapy apparatus can potentise water at any frequencies within its pass-band. Whether these are therapeutic, or stressful, depends upon the amplifier parameters selected.

If a sample of 'imprinted' water is placed on the left of the toroid and a sample of 'clean' water is placed equidistant to the right and successed, the latter acquires a phase-inverted 'imprint' of the water on the left. A practical application for this is for allergy therapy since it provides a way of converting stressful frequency components of allergens into therapeutic ones beyond the bandwidth of electrotherapy apparatus.

It is also possible to detect proton NMR in a water sample (2.13 kHz in a 50 μT geomagnetic field) and to confirm that it is and NMR, this resonance can be tuned through by moving a small magnet in the vicinity of the water.

It is also possible to use this method to detect resonances in series or parallel resonant inductance-capacitance (LC) circuits. The LC circuits **do** give octave spaced harmonics with the fundamental and all the other odd harmonics giving a stimulatory (counter-clockwise) pendulum rotation, while all the even harmonics give a depressive (clockwise rotation).

4.3. Erasing Water Memory

We now know much more about how to erase the 'water memory' albeit as the result of trying to measure erased samples. Heating water above 70°C will erase its memory, as will passing an amplifier off-set current greater than about a microamp through the water sample, or even the tube being handled by the 'wrong' person. But, most importantly, placing the water in a hypomagnetic container (e.g. a closed mu-metal box) erases not only water 'memory', but also tablets and liquid preparations of homoeopathic potencies and is very useful for producing "controls" by erasing a specimen after it has been measured. The threshold steady magnetic field necessary for the maintenance of water 'memory' in the environment is about 400 nT. At this field, the mean separation between magnetic flux quanta (2.07×10^{-15} Wb) would be 71 μm. A strong magnetic field >250 μT will suppress 'water memory', although in this case the original imprint is recoverable by succussion.

4.4 Coherence Frequencies in Living Systems

Work with a single cell system became possible during a visit to the laboratory of Dr. F-A Popp, (Technology Centre, Kaiserslautern, Germany) [24]. Here measurements were made at low frequencies and radio-frequencies on a single (10 cm long) filamentary cell of *Acetabularia.* The frequencies were observed to jump discontinuously by a small amount at reasonably regular intervals of a few minutes; the pattern repeated after about 25 minutes provided that the oscillator was only switched on while measurements were actually being made. When the oscillator was left on and tracked continuously on a 'stimulatory frequency', the periodicity dropped to about 2 minutes. While the oscillator was left on a 'depressing frequency' between measurements, no frequency jumping took place at all, even over several hours so long as the oscillator was left on that frequency.

This frequency jumping phenomenon applies equally to the human subject. Again, a depressing frequency stops all frequency jumping, while tracking the stimulating frequency speeds the frequency jumping from about a 2 hour cycle to a 6 minute cycle. Acupuncture points relate to specific body organs or systems. Electroacupuncture apparatus can be used to measure the frequencies on a meridian by first imprinting them into water in the output beaker, which can then be measured.

The empirical rule evolved for the therapy of electromagnetic hypersensitivity is to use an oscillator set to the highest stimulatory frequency or, to 'imprint' some, or all, of the stimulating frequencies into water or other substance tolerated by the patient. Thus, it seems that the therapy involves stimulating lethargic autonomic systems in the patient. The persistant presence of a depressing frequency in the environment can paralyse the natural frequency jumping of a living system, with most uncomfortable results.

5. Coherence or Wüst Waves

Specimens of potentised water seem to be well protected by wrapping the glass container in aluminium foil. The reason for this probably lies in the small critical angle (3.5°) for coherence waves passing from aluminium into water. It is theoretically possible to make lenses to focus coherence or Wüst waves. The critical angles do not alter measurably from at least 1 Hz to 100 kHz. Neither did there seem to be any appreciable dispersion when the velocities of coherence waves were measured by Fizeau's method [3], or in the case of stationary coherence waves on Lecher's wires. Therefore, coherence wave dispersion spectroscopy may not be available although, interference methods of spectroscopy remain a possibility. The 'imprint' of potentised water can be picked up by a light beam shone through it and this can be transferred into water or metal on which the beam falls by succussing it.

If an insulated wire is arranged to conduct a coherence wave as described previously [2] and is then passed through a hypomagnetic enclosure, it is found that is a critical magnetic field >740 nT within the enclosure for propagation beyond it. At this field, the mean separation between magnetic flux quanta would be 52 μm. A coherence wave will propagate through a 10A fuse wire (calculated diam. 322 μm), It will not propagate through a 3A fuse wire (calculated diam. 140 μm), there is what is probably a 1-D coherence in a 5A fuse wire (calculated diam. 200 μm), Coherence waves can be conducted through water in tube diameters >2.5 mm but, not in tubes <2mm diameter. These dimensions are all of the same order of magnitude.

The brass beakers that are used to contain samples whose signals are to be introduced into electroacupuncture apparatus provide a convenient alternative method of coupling a coherence wave into a wire. They are usually provided with more than one socket so that a pair of wires can be lead away, each carrying an identical coherence signal. If a beaker contains a sample of water excited at its imprinted frequency by a toroid, and if such a pair of wires is passed one either side of the end of an air-cored solenoid resting on end (e.g. 3000 turns/m, 50 mm diam.) through which a small steady current can be passed, this provides the basic arrangement for the Aharanov and Bohm experiment. It does give changes in the pendulum rotation sense proportional to the current in the solenoid for when the water in the beaker is excited at either the ESR or proton-NMR frequency in the ambient magnetic field.

The same result is also obtained if the water is contained in a circular plastic tube placed around the solenoid. This is in effect replacing the metal wires and the beaker containing the water sample with a water channel. The simultaneous application of both ESR and proton-NMR frequencies to water produces a phase inversion of the response.

The same changes are obtained for a metal sample (copper) placed in the beaker, but in this case only at the ESR frequency. The NMR for both Cu_{63} and Cu_{65} were detected, but, varying the current in the solenoid produced no effect at these resonances. Likewise, there was no effect at the NMR frequency of deuterium when an ampoule of pure D_2O was placed in the beaker. It has not been possible to try hydrogen and deuterium absorbed in palladium. This implies an absence of coherence wave phenomena for ions heavier than hydrogen, which just manages to be a quantum particle. Also, the 'imprint' in a metal must involve only electron coherence. The metal ions are already spatially coherent.

6. Potentisation by Contact

As already mentioned, water 'memory' can be imprinted with a frequency using an alternating magnetic field, or with an alternating magnetic vector potential plus either succussion or a magnetic field. However, the information will also slowly imprint by contact. For example, if a glass tube containing 'erased' water is placed in a beaker of water imprinted with a range of frequencies, the frequencies will slowly become imprinted, the higher frequencies imprint more quickly than the lower frequencies. This suggests that imprinting does not need energy other than thermal excitation, and that the imprinting awaits the random arrival of the correct frequency component in the thermal noise to provide a necessary bifurcation. The random statistic theoretical curve is close to the water-water curve. This is based on the uncertainty relations for the number of randomly arriving quanta within the coherence time which are sufficient to give an energy equal to thermal energy (kT) [11]. Chlorided-silver wire has no potential barrier with water at low frequencies and imprints the frequencies in times given by assuming the equivalent statistics for coherent quanta. The frequency imprinting times for copper and gold immersed in water come between these two limits. As Fröhlich often remarked, onset time delay is a hall-mark of coherence.

7. Discussion

The possibility of coherence in the ground state and the availability of the Josephson effect is supported by QED theory applied to water. For potentization and coherence wave propagation in water, a cluster of domains 1 - 2 mm in size (for a metal about 200 μm) is needed to have enough weak-link junctions to provide memory capability. If these coherent domains behave as superconducting regions, then magnetic flux will not penetrate them. The ambient magnetic flux provides a mechanism to prevent domain coalescence. If this falls below 740 nT (average magnetic flux line separation 52 μm) water memory is erased; this is the likely result of domain coalescence dissipating charge patterns giving rise to Josephson voltages between domains. In magnetic fields >250 μT, the flux may force the weak-link junctions too far apart for memory read-out. There is no dissipation of the charges, since memory is recoverable on succussion. Writing to memory with a frequency becomes a process of setting up internal charge patterns which give Josephson voltages between domains corresponding to the applied frequency. Read-out would be through the AC Josephson effect when a small supercurrent crosses a weak-link junction. In the *in vivo* state, redox reactions which work with electron-pairs, are the obvious source of read-out supercurrents. The process of succussion is then seen as re-distributing the coherence domains and their separating lines of magnetic flux. The diluent will only have noise in memory. The addition of a diluent to a potency/mother tincture is the addition of a larger memory capacity, realised on succussion, thereby permitting greater resolution and thus higher coherence. The signals in memory should increase in proportion to the number of potentizations, the noise from the diluent as their square-root.

8. Conclusion

In these lectures I have tried to present a picture of what can be found out about coherence in water and living systems by simple experimental techniques, many of which have had to be devised to meet circumstances as the results unfolded. The detailed mechanisms leading to electromagnetic hypersensitivity are still not determined, but it is clear that the patients are really sick. Those results obtained by subjective measurements obviously need repetition by objective physical techniques as soon as these can be designed and I trust that I have provided some useful design parameters. I can only plead that the good general first surveys the battle-field on foot.

9. References

[1] C.W. Smith and S. Best, 1989. *Electromagnetic Man: health and hazard in the electrical environment*. London: Dent, 1989,1990; New York: St. Martin's Press, 1989; Paris: Encre/Arys (Fr. Edn.), 1995.

[2] C.W. Smith, 1994. *Electromagnetic and magnetic vector potential bio-information and water*, in Endler P.C. and Schulte J. (Eds.) Ultra High Dilution: Physiology and physics, Dordrecht: Kluwer Academic, pp. 187-201.

[3] C.W. Smith, 1994. Coherence in living biological systems, *Neural Network World* 3: 379-388.

[4] H. Breithaupt, 1989. *Biological rhythms and communication*, in Popp F-A, Warnke U, König H L, Peschka W (eds.), Electromagnetic Bio-Information, 2nd Edn. Baltimore: Urban & Schwarzenberg, pp. 18-41.

[5] W. Ross Adey,1990 *Electromagnetic fields and the essence of living systems*, in Anderson J B (ed.), Modern Radio Science 1990. Oxford: University Press/URSI, pp.1-36.

[6] E. Aarholt, E.A. Flinn and C.W. Smith, 1982. "Magnetic fields affect the *lac operon* system" *Phys. Med. Biol.* **27**(4), 1982, 603-10.

[7] K. McLauchlan, 1992. "Are environmental magnetic fields dangerous?". *Phys. World* **5**(1), 41-45.

[8] H. Fröhlich (ed.), 1988. *Biological coherence and response to external stimuli.* New York: Springer-Verlag.

[9] F.R.W. Strafford, 1946. "The super-regenerative detector: an analytical and experimental investigation". *J. Inst. Elect. Engrs.* **93**(III), 1946, 23-28.

[10] P.E. Rapp, 1979. "An atlas of cellular oscillators". *J. exp. Biol.* **81**, 281-306.

[11] C.W. Smith, A-H. Jafary-Asl, R.Y.S. Choy and J.A. Monro, 1987. *"The emission of low intensity electromagnetic radiation from multiple allergy patients and other biological systems"*, in Jezowska-

Trzebiatowska, B., Kochel, B., Slawinski, J. and Strek, W. (eds.), Photon Emission from Biological Systems. Singapore: World Scientific, pp. 110-126.

[12] Hölzel R., Lamprecht I.: Electromagnetic fields around biological cells, *Neural Network World* **3**, 1994, 327-337.

[13] R.A. Wever, 1985. *The electromagnetic environment and circadian rhythm of human subjects*, in Grandolfo M, Michelson S. M, Rindi A (eds.) Static and ELF Electromagnetic Fields: biological effects and dosimetry. New York: Plenum Press, pp. 477-523.

[14] A.P. Dubrov, 1974. *The geomagnetic field and life.* New York: Plenum Press.

[15] P. Semm, T. Schneide and L. Vollrath, 1980. "Effects of earth strength magnetic field on the electrical activity of pineal cells". *Nature* **288**(5791), 607-608.

[16] R.J. Reiter, 1991. "The chemical expression of darkness". *Mol. C. Endoc.* 79(1-3), C153-158.

[17] R.J. Reiter, L.E. Anderson, R.C. Buschbom, et al. 1988. "Reduction of the nocturnal rise in pineal melatonin levels in rats exposed to 60 Hz electric fields in utero and for 23-days after birth". *Life Sci.* **42**(22), 2203-2206.

[18] R. McCraty, M. Atkinson, and G. Rein, 1993. "ECG spectra: the measurement of coherent amd incoherent frequencies and their relationship to mental and emotional states". *Proc. 3rd. Ann. Conf. Intl. Soc. for the Study of Subtle Energies and Energy Medicine*, Monterey CA. June 1993.

[19] G. Rein and R. McCraty, 1993. "Modulation of DNA by coherent heart frequencies". *Proc. 3rd. Ann. Conf. Intl. Soc. for the Study of Subtle Energies and Energy Medicine*, Monterey CA. June 1993.

[20] C.W. Smith, R.Y.S. Choy and J.A. Monro, 1989. "The diagnosis and therapy of electrical hypersensitivities". *Clin. Ecol.* **6**(4), 119-128.

[21] K.M. Quboa, 1990. *Environmental electric and magnetic fields: measurements and communications implications.* Ph.D. Thesis, Salford University, England.

[22] M-W. Ho, A. French, J. Haffegee and P.T. Saunders, 1994. *Can weak magnetic fields (or potentials) affect pattern formation?*, in Ho, M-W, Popp, F-A., Warnke, U., Bioelectrodynamics and Biocommunication, Singapore: World Scientific, pp. 195-212.

[23] C.W. Smith and P.C. Endler, 1994. *Resonance phenomena of ultra high dilution thyroxine - preliminary study*, in Endler P.C., Schulte J. (Eds.): Ultra High Dilution: Physiology and physics. Dordrecht: Kluwer Academic, pp. 203-207.

[24] C.W. Smith, 1995. "The electromagnetic aspects of biological cycles". *Environmental Medicine* **9**(3), 113-118.

ON THE STRUCTURE OF HIGH DILUTIONS ACCORDING TO THE CLATHRATE MODEL

G.S. ANAGNOSTATOS
Institute of Nuclear Physics, National Center for Scientific Research "Demokritos"
Aghia Paraskevi, Attiki, Greece

ABSTRACT

For the interpretation of high dilutions in water, the clathrate model is presented according to which round each molecule of the mother substance a clathrate is formed, made of water molecules interacting by hydrogen bonds. If forceful mechanical impulses are applied on the solution, the model assumes that a separation of the initial molecule from its clathrate is possible, a fact which leads to the formation of new clathrates round each of these two separated components. This effect is repeated again and again as we proceed in the preparation of a high dilution. Thus, eventually this dilution contains free molecules of the solvent (here water) and bound molecules of the solvent in the form of clathrates which are characteristic of the initial mother substance.

1. Introduction

High dilutions of a substance could be called those dilutions which go beyond the Avogadro number (i.e., $6.023 * 10^{23} \sim 10^{24}$). That is, they effectively contain no molecule of the initial substance. Thus, the chemical composition of such dilutions is identical to the chemical composition of the solvent, e.g., of water. Hence, any possible property of a high dilution which is not a property of the untreated relevant solvent cannot be due to a chemical action of the initial substance (which is not effectively present) and should be attributed to changes of the solvent made during the preparation of this dilution. Out of all possible changes, we are concerned with those referring to the structure of the solvent examined in the framework of the clathrate model discussed in section 3. Clathrate is the term for an inclusion compound made here strictly from molecules of the solvent arranged in such a way that a space is left in the interior [1]. For the special case where the solvent is water, the particular terminology clathrate hydrate is used. The interior space can be unoccupied (empty) or occupied by molecules of the initial substance (usually one) or by molecules of the solvent (one or more). When this space is occupied by molecules of the initial substance, then the clathrate is here called full. For a given initial substance in a specific solvent, a specific clathrate is formed. Thus, different substances diluted in the same solvent create clathrates which differ in form and volume of the interior space but also in the whole structure of the clathrate [1]. Such differences in the structure correspond to different amounts of energy (e.g. heat) necessary for the destruction of the corresponding clathrates. Also, different amount of energy is needed for the destruction of the same kind of clathrate when it is empty than that when it is full. For hydrophobic substances in the cases of clathrate hydrates (which in practice constitute the usual cases for high dilutions) the former amount is larger than the latter.

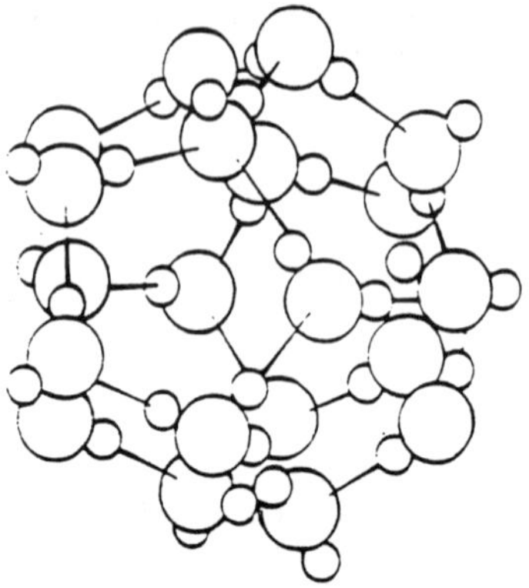

Fig. 1 Pentagonal dodecahedron made of 20 water molecules interacting by hydrogen bonds, called basic clathrate.

2. Building Blocks of Clathrate Hydrates

The basic building block in forming a clathrate hydrate involved in high dilutions is the pentagonal dodecahedron (Fig. 1) composed of 20 water molecules [2], i.e. $20H_2O$. Specifically, there are 20 oxygen atoms at the vertices of this polyhedron and 30 (out of 40) hydrogen atoms may be considered along its edges. The remaining 10 hydrogen atoms (directed outwards of the polyhedron) may be used for hydrogen bonding to adjacent water molecules thus acting as donators and 10 of the oxygen atoms (those whose both hydrogen atoms are considered along the edges of the polyhedron) can thus act as acceptors for hydrogen bonds from surrounding water molecules [3] (Fig. 2).

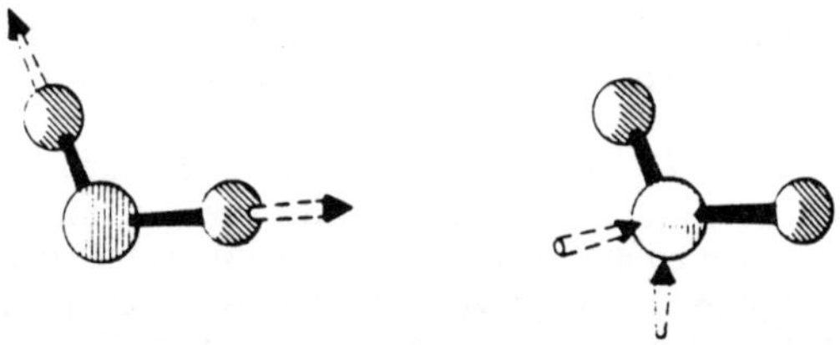

Fig. 2 (a) Hydrogen bonds donated; (b) hydrogen bonds accepted.

The internal angle of the pentagon is 108°, which is very close to the ideal hydrogen bond geometry for a water molecule [4] where the angle H-O-H is 104.6° (Fig. 3). This condition makes the pentagonal dodecahedron a tightly hydrogen-bonded entity and thus such a configuration could be seen as a very stable one. Another stable clathrate could be throught of as composed by several pentagonal dodecahedra sharing one or more faces [2] or by a framework of dodecahedra linked by larger polyhedral voids [4] (Fig. 4) or something else where again the dodecahedron plays a dominant role.

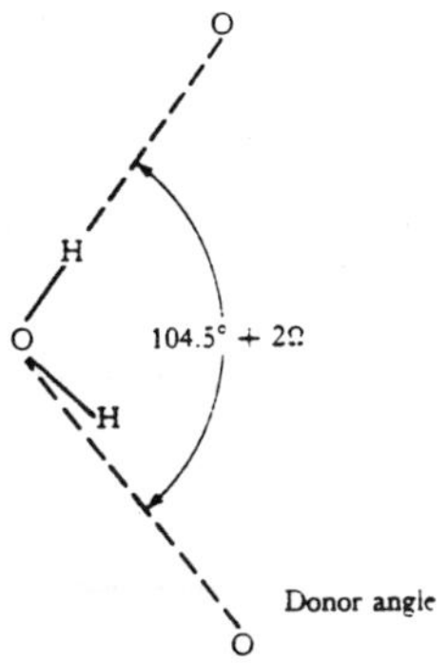

Fig. 3 Ideal hydrogen bond geometry for water molecules (104°.6) and its association to the donor angle $\alpha=108°$ in a pentagonal arrangement of oxygen atoms, where $\Omega=1/2(\alpha-104°.6)$.

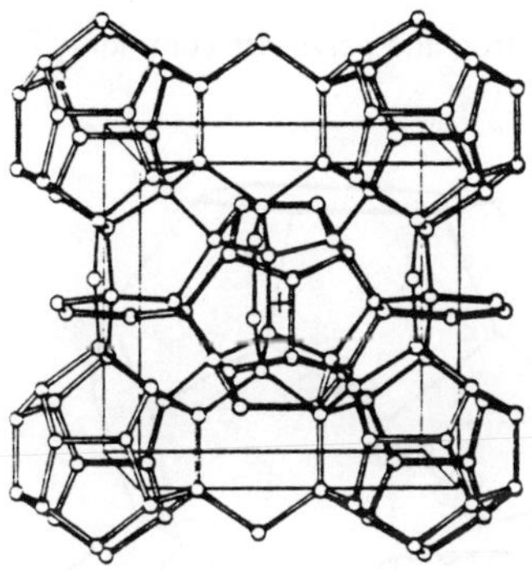

Fig. 4 Framework of dodecahedra linked by larger polyhedral voids.

In Fig. 5 the pentagonal dodecahedron and the polyhedra that are commonly found in conjunction with it in the clathrate hydrates are shown [1], while in Figs. 6 and 7 two specific cases of clathrate hydrates are shown referring to two particular solutes named in the relevant captions [1].

Although colloid stability is somewhat peripheral to the main subject pursued in the present work, we emphasize here the fact that structured water has been proposed to exist surrounding micro-sized colloidal particles. Similarly, we propose that structured water is formed around a solute in a water environment (solvent) as a necessary starting point for the preparation of a high dilution.

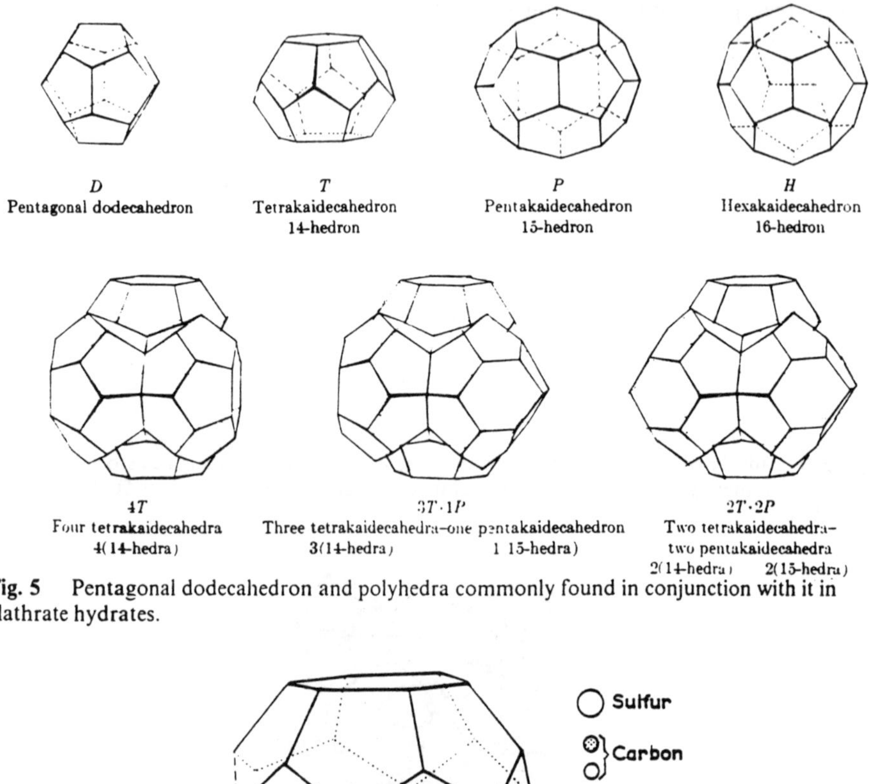

Fig. 5 Pentagonal dodecahedron and polyhedra commonly found in conjunction with it in clathrate hydrates.

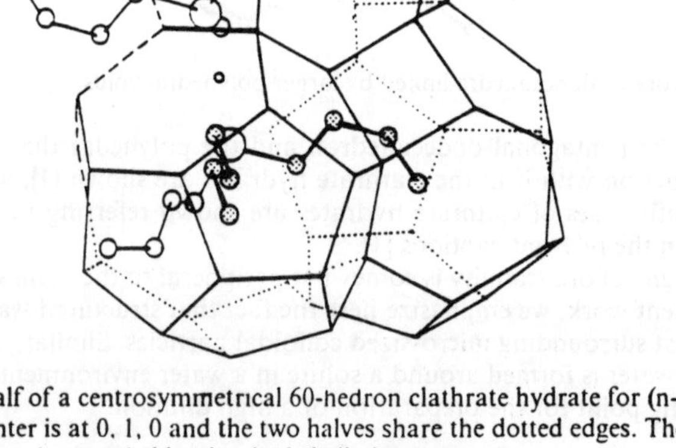

Fig. 6 One half of a centrosymmetrical 60-hedron clathrate hydrate for $(n\text{-}C_4H_9)_3\text{-}S^+F^- \bullet 23H_2O$. The center is at 0, 1, 0 and the two halves share the dotted edges. The half polyhedron shown is occupied by the shaded alkyl groups.

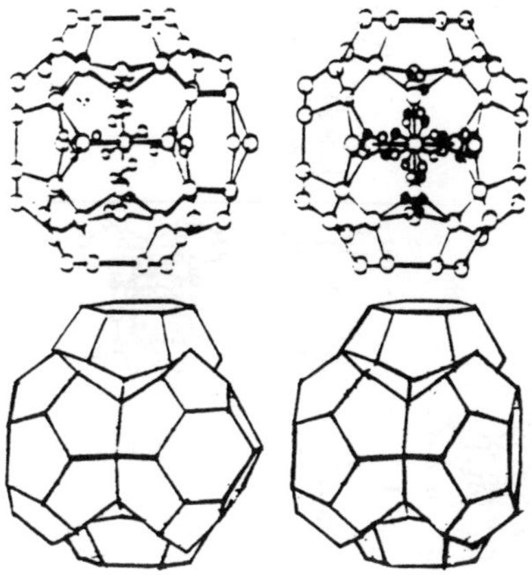

Fig. 7 Clathration of a $(n\text{-}C_4H_9)_4N$ cation within the complex polyhedra shown for two slightly different interior voids.

3. The Clathrate Model

In the present brief presentation of the clathrate model, the tetraisoanylammonium fluoride in water is taken as an example, because the stereochemical structure of its molecule and all details (form and size) of its corresponding clathrate hydrate are precisely known [1,5].

In Fig. 8a, b, c, all three stages in the preparation of a high dilution, according to the clathrate model, are demonstrated and are separated from each other by heavy horizontal lines. As stage one the molecule of the tetraisoanylammonium fluoride is shown at the center of Fig. 8a depicted by bold circles, while the water molecules of the corresponding clathrate hydrate are depicted by larger open circles.

It is apparent that the overall form of the clathrate shape resembles that of the molecule of the mother substance. Even more important is the remark that the three volumes comprising the clathrate are pentagonal dodecahedra which by two have a common face. This remark holds true for many clathrate hydrates, i.e., the building block for almost all cases of clathrate hydrates, as mentioned in the previous section, is the pentagonal dodecahedron (Fig. 5), mainly because of its basic property that its interior face angle between two adjacent edges of a pentagon (108°) very closely approximates the angle between the two hydrogen atoms in a water molecule (104.6^o). This near coincidence of angles substantially contributes to the necessary strength of the structure of clathrate hydrates made up from such dodecahedra [6-8].

In Fig. 8b the second stage in the preparation of a high dilution, according to the clathrate model, is demonstrated. During this stage the separation of the molecule of the mother substance from its clathrate hydrate, when a forceful mechanical impulse is applied to the dilution, takes place (Fig. $8b_2$-b_1). This separation is obtained due to the different inertial kinematic behavior between the (relatively large density) molecule of the mother substance and that of the (relatively small density, because of the much larger size) clathrate hydrate and also due to the repulsive forces between the water molecules and the hydrophobic molecule of the

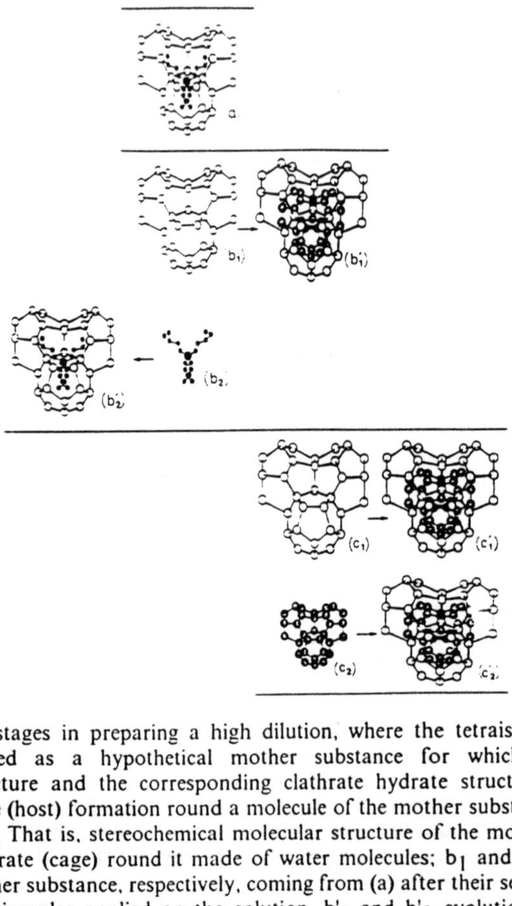

Fig. 8 The three stages in preparing a high dilution, where the tetraisoanyl-ammonium fluoride is employed as a hypothetical mother substance for which the molecular stereochemical structure and the corresponding clathrate hydrate structure are precisely known; (a) Clathrate (host) formation round a molecule of the mother substance (guest) after its dilution in water. That is, stereochemical molecular structure of the mother substance at the center and clathrate (cage) round it made of water molecules; b_1 and b_2 clathrate and molecule of the mother substance, respectively, coming from (a) after their separation due to a forceful mechanical impulse applied on the solution. b'_1 and b'_2 evolution of b_1 and b_2, respectively, shortly after having applied the mechanical impulse. Specifically, the mother molecule in b_2 interacts with surrounding water molecules and, as a result another clathrate hydrate is formed leading to b'_2 which is identical to (a). At the same time, b_1 shrinks (due to absence of hydrophobic forces with the molecule of the mother substance) and interacts with surrounding water molecules leading to the formation of another (mantle) clathrate round the previous (core) clathrate as shown in b'_1. c_1 and c_2 mantle and core clathrate, respectively, coming from b'_1 after their separation due to a forceful mechanical impulse applied on the solution. c'_1 and c'_2 evolution of c_1 and c_2, respectively, shortly after having applied the mechanical impulse. Specifically, the core clathrate in c_2 interacts with surrounding water molecules and, as a result, a new mantle clathrate is formed leading to c'_2, which is identical to b'_1. At the same time, c_1 shrinks (due to absence of repulsive forces between hydrogen atoms along parallel edges of the mantle and the core clathrates in b'_1) and interacts with surrounding water molecules leading to the formation of another new (mantle) clathrate as shown in c'_1 which is identical to c'_2 and b'_1.

mother substance. The inertia of the clathrate hydrate is additionally increased due to its connections, via hydrogen bonds, with water molecules of the water layers immediately after the clathrate hydrate.

The molecule of the mother substance at its new, relocated position (Fig. $8b_2$) interacts with water molecules and forms a new clathrate hydrate identical to the initial one (Fig. $8b'_2$). At the same time, the left-over empty clathrate (Fig. $8b_1$) acquires a more compact structure than before (due to the absence of the hydrophobic repulsive forces between mother substance and water molecules of the clathrate). Simultaneously, the water molecules of the water layer immediately after the clathrate are affected by the symmetry of the shrunken clathrate (now called core clathrate) and form another clathrate (called mantle clathrate) imitating the shape of the previous clathrate vertex by vertex (Fig. $8b'_1$). The mantle clathrate is apparently looser than the core clathrate due to the apparently larger distances between its oxygen atoms located at the mantle-clathrate corners forming hydrogen bonds with hydrogen atoms located along its edges. The size of these distances is further increased due to the repulsive forces between hydrogen atoms of the two clathrates, especially between pairs of hydrogen atoms along pairs of parallel edges of the core and mantle clathrates.

In Fig. 8c the third stage in the preparation of a high dilution is presented, where the separation of the core clathrate from the mantle clathrate is obtained, when a forceful mechanical impulse is applied on the dilution (Fig. $8c_2$ and c_1). This is possible due to the different inertial kinematic behavior between the compact core clathrate (behaving like a solid) and the loose mantle clathrate (behaving like a liquid with an increased viscosity), whose inertia is additionally increased due to its connections, via hydrogen bonds, with water molecules of the water layers immediately after the mantle clathrate.

The core clathrate at its new, relocated position (Fig. $8c_2$) interacts with water molecules, and forms a new mantle clathrate identical to the initial mantle clathrate (Fig. $8c'_2$). The left-over empty clathrate (Fig. $8c_1$) acquires a more compact structure than before, since, due to the absence of the core clathrate, shorter distances between oxygen atoms at the vertices of the mantle clathrate are permitted. Simultaneously the water molecules of the water layer immediately after the clathrate are affected by the symmetry of the shrunken clathrate (now becoming core clathrate) and form another (mantle) clathrate imitating the shape of the previous clathrate vertex by vertex (Fig. $8c'_1$). This new mantle clathrate is identical to the mantle clathrates of Fig. $8b'_1$-c'_2.

An essential feature of the model is the assumed stability of clathrates involved (full or empty). Theoretical and experimental [6-8] studies, however, lend support to this assumption. It is finally going to be tested simultaneously with the experimental efforts to detect possible physico-chemical properties of high dilutions discussed in the next paper [9].

4. Conclusions

In the framework of the clathrate model for an understanding of water high dilutions, the formation of clathrate hydrates around the molecules of the mother substance in water is essential. In general, a specific clathrate hydrate is formed around each molecule of a given mother substance, which is different for each different substance. Substances which do not form such a structured water around the solute, according to the model, cannot form high dilutions as well. Also, in obtaining such dilutions the application of forceful mechanical impulses on the dilution is absolutely necessary. Thus, both formation of clathrate hydrates and mechanical impulses above a certain threshold of strength, characteristic of the specific substance, are required. Hence, in the clathrate model, a high dilution in water possesses a structure that of the relevant clathrate hydrate or of a net (via hydrogen bonds) of such clathrates.

5. References

[1] G.A. Jeffrey, R.K. McMullan, "The clathrate hydrates", and references therein. *Progress in Inorganic Chemistry* **8**, 43-107, 1967.

[2] A.W. Castleman, Jr., "Probing the structure and dynamics of cage-like clusters: from water to Met-Cars". In *Atomic and Nuclear Clusters*, pp 160-166. Eds G.S. Anagnostatos and W. von Oertzen (Springer-Verlag, Heidelberg, 1995, ISBN 3-540-59160-5).

[3] I. Olovsson and P-G Johnsson, "X-ray and neutron diffraction studies of hydrogen bonded systems". In *The Hydrogen Bond*, Chapter 8, p. 413. Eds. P. Schuster, G. Zundel, and C. Sandorfy (North-Holland Publishing Co., Amsterdam, 1976, ISBN 0-7204-0315-4).

[4] W.C. Hamilton and J.A. Ibers, *Hydrogen Bonding in Solids*, pp. 196, 198 (W.A. Benjamin, Inc., New York, 1968).

[5] G.S. Anagnostatos, In *1992 Omeomed Proceedings Book*, pp. 77-86. Ed. C. Bornoroni (Editrice Compositori, University of Urbino (PS), 1992).

[6] R.J. Stanley, M. Cook, A.W. Castleman, Jr., "Unimolecular dissociation rate constants; chlorobenzene cations revisited by using a new method". *Journal of Physical Chemistry*, **94**, 3668-3674, 1990.

[7] X. Yang and A.W. Castleman, Jr., "Large protonated water clusters $H^+(H_2O)_n$ (1≤h>60); the production and reactivity of clathrate-like structures under thermal conditions". *Journal of the American Chemical Society*, **111**, 6845-6846, 1989.

[8] X. Yang and A.W. Castleman, Jr., "Production and magic numbers of large hydrated anion clusters $X^-(H_2O)_{n=0.59}$ (X=OH, O, O_2, and O_3 under thermal conditions". *Journal of Physical Chemistry*, **94**, 8500-8502, 1990.

[9] G.S. Anagnostatos, "Experimental detection of the existence of structures in high dilutions". See these lectures.

STRUCTURE-FUNCTION PROPERTIES OF WATER CLUSTERS IN PROTEINS

DORIT ARAD*, KEREN MOSS, YUVAL ELIAS and GAI ANBAR
Department of Molecular Microbiology and Biotechnology, Tel Aviv University
Ramat Aviv, Israel

Introduction

The study of water clusters is an ever growing interdisciplinary area of interest in physics, chemistry and biology. The properties of clusters in the gas phase have been thoroughly studied, giving rise to the existence of distinct water clusters as a new kind of chemical moiety (1). The structures geometries, and the energetics that relate to small water clusters have been calculated by high level ab-initio calculations and compared to experimental values. Some of the calculations are shortly reviewed here to demonstrate the complimentarity between the gas phase and calculation results to the types of water cluster structures observed in protein structures.

In recent years a bulk of scientific evidence has been accumulated indicating that liquid water is not an inert solvent, but rather alters the structural properties of small molecules and proteins (2). Furthermore, the structure of water itself is not amorphous, and appears to consist partially of clusters. (3). By considering water a structured moiety, light may be shed on many physical chemical and biological phenomena presently unsolved.

The tetrahedral SP_3 oxygen of water resembles the tetrahedral SP_3 carbon (Fig. 1). The change in the chemical features between the two is mainly due to the change in electronegativity, and the fact that carbon forms four covalent bonds, in contrast to only two bonds formed by oxygen. The weaker hydrogen bonds, formed by the two lone pairs of oxygen, however, provide the possibility to the formation of more labile structures.

*To whom correspondence should be addressed

The water oxygen, thus, may be regarded as a "virtual carbon" which nature has formed in order to transfer information.

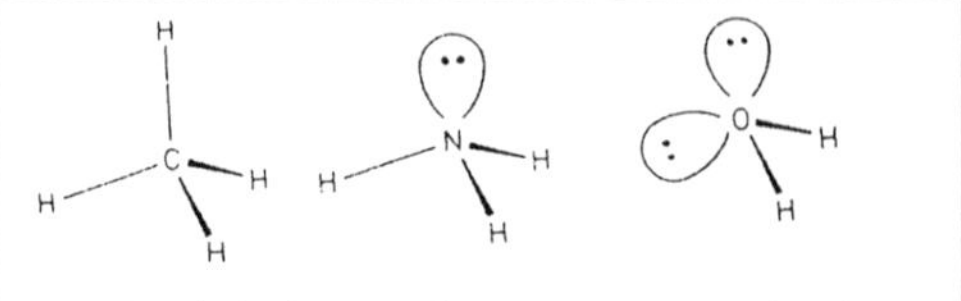

Fig. 1: *SP_3 hybridization of oxygen, nitrogen and carbon*

Structures built from several water molecules clustered into a distinct form have been observed in ice, as well as in the gas phase (1). In many cases water is organized in symmetrical shapes general to other tetrahedral species. Water clusters feature spatial geometrical properties similar to those of organic molecules. This is illustrated by the adamantane-like structure of water in ice (Fig. 2).

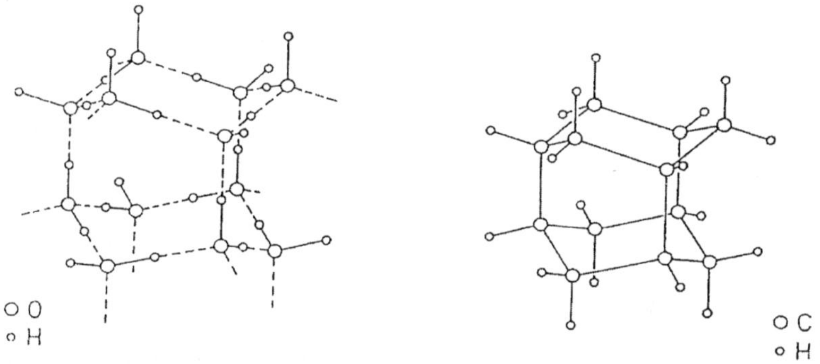

Fig. 2: *adamantane, and adamantane-like structure of water in ice.*

Calculations and experiments

The predicted geometries and energetics of small water clusters have been calculated by ab-initio calculations (5) and compared to experimental gas-phase values. Laasonen et al. (6) used Car-Parrinello method which enables the treatment of large systems, using ab-initio molecular dynamics. They used the gradient corrected density function, which was proved adequate for describing hydrogen bonding. The most stable conformation for $(H_2O)_3$ was shown to be a slightly distorted equilateral triangle. This result is in agreement with other calculations using Gaussian and the Hartree Fock (HF) approximation (7) (see Fig 3). The preferred geometry for $(H_2O)_4$ was found to be a flat ring. Hexamers are found in several equilibrium geometries, similar in energy. The lowest energy species has a chair-like ring structure, although the number of possible conformers increases considerably. The structures were grouped into classes according to their ring size. The lowest energy structures are composed of five four membered rings. Excluding this structure, the rest of the conformations were close in energy. Thus it was concluded that several structures undergo interconversion at room temperature (8).

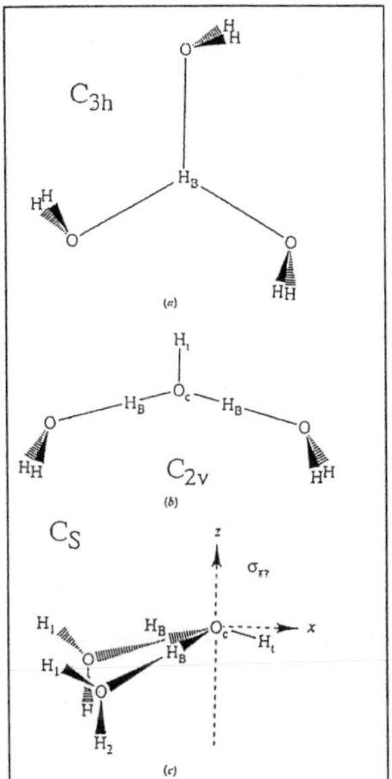

Fig. 3: *Possible structures of water trimers. The cyclic structure is preferred.*

Burke et al. (9) used the HF wave function using 3-21G, 6-311G** basis sets to study $(H_2O)_5$ clusters. Four structures which correspond to a single ring and bipyramidal structures were obtained. The enthalpy difference was at least one Kcal/mol, favoring the single ring system. Entropy favored, however, the bipyramidal structures.

For $(H_2O)_8$ the three dimensional arrangement was more stable than the ring structure (10). A cubic structure was found to have the lowest energy.

The existence of 'magic numbers' for cluster size is known from noble gas chemistry, where distinct 'magic numbers' of Argon atoms, Helium atoms, etc. have been observed (11). In the case of water, a cluster of 20 water molecules, $(H_2O)_{20}$ has been detected. It has been predicted that the cluster is arranged in a cage-like conformation, namely a clathrate (12). A dodecahedron clathrate consists of a regular dodecahedron with one water molecule at each of the 20 vertices and an additional H_2O in the center.

Jurema et al. (13) studied the structure, energetics and hydrogen bond pattern of the $(H_2O)_{20}$ cluster, which was assumed to have a clathrate structure, with and without

H_3O^+. They used Mopac semi-empirical calculations, including correlation. The initial geometry was a pentagonal dodecahedron, placing the ion caged in different regions within the sphere. $(H_2O)_{20}$ was found to have a minimal energy structure of a symmetrical clathrate, with 30 hydrogen bonds. For $(H_2O)_{20}H^+$, however, 6 minima were found, representing various distortions from the pentagonal dodecahedron. More recent electron diffraction experiments (14) failed to reveal the predicted dodecahedron structure experimentally. Clusters around n=20 did not show the clathrate structure. Electron diffraction data agreed best with a model of amorphous ice clusters, which are composed of a network of distorted rings with three to six hydrogen bonded molecules.

Sukhodub et al. (15) studied the thermodynamic properties of small water clusters in the gas phase, using temperature dependent field mass-spectrometry. For the trimer, the predicted geometry is that of a strained cyclic structure. For larger clusters the energy minima were also those of planar cyclic structures, in which each water molecule acts as a proton donor and acceptor. The equilibrium structure of water is dynamic, however, due to the numerous degenerative structures.

In a recent study published in Science (16), it was demonstrated that water, in the gas phase, can form asymmetric clusters that consist of three or more molecules in a cyclic arrangement, similar to the most stable calculated three-membered water cluster (fig. 4). The chiral structures are interconverted from one mirror image to the other via the tunneling effect. The interaction of such chiral structures with chiral molecules enforces chiral clusters, whereas the interaction with a racemic mixture destroys them.

Structured water in proteins

Water may play a role in protein folding, in determining the tertiary conformation, stabilizing active conformations, and participation in the mechanism of action of proteins (17).

It has been estimated from the location of the oxygen atoms that more than 95% of water actively participates in constructing the protein geometry, by forming hydrogen bonds between themselves and to the protein, and by arranging in distinct forms around hydrophobic residues.

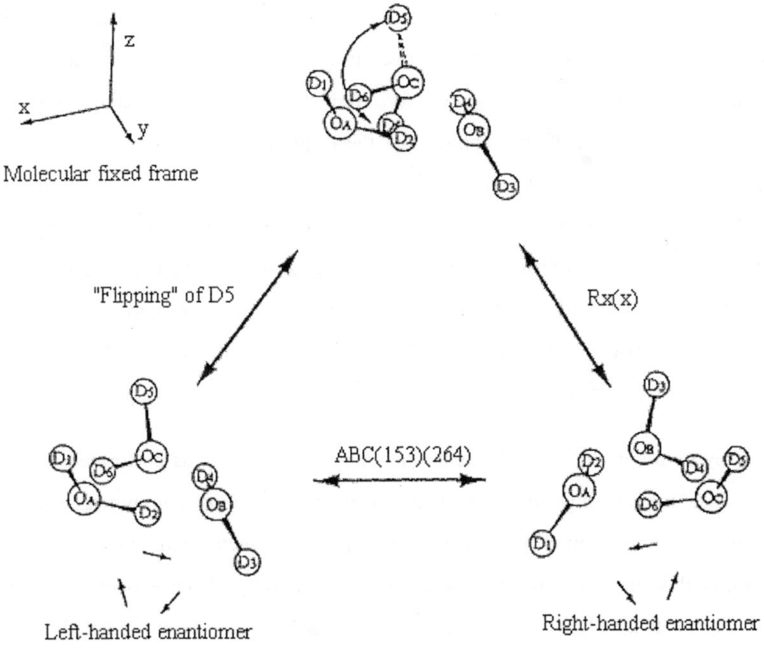

Fig. 4: *Tunneling interconversion between chiral water clusters.*

Proteins usually contain 25% to 90% water in their structure, also in their crystal form. From X-ray analysis the positions of the oxygen atoms of water are being detected. A fundamental question arises when considering water in crystallized forms of proteins: how many of the water molecules conserve their positions in the proteins, and persist in free solution? NMR spectroscopy, a very rapidly developing method for protein structure analysis in solution, has detected, in several cases, internal water molecules seen crystallographically (18).

Information about the structure of water in solution is currently very limited. Although protein suspensions are not considered as bulk solution, they may provide some insight to the structure of liquid water because of their high water content and due to their high degree of functionality to the protein.

We have raised a number of questions regarding the aqueous solvent molecules, which constitute such a considerable percentage of protein mass in both crystal and solution forms. Are these solvent molecules arranged in clusters? If they are, what are the most common forms and sizes of these clusters? And do these clusters interact with groups belonging to the protein? We approached these questions by utilizing protein three-dimensional structure data, collected by X-ray and NMR structures, from the Brookhaven crystallographic database. The tool we used was to statistically locate water clusters in a set of different proteins for which structures are available and analyze the common ones according to their shape and size. This information may provide a clue to the question whether distinct water clusters that mimic the original shape and some physical properties of an organic molecule exist in solution and affect protein structure.

A survey of water clusters in proteins including 40 proteins was performed. We have found that trigonal rings of water clusters constitute the most common geometry in proteins. They are found as single or multiple three-membered clusters, including bipyramidal forms. The five-membered ring also show preference over other forms. Interestingly, these observed geometries in proteins have also been predicted to be the most stable ones by calculations and preferred over linear and opened structures (19) (see above).

As mentioned above, cyclic trimers of water have been the subject of study in the tunneling shift between chiral water conformers (16). The tunneling effect is believed to be quenched when the cluster interacts with an achiral molecule, and to enforce when interaction with chiral molecules occurs. Since the interaction of water clusters with the protein moiety is mostly of chiral nature, the chiral clustering water observed in our statistical survey may induce chirality upon interaction and may therefore be involved in the process of protein folding. The results may imply that the interaction of water clusters with proteins is of a different nature than the interaction with single water molecules. One of the unique properties of clusters in general in relation to a bulk of atoms or molecules is the large extent in which the molecules of the cluster are exposed to the surface. This property imposes a new chemistry (20). For example, both hydrogen and ammonia molecules react when they strike clusters of iron atoms. They do not interact with bulk iron.

Water clusters in proteins, thus, especially if chiral, may be regarded as a unique kind of molecular form, capable of affecting protein structure and function, through cluster formation, destruction and specific interaction. A good example to this "cluster machinery" is presented in fig. 6.

An interesting structure of water clusters was found in a neutron diffraction study of vitamin B12. Neutron diffraction enables the positioning of the hydrogen atoms of water since the resolution in this method is less than 1.0.. In regular X-ray crystallography of proteins only heavy atoms are observed in a resolution of 1.4-3.0.. Vitamin B12 has a pocket and a channel completely filled with water molecules, forming a continuous water layer. The water molecules detected in this study are completely clustered, comprising of an array of 4, 5 and 6 membered rings. In addition, partial clathrate cages are observed in the pocket region. Clathrate cages usually form around apolar groups when the water molecules tend to maximize the number of hydrogen bonds that are being formed (21).

The most exciting insight on water arrangement in proteins is gained by observing clusters of water molecules at the active site of enzymes, the structures of which have been determined by X-ray crystallography.

Most proteins, including enzymes, transport proteins and immunoglobolins, manifest biological activity by binding to other molecules or ions. The binding site often takes the form of a cleft or depression in the protein surface and is thus likely to provide a favorable environment for bound solvent molecules. Most importantly, the water molecules in the active site may be of profound importance to protein function, either due to their participation in binding or catalysis, or because they must be displaced by an incoming substrate which exhibits a tighter binding (22).

An ordered solvent has been found in the active site of many enzymes. The bacterial serine protease SGPA, whose active site contains 24 water molecules (!) provides an excellent example (23). The most striking feature in this case is the near-perfect match in the electron density patterns, as detected by X-ray diffraction, between the solvent molecules and the polar groups of the bound tetrapeptide substrate (23). A chain of clustered water molecules (i.e., connected within themselves in a specific pattern) is ordered approximately parallel to the polypeptide chain that forms the S1 to S3 binding sites, and another cluster occupies a hydrophobic pocket that binds the product's benzyl group.

A very similar pattern is seen in lysosyme, where the active sites of both the human and turkey egg white enzyme contain similar distributions of bound water molecules (clusters of n = 12 - 15 water molecules) (24). Comparison with the hexasaccharide binding models show that these bound water molecules mark the positions that will be occupied by the polar groups of the bound substrate.

Another striking example is found in the biotin-avidin system. Avidin is a tetrameric protein that binds biotin with exceptionally high affinity. This system is one of the strongest natural binding systems known in proteins. It was observed that avidin alone (without the substrate) exhibits binding activity similar to that observed with biotin.

It was detected by X-ray crystallography that five water molecules occupy the binding site region of biotin, comprising an electron density map similar to that of biotin in the enzyme-substrate complex (25). The five water molecules are thus distributed in the active site in a similar way to the arrangement of biotin (fig. 5). It was concluded that the water molecules serve, in this case, to enhance the affinity between biotin and avidin, by forming a pre-organized structure. The local conformational changes and entropy loss accompanying regular binding processes are thus unnecessary when biotin binds avidin.

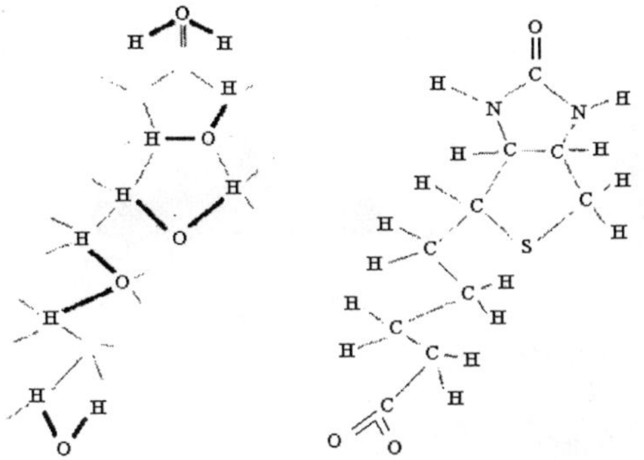

Fig. 5: *Biotin molecule, and its water cluster counterpart.*

Several other proteins, such as arabinose binding protein, show similar behavior. The interesting feature in these, is that the cluster of water molecules at the active site mimics not only the active site arrangement, but also the exact shape of the sugar moiety.

The role of water clusters in protein function

The role of water clusters in protein function, i.e. in the biological and chemical reactions that occur, is a subject still under investigation. Kong et al. (26) studied a possible error mechanism in the genetic code replication as a consequence of tautomer formation. They proposed that the proton transfer that causes tautomerization takes place through the mediation of water clusters. They found that the mutagenesis probably occurs in a double proton transfer mechanism. Krauss et al (27) have studied the energetics of "direct" as oppose to "through water" binding of the cis-diaminoplatinum adduct to DNA, namely the $Pt(NH_3)_2^{2+}$ - H_2O - $H_2PO_4^-$ cluster. They used molecular mechanics calculations and found the two conformations (direct and through water) to be energetically competitive. They concluded therefore that mediation of the drug through water clusters is a possible mechanism.

Krauss and Stevens (28) later performed ab-initio calculations of the binding energy of metal cations to octahedral clusters of water. They used the conformation of a single shell cluster to emulate the first coordination sphere in water or a protein. Using this model, they found the cation-binding selectivity of the protein to be correlated with the calculated energies required for transferring different cations from a water cluster to the protein model. This result, based on a reliable calculation method indicates the direct effect of water clusters on the binding properties of proteins.

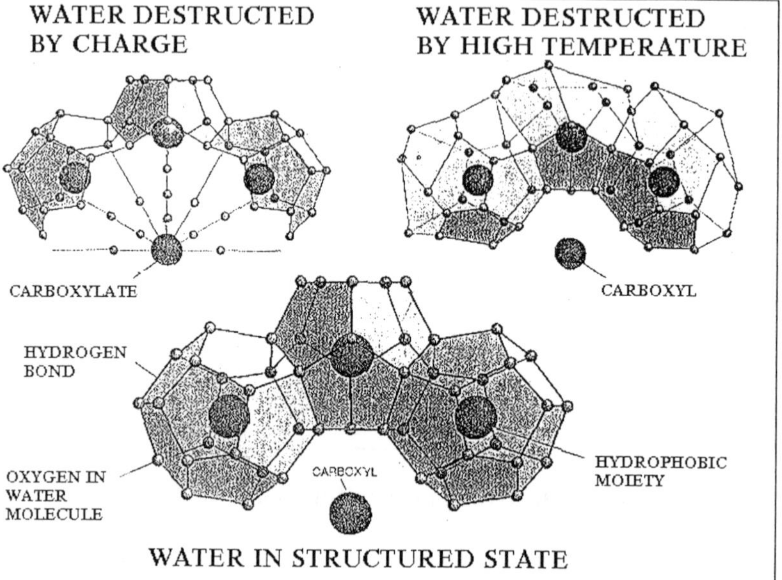

Fig. 6: *water cluster machinery for expanding-contracting mechanism in elastin.*

Urry (29) studied the behavior of proteins which have the ability to stretch and contract such as elastin, which is found in the connective tissue of mammals. He noticed the connection between the water content in the protein and its ability to contract efficiently as a function of the temperature rise. He explained that the water around hydrophobic amino acids is arranged in a series of pentagonal structures of water molecules. These structures help keeping the polymer in an extended shape. As the temperature rises the pentagonal structure is disrupted and becomes amorphous. The protein can now accommodate a folded shape and contracts. He showed that this change of the amount of "ordered" water in the protein can be triggered by different stimuli, which resemble biological stimuli, such as pH change and electric fields induced. If these factors affect both water clustering and protein function it is reasonable that water clusters have a role in affecting protein structure-function relationship.

The role of water in active sites

Water molecules in well organized forms that mimic the shape of the real ligand occupy the active sites of proteins - where chemical reactions occur. The question is what role do these water molecules play at the active site. Is it solely pre-organization, that allows the native substrate to enter the active site more easily, or can they also replace in some conditions the real molecules - introducing inhibitory effects? The work of Burlakova (30) hints towards the second possibility. Burlakova has found that a series of selected inhibitors for PKC inhibit the enzyme activity in very low concentration of the substrates (10^{-15}), i.e., the numbers of real molecules in each sample is very low. Only a group of inhibitors from one family for PKC exhibits this phenomenon. It is attractive to speculate that the inhibition effect is caused by water clusters that arrange in structures similar to the original substrate molecule. This speculation is supported by the numerous cases of structural information showing substrate-like structures of water molecules in active sites of enzymes. Open questions, however, remain regarding the limited stability and high mobility of water clusters in solution. Can the clusters be seeded by a real molecule, and undergo subsequent transfer to a remote protein site? Can the protein molecule exhibit some degree of functionality when water, instead of a "real" substrate enters its active site? The current information about water clusters in solution has not yet developed to an extent where we can state such an hypothesis as a realm. Yet, facing a scientific period in which imaginary processes are actually being observed by sophisticated physical methods, and in which new molecular structures, previously not considered classical, are defined as new molecular types, we expect these questions to obtain a serious response in the near future.

References:

1. Clusters of Atoms and Molecules, H. Haberland, Ed., Springer Verlag, Berlin, 1994.
2. Water and Biological Macromolecules, E. Wedthof , Ed., Topics in Molecular and Structural Biology, **Vol. 17**, Macmillian Press, France, 1992.

3. H.F.J. Sabage, ref. 2, pp. 3.

4. W.C. Hamilton, B. Kamb, S. Laplaca and A Prakasb, "Structural chemistry and molecular biology". Freeman, San Francisco, pp. 466., 1969.

5. L.S. Sremaniak, L. Perera, M.L. Berkowitz, *Chem. Phys. Lett.*, **218**, 377-382 (1994).

6. K. Laasonen, M. Parrinello, R. Car, L. Changoil and D.Vanderbilt, *Chem. Phys. Lett.*, **207**, 208-213 (1993).

7. E.P.F. Lee and J.M. Dyke, *Molec. Phys.*, **73** (2), 3755-405 (1991).

8. C. J. Tsai, K.D. Jordan, *Chem. Phys. Lett.*, **213**, 181-187 (1993)

9. L.A. Burke, J.O. Jensen, P.N. Kirshnan, *Chem. Phys. Lett.*, **206**, 293-296 (1993).

10. K. Laasonen, M. Parrinello, R. Car, L. Changoil and D.Vanderbilt, *Chem. Phys. Lett.*, 207, 208-213 (1993).

11. J. Yortner D. Scharf U. Landmann in: "Elemental and molecular clusters", G. Benedek, T.P. Martin, G. Pacchioni Eds., pp148, Springer, Berlin, (1988).

12. G.A. Jeffrey, *Acc. Chem. Res.*, **2**, 344-352 (1969).

13. M.W. Jurema, K.N. Kirshner and G.C. Shields, *J. comp. chem.*, **14**, 1326-1332 ((1994).

14. J. Farges, M.F. de Feraudy, B.Rault and G. Torchet, *Adv. Chem. Phys.*, **70**, part 2, 45 (1988).

15. L.F. Sukhodub, Y.V. Telezhenko and V.S. Shelkovskii, *Zh. Strukt. Khim.*, **29** (5), 89094 (1988).

16. N. Pugliano and R.S. Saykally, *Science* , **257**, 1937-1940 (1992).

17. "Protein solvent interactions" R.B. Gregory, ed., Marcel Dekker Inc., NY (1995).

18. K. Wuthrich, *J. Biol. chem.*, **265**, 22059-22062 (1990).

19. P.A Karplus and G.E. Shultz, *J. Mol. Biol.*, **195**, 701-729 (1987).

20. Ref. 1, pp. 3-15.

21. H.F.J. Savage, Nature, **322**, 717-720 (1986).

22. Ref. 17, E.N. Baker, Solvent interactions revealed by X-ray crystallography, pp. 143-189 (1995).

23. G.N.M. James, A.R. Sielecki, G.D. Brayer, L.T.J. Delbaere and C.A. bauer *J. Mol. Biol.* 200, 523-551 (1980).

24. C.C.F. Blake, W.C.A Pulford and P.J. Artymiuk, *J. Mol. Biol.*, **167**, 693-723 (1983).

25. Coordinates available from O. Livne, private communication. Ref: O.Livne, E.A. Bayer M. Wilchek and J.L. Sussman, *Proc. Natl. Acad. Sci. USA*, **90**, 55076-5080 (1993).

26. Y.S. Kong, M.S. John and P.O. Lowdin, *J. Quant. Chem.*, Quantum Biol., Symp., **14**,189-209 (1987).

27. M. Krauss H. Basch, K.J. Miller, *J. Am. Chem. Soc.*, **110**, 4517-4521 (1988).

28. M. Krauss and W.J. Stevens, *J. Am. Chem. Soc.*, **112,** 1460-1466 (1990).

29. D.W. Urry, *Scientific American*, **272**, 44-49 (1995).

30. N.P. Pal'mina, E.L. Mal'tseva, N.V. Kurnakova and E.B. Burlakova, *Biochemistry (Moscow)*, **59** (2), 135-139 (1994).

EXPERIMENTAL DETECTION OF THE EXISTENCE OF STRUCTURES IN HIGH DILUTIONS

G.S. ANAGNOSTATOS
Institute of Nuclear Physics, National Center for Scientific Research "Demokritos"
Aghia Paraskevi, Attiki, Greece

ABSTRACT

For a verification of the clathrate model, introduced in the previous paper, the dielectric and the differential scanning calorimetric methods are applied. The first method is capable of verifying that a high dilution contains free and bound (e.g., in the form of clathrates) molecules and the second that a change of phase occurs when the clathrates are destroyed (freeing their bound molecules) at a certain temperature. This temperature depends on the specific clathrate involved which is characteristic of the specific mother substance. Thus, while many more methods should be applied before a final conclusion is drawn, it is encouraging that the results of both methods applied here support the predictions of the clathrate model.

1. Introduction

By following the procedure described in the previous paper [1] (i.e., that of sequential dilutions and application of mechanical impulses on each dilution) we can reach to any high dilution we may wish, where the solution could be thought of as effectively composed of only empty clathrates and of molecules of the solvent. In this respect, since the clathrates are also made of molecules of the solvent, high dilutions could be thought of as comprised of two phases of the solvent, i.e., the one of the free molecules of the solvent and the other of the bound molecules of the solvent in the form of clathrates. Thus, experimental techniques distinguishing free from bound molecules of the solvent (e.g., dielectric measurements [2-4] (DRS)) or techniques detecting change of phase occurring, for example, during heating of the solution (e.g., differential scanning calorimetric (DSC) measurements [5]) are appropriate for testing the predictions of the clathrate model [1]. Both techniques have been applied and others are planned. The results are really in favor of the clathrate model. Indeed, by performing experiments by DSC on high dilutions of several substances in water, we observed that at a certain temperature (different for each substance and slightly different for each dilution of the same substance) of the solution a peak appears characteristic of a change of phase. That is, at this temperature the structures of the clathrates break and thus the molecules of the solvent bound in the form of clathrates become free molecules like the remaining molecules. Also, our repeated DRS experiments show the existence of bound and free molecules of the solvent in high dilutions in consistency with our DSC experiments. However, many more experiments using the above and additional techniques should be performed before an unambiguous conclusion is drawn on this extremely sensitive matter of high dilutions.

What we have obtained so far and we expect to verify again in the future is that high dilutions, while chemically no different from the initial solvent, have physically detectable properties due to local changes of the structure of the solvent obtained during the procedure applied for the preparation of these dilutions. It is not included in the purpose of our research to investigate, e.g., whether the human organism is sensitive to the above properties of high dilutions or, even if so, if they have any therapeutic value.

2. Dielectric Measurements

Dielectric methods are very powerful in hydration studies. They make use of the different dielectric relaxation time τ of hydration water molecules compared to that of free water molecules due to disturbance of their microdynamics by the solute molecules.

It has been pointed out in a previous paper [2] that one may avoid problems in the analysis of the measurements by using the depolarization thermocurrent (DTC) method. In the DTC method the measurements have to be carried out in frozen solutions at subzero temperatures. For interpretation, the results of these measurements have to be compared with those of DTC measurements in pure ice.

The DTC method is as follows [3,4]. The sample is polarized by an applied electric field E_p at a temperature T_p. This polarization is subsequently frozen in by cooling the sample to a temperature T_0 sufficiently low to prevent depolarization by thermal energy. The field is then switched off and the sample is warmed at a constant rate b, while the depolarization current is detected by an electrometer. In the case of a single relaxation process obeing the Arrhenius equation $\tau(T) = \tau_o \times \exp(W / k_B T)$, the depolarization current density J(T) is given by Eq. 1

$$J(T) = (P_o / \tau_o) \exp(-W / k_B T) \times \exp\left(-\frac{1}{b\tau_o} \int_{T_0}^{T} \exp(-W / k_B T') dT' \right), \qquad (1)$$

where τ is the relaxation time, W the activation energy of the relaxation, τ_0 the preexponential factor, T the absolute temperature, k_B Boltzmann's constant, and P_O the initial polarization. The analysis of the shape of this curve makes it possible to obtain the activation energy W, the pre-exponential factor τ_0, and the contribution $\Delta\varepsilon$ of the peak to the static permittivity [3,4].

The DTC apparatus (of the National Technical University of Athens) used in the present study consisted of a cryostat, in which the sample temperature could be varied from 77 to 400 K, at heating rates between 0.8 and 15.0 K min^{-1}. The measuring capacitor was made of brass. The sample temperature was measured by a copper-constantan thermocouple attached to the upper (grounded) electrode. The current was measured with a Keithley 610C electrometer and recorded, simultaneously with the temperature of the sample, with a YEW 3083 XY recorder. The temperature was controlled by a Barber-Colman 520 controller.

We repeatedly performed dielectric measurements on several samples rather shortly after their preparation. The samples were chamomile extract prepared by exclusive use of de-ionized, double distilled water at room temperature (i.e., no alcohol was used at any stage of their preparation).The dilutions employed were 0c, 1c, 6c, 12c, and 30c, where c denotes that centesimal dilutions were used (i.e., each new dilution was obtained by adding one drop of the just previous dilution into 99 drops of double distilled, de-ionized water and then applying a number of forceful mechanical impulses) and 0,1,6,12,30 stand for an identification of the mother tincture (here chamomile extract) and the first, sixth, twelveth, thirtieth dilution.

The results of dielectric measurements were positive for all samples in the sense that a substantial percentage of bound water molecules was detected in all cases as shown in Figs 1 and 2, and discussed below.

Figure 1 presents three curves labelled H_2O, 0_c and 30c. The H_2O curve stands for DTC measurements of pure water and its characteristic peak at about 130°K is used as a reference for the other two curves and, in general, for any other curve derived from such measurements as, e.g., for those shown in Fig. 2.

The 0_c curve in Fig. 1 stands for DTC measurements of the mother tincture (here chamomile extract). It is noticeable that its peak at about 130°C presenting unbound water (since its temperature about coincides with that of the reference peak for unbound, pure water in curve labelled H_2O) is shorter (it corresponds to less pA) than the second peak (at about 150°K presenting bound water (since it does not appear at all in the reference curve H_2O for unbound water). This difference in the height of the two peaks means that a greater percentage of water molecules are

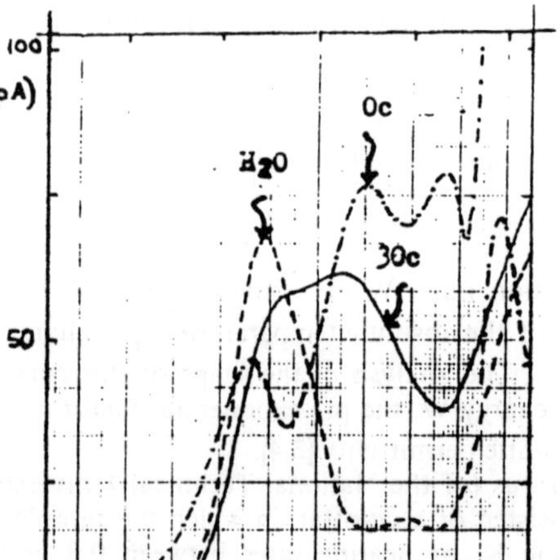

Fig. 1 Dielectric measurements plots for chamomile-extract water solutions of 0c and 30c potencies together with the reference plot for water.

bound than unbound in the extract of the mother tincture. This, of course, is not surprising since in our dense chamomile extract many water molecules, e.g., in the form of clathrate hydrates are bound (i.e., they have restricted motion with respect to molecules far apart from "impurity" - chamomile origin - molecules).

The 30c curve stands for similar measurements of the homonymous dilution. Again, like in 0c curve, two peaks appear at about 130°K and 150°K. The first stands for unbound water and the second for bound water in slightly higher percentage. Additional peaks at different temperatures in all three parts of Fig. 1 are not meaningful for the present discussion. The basic meaning of Fig. 1 is that a large percentage of bound to free water molecules exist in the dense 0c solution, where they are many "impurity" molecules coming from chamomile, but also in the 30c solution where effectively there is not even one "impurity" molecule. This important conclusion is consistent with the predictions of the clathrate model[1], where the "impurity" molecule and the core clathrate can form new clathrates during the application of mechanical impulses in the solution [1].

Figure 2 presents also three curves labelled 6c, 12c, and 30c whose notations have already been explained. Two main remarks could be drawn from this figure. The one is that the peak of bound water, relatively to the peak of free (unbound) water, is more pronounced for the lower dilutions in such a way that the free water peak hardly can be seen in the curve 6c. This could be thought of as related to the presence of molecules coming from chamomile which becomes more and more as we go to lower and lower dilutions. The presence of such molecules in lower dilutions

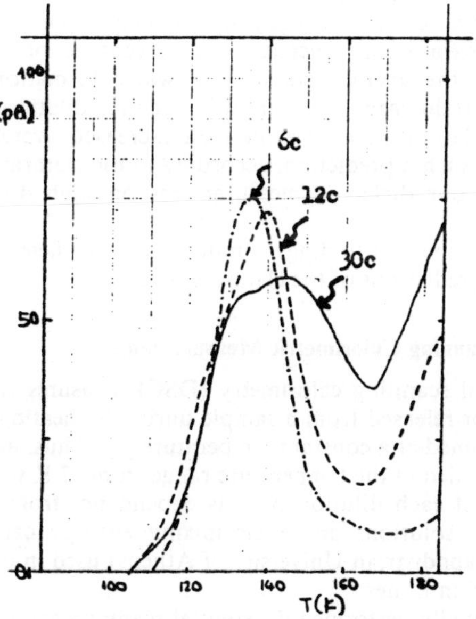

Fig. 2 Dielectric measurements plots for chamomile-extract water solutions of 6c, 12c and 30c potencies. Going from lower to higher potencies, a shift of the plot to higher temperature is noticeable.

seems that makes the creation of new water formation, during the application of forceful mechanical impulses, on each solution more and more efficient than the absence of such molecules in higher dilutions. Of course, more and more water formations (e.g., in the form of clathrates) means more and more bound water molecules. The reason for observing bound water molecules in the dilutions 12c and 30c apparently has no relationship to molecules coming from the chamomile extract which, for these dilutions, are practically zero. The appearance of bound water in these dilutions, however, is explained by the clathrate model [1].

The second remark is that the peak of bound water shifts more and more to the right (i.e., to higher and higher temperature) as we go to higher and higher dilutions. This could be attributed to the formation of stronger and stronger structure of water formations, made of bound water molecules for example in the form of clathrates, as we go to higher and higher solutions. Such possibility is permissible in the framework of the clathrate model as has been discussed in Ref. [1].

By assuming hydrophobic forces between water and molecules coming from chamomile, water molecules are bound among themselves via hydrogen bonds and form a filled clathrate around each chamomile molecule (like that presented by Fig. 1 a-b'_2 of Ref. [1]), which is energetically weaker than an empty clathrate formed by water molecules alone (like that presented by Fig. $1c_2$ of Ref. [1]). This is reasonable, since the absence of hydrophobic forces in the case of an empty clathrate makes the corresponding structure

However, another possible reason for an increased stability of water formations in high dilutions is the presence of a huge number of clathrates in these solutions. Indeed, when the number of clathrates is substantial with respect to the number of free water molecules, extra hydrogen bonds may occur among the clathrates themselves. Such extra bonds increase the percentage of bound to free water molecules and also the average stability of water formations. Thus, a solution resulting from a certain large number of sequential dilutions (which has a huge number of empty clathrates) may exhibit an increased average stability of water formations. Indeed, such a prediction, according to the clathrate model, is consistent with the results of our dielectric measurements on high dilutions of chamomile extract.

Hence, all experimental findings coming from our dielectric measurements are consistent with the predictions of the clathrate model.

3. Differential Scanning Calorimetric Measurements

The differential scanning calorimetry (DSC) measures the amount of energy which is absorbed or released from a sample during its heating or cooling, or when the sample is maintained at a constant temperature [5]. Thus, any change of phase of a material can be studied in the temperature range from 77 K to 900 K. In the case of solutions, the heat of each dilution ΔH, its dependence from the temperature, the thermal capacity of solutions and their mixing entropy can be calculated. The apparatus (of the Kapodistrian University of Athens) used in our experiments was a DSC-4 model of Perkin Elmer.

We have repeatedly performed differential scanning calorimetric measurements on the same samples with those used in dielectric measurements. That is, we used again chamomile extract of 0c, 1c, 6c, 12c, and 30c dilutions. The results of our DSC measurements were positive for all samples in the sense that a change of phase was observed in all cases (Fig. 3). This change of phase, according to the clathrate model, implies the destruction of the clathrate structures at about a certain temperature. It is noticeable that this temperature shows a small shift to higher values as we proceed from low to high dilutions. This shift reminds one of the similar shift observed in the dielectric measurements discussed above and, according to the clathrate model, has the same explanation.

By using the DSC technique the following two additional series of experiments were performed. The one was a repetition of the previous experiments on the same sequential dilutions of chamomile extract but without applying any mechanical impulse on any dilution. For all dilutions no detectable effect was observed. Thus, mechanical impulses are absolutely necessary for the appearance of a measurable physicochemical effect. Of course, up to 12c dilutions some molecules (and their clathrates) coming from the chamomile extract are present, but there are not enough of them for a sizeable (detectable) effect by the DSC technique. The mechanical impulses, according to the clathrate model, multiply the number of clathrates and make their effect detectable.

The other additional series was a repetition of the same experiments but using de-ionized, double distilled water as mother substance in the place of chamomile extract and applying again a number of mechanical impulses on each dilution. That is, the 1c dilution is obtained by adding one drop of de-ionized, double distilled

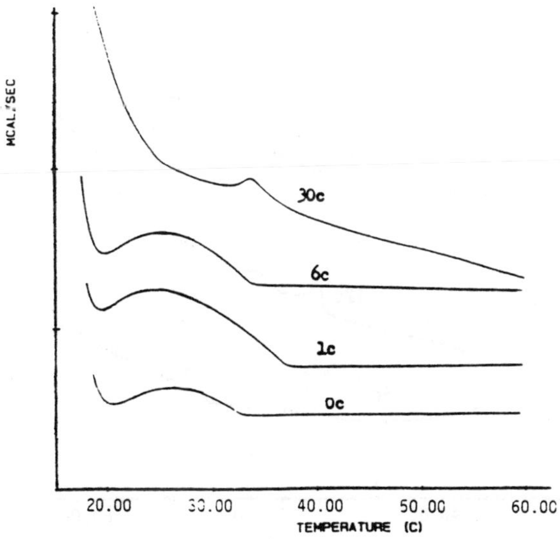

Fig. 3 Differential Scanning Calorimetric plots for chamomile-extract water solutions of 0c, 1c, 6c, and 30c potencies.

water into 99 drops of the same water and then applying a number of mechanical impulses (the same as previously) on this dilution. The higher dilutions are also obtained by following the standard approach, that is each diution is obtained by taking one drop of the previous dilution and adding this drop into 99 drops of double distilled water and then applying the same as before number of mechanical impulses. For all dilutions (i.e., 0c, 1c, 6c, 12c, and 30c) again no effect was observed. Thus, the presence of a mother substance for the preparation of a high dilution is essential, a fact which is also consistent with the predictions of the clathrate model, according to which the initial (full) clathrates are formed round the molecules of the mother substance (see Fig. 1a-b'$_2$ of Ref. [1]), which specifies the kind of clathrates formed. Without these initial clathrates no further clathrates can be formed.

Another series of DSC experiments was performed by employing methanol (CH_3-OH) as mother substance [6] in the place of chamomile extract. All other parameters were kept the same as for the experiments described above. Again the results were positive with respect to the predictions of the clathrate model. That is, again a change of phase is observed at a certain temperature, where according to the model a destruction of the clathrate structures occurs (Fig. 4). However, it is not clear here if an upward shift of this characteristic temperature occurs by going from lower to higher dilutions, as it has clearly been noticed in the chamomile DSC experiments.

What is important in this series of experiments is the difference, with respect to

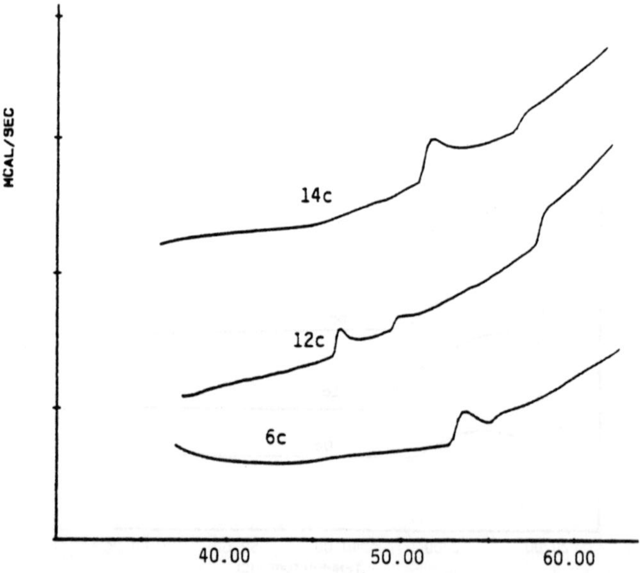

Fig. 4 Differential Scanning Calorimetric plots for methanol (CH_3-OH) water solutions of 6c, 12c, and 14c dilutions.

the chamomile experiments, in the characteristic temperature where we observe the change of phase. That is, while this temperature is about 24°C for the chamomile extract, it is about 52°C for the methanol. This difference in temperature constitutes a further justification of the clathrate model. Indeed, for a different mother substance, a different clathrate is expected which, in general, requires a different amount of heat in order to be destroyed. Since all masses involved in the experiment are kept constant (i.e., the same masses for the chamomile extract and for the methanol experiments), the different amount of heat necessary for the destruction of the different clathrates involved corresponds to different temperature.

In Fig. 4 we observe two instead of one peak for the 12c dilution. While many other explanations could exist, the following is a plausible one. That is, that the initial water formation (clathrate) is transformed during the procedure of preparation or of heating into two simpler (and thus probably thermally weaker) clathrate structures. Each of these two structures is sensitive at its own temperature.

Finally, a preliminary experiment was done for another 1c dilution by using agaricus muscaricus as mother tincture. Again a change of phase is observed and this time at a temperature which is different than both characteristic temperatures observed above for the previously described chamomile and methanol experiments. Indeed, this characteristic temperature of change of phase constitutes a way to differentiate among high dilutions of different substances.

Hence, all differential scanning calorimetric findings are consistent with the predictions of the clathrate model and also with the previous findings of dielectric measurements.

4. Conclusions

All experiments performed so far either by applying Dielectric Measurements or Differential Scanning Calorimetric Measurements are in favor of the predictions of the Clathrate Model. That is, that high dilutions in water, before and after Avogadro number, have a clathrate-hydrate structure which is different mainly for different mother substances and slightly different from dilution to dilution of the same substance. It is this structure which makes the high dilutions meaningful. That is, while these dilutions of a substance have identical chemical composition with that of the relevant solvent (here water), they have physically detectable properties, in addition to the properties of the solvent, revealable by following standard scientific methodology.

However, despite the experimental successes described here and elsewhere [7], due to the very delicate character of the matter under investigation, we emphasize that many more series of experiments using different mother substances should be performed by employing the present and other experimental techniques before a final conclusion is drawn.

5. References

[1] G.S. Anagnostatos, "On the structure of high dilutions according to the Clathrate Model". See these lectures.

[2] P. Pissis, D. Diamanti, and G. Boudouris, "Depolarisation thermocurrents in frozen aqueous solutions of glucose". *Journal of Physics* **D16**, 1311-1322, 1983.

[3] C. Bucci, R. Fieschi, and G. Guidi, "Ionic thermocurrents in dielectrics". *Physical Review* **148**, 816-823, 1966.

[4] J. Vanderschueren and J. Gasiot, "Field-induced thermally stimulated currents". In *Topics in Applied Physics*, **37**, ch. 4, pp 135-223; "thermally stimulated relaxation in solids". Ed. P. Braunlich (Springer, Berlin, 1979).

[5] See references in: *Inclusion Compounds*. Eds. J.L. Atwood, J.E.D. Davies, and D.D. MacNicol (Oxcord University Press, Oxford, 1991).

[6] The experiments on methanol constitute an unpublished work. These experiments and all previous work on chamomile extract have been performed as a team project in cooperation with Professors K. Viras (Kapodistrian University of Athens) and P. Pissis (Technical University of Athens). Specifically first experiments on high dilutions were performed using dielectric measurements on chamomile solutions by Prof. P. Pissis in 1991, while experiments by using differential scanning calorimetric measurements on the same solutions and on methanol solutions were performed by Prof. K. Viras and his assistant Ms Maria Soudgidou in 1992 and 1995, respectively.

[7] G.S. Anagnostatos, P. Pissis, and K. Viras, "Possible water cluster formation by dilution and succussions". In *Atomic and Nuclear Clusters*, pp. 215-217. Eds. G.S. Anagnostatos and W. von Oertzen, (Springer-Verlag, Heidelberg, 1995, ISBN 3-540-59160-5).

A NEW PARADIGM APPLIED TO HIGH DILUTION EFFECTS ON THE LIVING BODY

MADELEINE BASTIDE
Laboratory of Immunology, Faculty of Pharmacy,
University of Montpellier 1, France.

and

AGNES LAGACHE
Lycée Carnot, 145 Bd Malesherbes
75017 Paris, France

ABSTRACT

We can present the paradigm of signifiers as following: living beings communicate with their world in a non verbal way, whether on a somatic or psychological level. This paradigm takes place within the framework of the logic of analogy. The signifier is the semantic object which materially designates information to be transmitted and dealt with: the homeopathic remedy is the mimetic representation of the disease. Different levels of information organize the spread of signifiers; each level is the regulation and integration of the previous one. The living self is the never-ending process whereby levels of information are synthetized, in the face of the informing environment. Such structures meet each other in the communication between the patient and the physician-remedy system. The medical device has to reinform the patient and makes his symptoms move on towards a higher level of integration. The dilution of the remedy permits us to read it as information about disease. Symptoms can be recognized as an erroneous adaptation and the organism is engaged in a process of paradoxical negation. The action of the remedies consists of a dynamic analogy between pieces of information. The paradigm of signifiers gives therefore a new possibility in the exploration of informative therapeutics.

1. Introduction

The mechanistic paradigm has been so fruitful that we think it is the only law of rationality. However, other facts necessitate the creation of new ways of thinking but we still need to be rational. Each kind of interaction has its own logical modalities. Just as the mechanistic paradigm seizes the material interactions between objects add the symbolic paradigm assumes linguistic acts, we need a paradigm that allows the

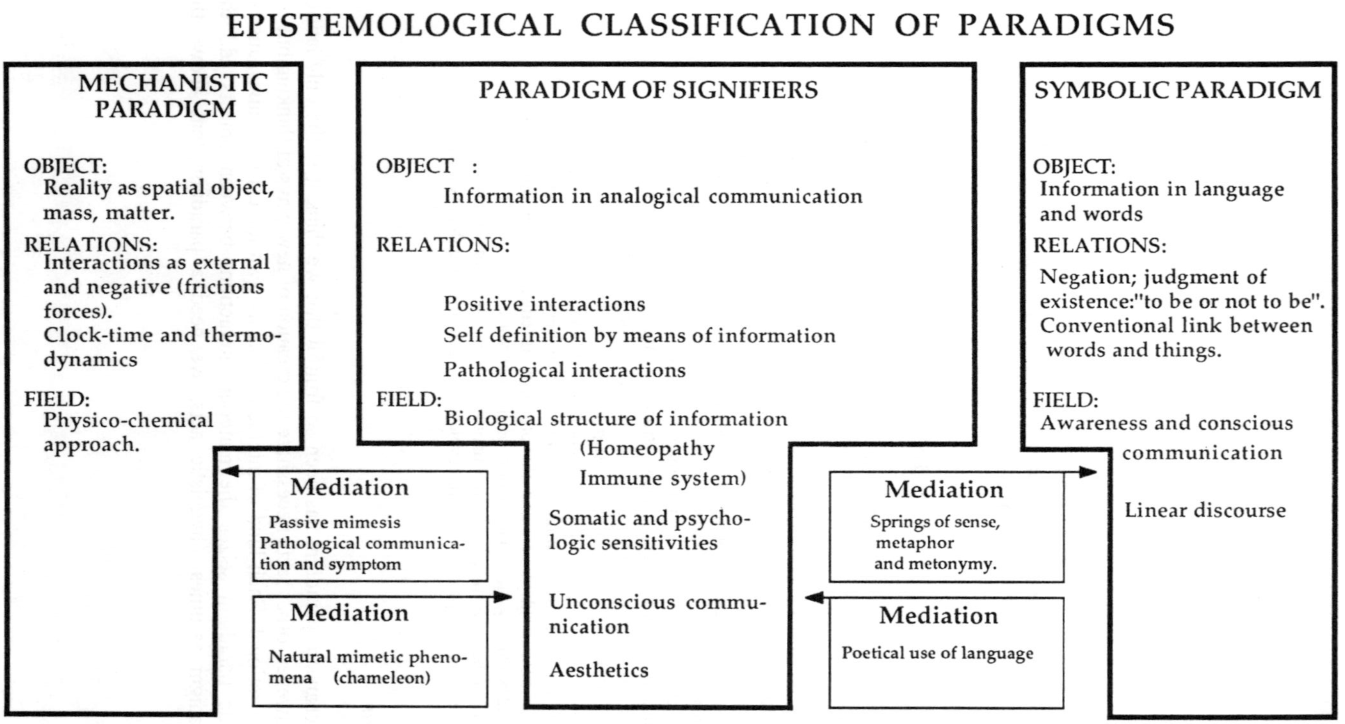

Figure 1 : Relationships between the paradigm of signifiers and the mechanistic and symbolic paradigms [1]

understanding or the structuration of new concepts applied in concrete experiments or in therapeutic applications. As such, we can also present the paradigm of signifiers [1, 2, 3, 4]: living beings communicate with their world in a non verbal way, whether on a somatic or psychological level. This paradigm takes place within the framework of the logic of analogy. Its application can be used to interprete homeopathic experimentation or the therapeutic effects of homeopathy. It is also very useful in understanding immunological concepts. We will limit this paper to the domain of homeopathic interpretation in the framework of this paradigm.

2. The living body is not an object

Our way of thinking is inherited from positivism:"everything is matter, according to the very restricted definition of the matter provided by the mechanistic paradigm. Consequently, it was thought that every problem (or every object) was solvable by reducing its terms (or itself) into simple material elements. Matter rules over everything."[5]. If we consider the properties of the object according to the mechanistic paradigm, it is homogenous and it may be separated into its different elements; it does not change except by wear of time; it may be different in terms of its spatial position and it is characterized by autarky. The object responds to thermodynamic laws and is, by approximation, not related to its environment; it is a physico-chemical material [2]. These characteristics are not acceptable for the living body.

On the contrary, the living body is heterogenous and cannot be separated into its different elements as it functions as a whole. This represents a major difficulty for biological science. Interrelationships between the different biological systems of the body are so strong that the body must be considered as a whole. However, scientists traditionally have always dealt the problem of individual biological systems. Moreover, the living body is different before and after the events that it comes accross and it is continuously modified as a function of time. The body has a psychological memory but a physical one too as shown by the immune system: it learns how to fight against agressions and remember them. The body is in a continual and irreversible learning process. It shows interactions at every level from psychic to physical as demonstrated for example in psycho-neuro-immunology. It is never independent from its environment and is in continuous relationship with the external world [2]. Therefore, it is able to receive *information*. The living body is not an object and needs to communicate with the environment [6].

3. What is information ?

According to the Bateson's definition, information is "a difference which makes a difference" [7]. A piece of information is not an object, whatever the medium used to transfer it. "A lost jigsaw piece in a children's room has no meaning, except to make the difference between a tidy and untidy home. Putting it in its jigsaw's frame with the other pieces makes a new meaning arise: it was this pretty star which was lacking in the sky in the left corner.... Sense is an emigrant, the new and impalpable dimension which arises not from objects, but from the network of relationships woven around objects" [6]. Information is not an object even though a carrier of that information may exist. The result of information is only observed in the receiver and depends on the capacity of the receiver to read and treat this information: the receiver is usually active. Information is only read in a context which gives a sense to the information which therefore becomes significant to the receiver. These pieces of information which have a sense are designed as signifiers: when this information is perceived by the living body, we can then call them *semantic objects* (or physical signifiers): they will transfer information to the living body. The body will then deal with this information and modify its behavior accordingly. This communication is within the framework of *mediation.*

4. The notion of mediation

This notion concerns two different things whose relationship has traditionally been ignored. For example, in modern science, body and mind are two separate things, each one isolated in its own camp. Some attemps to join them together exist, but the symbolic paradigm is unable to explain objects (function of molecules, biochemistry of physiological regulations, etc...) The mechanistic paradigm would like to explain mind, memory and behavior exclusively in terms of interactions between molecules and receptors. A good example is given by psycho-neuro-immunology where one part of the research analyses the interactions of neuropeptides with the immune system and another part studies immunological products like cytokines and their interactions with the nervous sytem or on behavior. It is quite difficult to approach the real synthesis between these events.

On the contrary, mediation creates a link which discriminates both elements."We need something which links mind and body, creating a bridge between physical and semantic facts, whilst maintaining the distance and difference between them: that is a mediation"[5]. Mediation takes place before the objects that arise from it: it is a creative element which unables different levels of reality to exist according to their own identity, while working together in nature. This concept of mediation provides a way to understand creative organisations and temporal regulations [5]. The high dilution informative function read and treated by an organism is a mediation. Examples of mediations are given by psychoanalysis and freudian concepts which reveal the somatisation of psychological facts. Other examples of mediation which show *semantic objects* acting in the high dilution model are given in section 9 .

5. Signifiers for the body: semantic objects

The semantic object materially designates information to be transmitted as a signifier for the body to deal with. For example, according to the similia principle, the homeopathic remedy is the mimetic representation of disease.This representation is understood by the body. Another example is given by aesthetics in which concrete phenomena (music, painting) are meaningful to the body. An extreme example is given by Aldrige who uses music in therapy applied to coma patients or in Alzheimer's disease [8, 9, 10]. In this approach, the body receives musical information within a rhythmic context of communication which enables healing to take place. In the above cases, the musical components are fundamental to the body's communication and these play a role by their rhythmicity. According to the level of the pathology, different levels of rhythmicity as carriers of information may act upon the body and represent different levels of semantic objects.

This definition belongs neither to the mechanistic paradigm concerning spatial objects and the physico-chemical approach nor to the symbolic paradigm which implies information by language, words and conscious communication.The living body may follow other rules which follow the principles given by the paradigm of signifiers. We design semantic objects as signifiers like objects carrying information to the body.This information may be carried by successed homeopathic dilutions for example: the information of the remedy is not related to the presence of the molecules but to the passage of information given by the molecules after dilution and succussion. Another example is provided by the immune system: for example, anti-idiotypic antibodies considered as the internal image of the antigen are informing proteins able to give to the body the notion of the tumorigenic effect of a chemical as shown by Faiderbe et al [11].

The *semantic object is a concrete element, neither symbolic nor linguistic but treated by the body as a piece of information.* It fulfils the function of a mediation. The semantic object may induce a learning process which may be memorized by the living organism.

6. The particularity of the homeopathic model

Basic scientific experimentation or clinical applications using homeopathic cures are characterized by the technique used to prepare solutions (potentized solutions prepared by serial dilutions which are successed vigorously in a dynamisation process). These solutions may be diluted until they contain no more molecule and are designated as high dilutions. The second originality of the product tested in the experimental model or chosen as a remedy lies in its choice. It always needs to be "recognized" by the receiver (cell, animal or patient) according to either the law of identity, or the similia principle or as an endogenous molecule [4].

7. Semantic objects induce a learning process

The logic of exchanging information does not follow the same rules as an exchange of objects where one loses and the other gains; the exchange of information creates a new situation which is not the sum of exchanges."It is a dynamic process, through which the partners change and build themselves by dealing with information: the very mediation creates what it mediates. Reasoning in terms of mechanistic elements is not enough to understand life, but we assume that: -first, living structures imply mechanical organization combined with semantic facts; - secondly, mediations are the dynamic elements which make living structures work as a whole"[5]. We can therefore consider semantic objects as pieces of information which function as a whole and cannot be reduced into its elements.

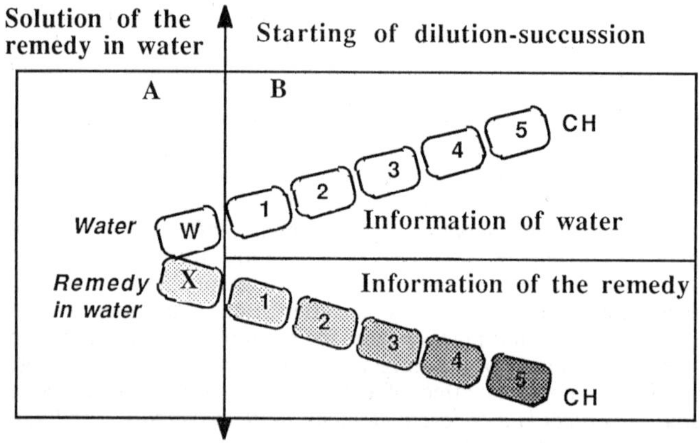

Figure 2: The information given by the remedy is not the sum of the [information of Water] + [information of the remedy]. The sources of information are different.
Part A: Water is the control of the remedy
Part B: Succussed water is not the control of the remedy

An example may be given by the choice of the "control" in experimentation on high dilution effects. Reasoning mechanistically, in order to assess that the effect is the difference between succussed remedy and succussed solvent, the control should be the succussed solvent . But what is the succussed solvent in reality ? it is an autonomous substance which gains properties by succussion and cannot be compared to the succussed remedy. This remedy (designated as substance X) becomes after dilution and succussion the informative preparation of X (fig.2). The only true controls are either unsuccussed solvent or still better the unsuccussed diluted substance X.

8. Consequences of the treatment of information by living organisms

In order to communicate, the living organism needs to establish a representation of the world. The most simple way for it to do this is by using analogical model designated by mimesis. This means that the image of the object is the easiest representation and only possible one. We consider for example that a diluted succussed solution prepared with a chemical carries the mimetic representation of the effects of that chemical on the healthy body: this effect on the healthy body is described as pathogenetic symptoms used in similia principle in homeopathy. When the organism receives the information that corresponds to its symptomatic manifestation, it recognizes it and deals with it.

Consequently, regulatory effects will occur by modifying the symptoms of the body, if the representation brought by the successed remedy is relevant to the symptoms, i.e. if the similarity is good. This phenomenon is named *active mimesis*, which means that the organism was able to treat the information and to modify itself by switching to "new" normal state. This process is designated by paradoxical negation which means that the symptoms disappear once the body receives and deals with information. The method used to reach this point is related to the level of information.

When the organism is able to receive and recognize the information but is not able to treat it, we observe a *passive mimesis* which looks as if the symptoms have increased rather than being corrected (worsening effect). Examples of worsening are currently observed in experimental models based on hormesis or the activity of endogenous molecules.

These points have raised the question of the regulatory effect of active mimesis which is organized according to the level of information. This hierarchy is organized according to the level of complexity of information (see examples in the chapter 9).

9. Examples of successed high dilutions as semantic objects

We will validate the use of the paradigm of signifiers by demonstrating the effects of semantic objects in experimental models described in an other paper [4].

Hormesis and the law of identity: the most simple learning process is the hormetic model designed traditionally by "mithridatization". This phenomenon implies an opposite effect of high concentrations (toxic effect) and low concentrations (stimulating effect) (figure 3). It is the most classical model allowing to demonstrate the biological effect of high dilutions [4]. Pretreatment either by low doses or high dilutions prevents the toxic effect. This situation is typically a learning process with or without a short memory. The living organism (from procaryots to insects and to mammals) prepares its defence according to the nature of the aggression: a heavy metal needs heat shock proteins as specific defence effect[12, 13, 14] and increases the vitality of the organisms or the cells [14].

In the same way, a pharmacological compound like aspirin will induce a specific effect when received in allopathic concentration. This effect is received as an aggression by the body: the Arndt Schultze law is an observation related to hormesis taken in its general meaning. Logically, according to the notion of information, high dilutions of aspirin will induce the opposite effect allowing the defence of the organism against the aggression of ponderal concentrations of the remedy, whatever their therapeutic effect. Effectively, it has been observed [15] that high dilutions of aspirin had a statistically significant thrombogenic effect which corresponds to the opposite classical therapeutic effects which is anti-thombogenic when used in allopathy. This understanding of hormesis in the framework of the paradigm of signifiers may be represented as shown in figures 3 and 4.

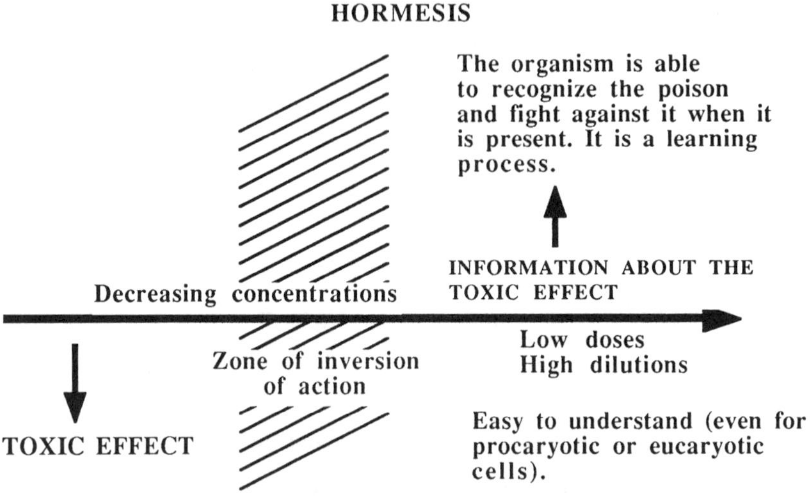

Figure 3: Hormesis interpreted in the framework of the paradigm of signifiers. The information about the poison (or all ponderal administration) induces an opposite effect to counteract the toxicity with the appropriate tools. This opposite effect is a function of concentration.
The pretreatment by low doses or high dilutions prepares the organism by informing it of the danger before it arrives. It has a very short memory.

Endogenous molecules: some experimental models are performed by using high dilutions of endogenous molecules, belonging to the organism itself (figure 4). The information is automatically read thanks to genomic structure. Different models have been tested: histamine on isolated heart [16], thymulin in immunodeficient mice or bursin in bursectomized chickens [17, 18], and thyroxin during the frog metamorphosis. The results are significant and have been reproduced. In these three models, the information was clearly read by the receiver and the living structure was able to treat this information. The possibility of a strong worsening effect exists if there is a misunderstanding which will causes a passive mimesis. The clearer the information, the better the correcting effect. An organism that is too weak or a piece of information that is too strong can provoke irreversible worsening effects.

Homeopathy and the similia principle: The diluted succussed remedy chosen according to the pathogenesy of the remedy (which describes the symptoms observed after administration of this remedy to a healthy patient) is prescribed according to the similia principle (figure 4). The more similar the symptoms of the patient and the remedy, the more effective the remedy . The medical device has to re-inform the patient and makes his symptoms move on towards a higher level of integration. The dilution of the remedy permits to read it as information about disease. Symptoms can be recognized as an erroneous adaptation and the organism is engaged in a process of paradoxical negation. The action of the remedies consists of a dynamic analogy between pieces of information [1, 2, 3] .

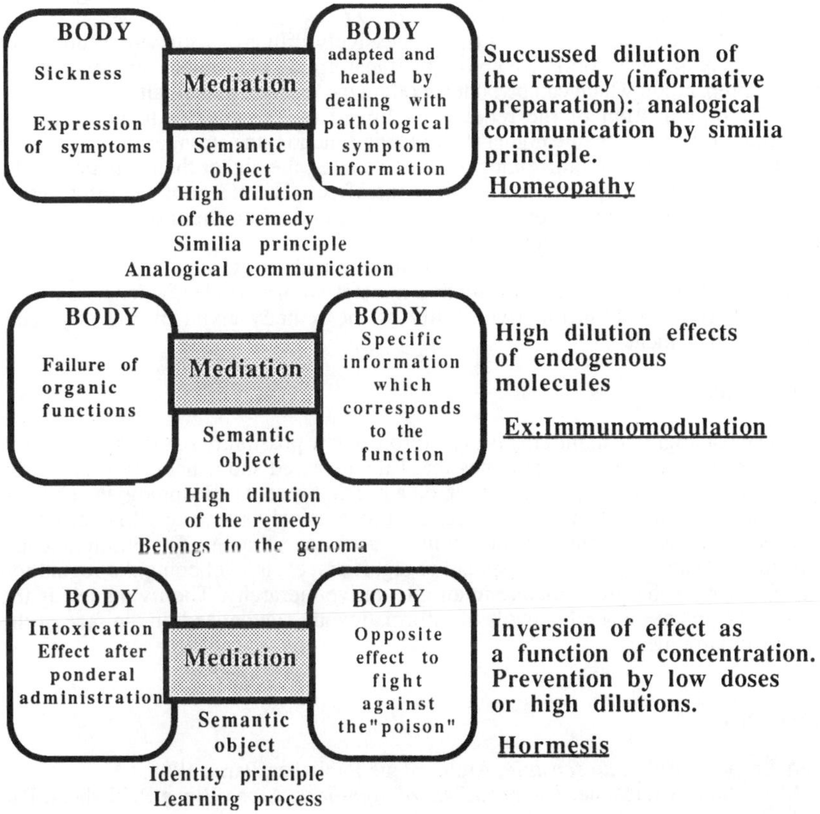

Figure 4: Examples of semantic objects in high dilution effects

10. The hierarchy of semantic objects as informative structures

The application of the paradigm of signifiers to the homeopathic model is determined by the relationship between several parameters due to the informative ability of the remedy.

The power of the homeopathic remedy is related to its capacity to be recognized as a signifier by the receiver in a mimetic function; an exogenous remedy needs to be evaluated by a pathogenesy (effect of ponderal doses on a healthy subject) in order to establish the mimetic function of the remedy. However an endogenous remedy (which belongs to the self molecules) will be automatically recognized by the receiver.

The dilutions which contain the successively diluted successed solutions of molecules increase the informative ability by transforming the "object-remedy" into an informing preparation which can be understood by the receiver as a signifier.

The organization of the receiver is linked to its capacity to perceive the information of the remedy: the higher the mimetic function, the stronger the effect as in the "sensitive types"; regarding endogenous molecules, the higher their function in the hierarchy of the living system, the stronger the effect. The effect of the information given by endogenous molecules can be misunderstood when the information is not targetted, and a worsening effect may occur.

The regulating effect of the homeopathic remedy prescribed according to the similia principle is linked to an informative transfer whose efficiency depends on the difference between the informative ability of the remedy given by the physician compared to the patient.

10. Conclusion

The paradigm of signifiers gives therefore a new possibility in the exploration of informative therapeutics. One imagines that Life appeared spontaneously four billion years ago, from or near the archaïc ocean, on a beach. Since the beginning, the ebb and flow of the ocean was an informative carrier progressively allowing all structures to interact with the environment in natural diluted-successed solutions. The different levels of information have organized the spread of signifiers, each level being the regulation and integration of the previous one in an integrative hierarchy. The living self is the never-ending process whereby levels of information are synthetized, in the face of the informing environment.

References

[1] A.Lagache, *Echos du sensible*, Alpha Bleue Publisher, Paris, 1988.
[2] M.Bastide, A.Lagache, *The paradigm of Signifiers,* Alpha Bleue Publishers, Paris, 1992.
[3] M.Bastide, A.Lagache, C.Lemaire-Misonne, "Le paradigme des signifiants: schème d'information applicable à l'Immunologie et à l'Homéopathie", *Revue Intern. Systémique*, 9, pp 237-249, 1995.

[4] M.Bastide, "Basic research on high dilution effects", see the corresponding chapter in this book.
[5] A.Lagache,"Notes on the conceptual basis of Science", *Signal and Images*, M.Bastide Ed, Kluwer Academic Publisher, Dordrecht, in press.
[6] A.Lagache,"What is Information", *Signal and Images*, M.Bastide Ed, Kluwer Academic Publisher, Dordrecht, in press.
[7] G.Bateson, *Steps to an ecology of mind*, Chandler Publisher, New-York, 1972.
[8] D.Aldridge,"Music, Communication and medicine: discussion paper", *J. Royal Soc. Med.*, 82, pp 743-746, 1989.
[9] D.Aldridge, G.Gustorff, H.J.Hannich, "Music therapy applied to coma patients", *J. Royal Soc. Med.*, 83, pp 345-346, 1990.
[10] D.Aldridge, "Music and Alzheimer disease- assessment and therapy: discussion paper", *J. Royal Soc. Med.*, 86, pp 93-95, 1993
[11] S.Faiderbe, J.L.Chagnaud, M.Bastide, P.Dorfman, M.Geffard, "Preventive effect of a monoclonal anti-idiotypic antibody compared to a hormetic model on rat sarcomas", *Signal and Images*, M.Bastide Ed, Kluwer Academic Publisher, Dordrecht, in press.
[12] M.Oberbaum and J.Cambar, "Hormesis; dose dependent reverse effects of low and very low doses", pp 5-19, in *Ultra High Dilution, Physiology and Physics,* Endler and Schulte Eds, Kluwer Academic Publisher, Dordrecht, 1994.
[13] R.Van Wijk, M.Welters, J.A.Souren, H.Ovelgonne, F.A.Wiegant. "Serum-stimulated cell cycle progression and stress protein synthesis in C3H10T1/2 fibroblasts treated with sodium arsenite", *J.Cell.Physiology*, 155, pp 265-272, 1993.
[14] A.Delbancut, "Contribution à l'étude des effets de hautes dilutions de métaux vis-à-vis de la cytotoxicité du Cadmium sur des cultures de cellules tubulaires rénales", *Thèse Université Bordeaux II, France*, Juillet 1994.
[15] C.Doutremepuich, O.Aguejouf, D.Pintigny, M.N.Sertillanges and O. De Seze, "Thrombogenic properties of ultra-low-doses of acetylsalicylic acid in a vessel model of laser-induced thrombus formation", *Thrombosis Reseach*, 76, pp 225-229, 1994.
[16] L.Hadji, B.Arnoux and J.Benveniste, "Effect of dilute histamine on coronary flow of guinea-pig isolated heart. Inhibition by a magnetic field". *FASEB*, n°7040, 1992.
[17] B.J.Youbicier-Simo, F.Boudard., M.Mekaouche, M.Bastide., J.D.Baylé. "Effects of embryonic bursectomy and *in ovo* administration of highly diluted bursin on adrenocorticotropic and immune response of chickens", *Int.J.Immunotherapy,* 9, pp 169-180, 1993.
[18] M.Bastide, "High dilutions as a tool of immunomodulation", see the corresponding chapter in this book.
[19] P.C.Endler, W.Pongratz, G.Kastberger, F.A.C.Wiegant, J.Schulte, "The effect of highly diluted agitated thyroxine on the climbing activity of frogs", *Vet.Hum.Tox.*, 36, pp 56-59, 1994.

PHARMACOLOGICAL FREQUENCY TRANSFER

M. CITRO
Istituto di Ricerca "A. Sorti", IDRAS, via Passalacqua, 19, Torino, Italia

1. Introduction

Our research is based on previous experiments concerning biological effects produced by high dilution preparations [1-4] and resonance phenomena in the medical practice of electroacupuncture [7]. A homeopathic remedy connected to an EAV (Electroacupuncture according to Voll) apparatus enters in resonance frequency with the acupuncture spots of the patient and modify its lectures. The homeopathic remedies seem to emit storable signals. They can be picked up by an appropriate device either in contact or near (≤ 5 mm) the preparations and transmitted by cable or radio [6]. F. Morell and E. Rasche have built an apparatus named MORA (see below). It is possible with this device to transfer by cable to a patient the RF emitted by a homeopathic remedy. Until 1989, this transfer seemed possible only with the homeopathic preparations [5]. In 1990, we abandoned the restricted field of homeopathy and for the first time we transferred electronically the therapeutical properties of drugs. Our hypothesis is that each drug emits continuously and spontaneously specific signals bearing a therapeutical information (MIS: Metamolecular Informed Signals) [3]. The following sections describe the essential of our finding of the effects of Pharmacological Frequency Transfer (TFF) on various life forms, *i.e.* humans, animals and vegetals.

2. Materials and Methods

The figure 1 illustrates the electronic circuitry as well as the method's scheme. An amplifier (MORA III 200, Med Tronik, Friesenheim, Germany) was used. It has a gain from 1 to 40 and a post-amplifier x 12 (total gain from 1 to 480). It has a minimum frequency distortion fed by accumulators. It intercepts both entry background noise and signals coming from the drug, amplifies the resonance frequencies and transfers them with selectable periods; we used periods of 7 s amplification with pauses from 3 s for a total of 15 min. The input is connected to a brass support where the drug to be transferred is placed, and the output is connected either to metallic supports with bottles containing various preparations (*e.g.* aqueous, ethanolic or physiological solutions), or to brass electrodes placed in contact with the patient. The information is transmitted along insulated coaxial cables.

The patients can receive the therapeutic frequencies directly from the exit electrode, or indirectly by sublingual or parenteral administration of activated water. The choice of the correct drug to be given for curing a particular disease is based on the conventional medical investigation. Once the right drug was found and gave positive results, the conventional treatment was stopped and substituted by the TFF of the same drug.

The animals chosen were amphibian. The parameter of choice were their phase development. Their aqueous environment was treated with TFF of an hormone. The vegetals were simply treated with water containing herbicide TFF information. The parameter chosen was the length of the sprouts. All tests were made with references that were in identical conditions of temperature, humidity and light (see also the report by Vinattieri, Picard and Citro, this volume).

This method, that we have conceived and patented, is called Pharmacological Frequency Transfer or TFF (Trasferimento Farmacologico Frequenziale). In our work the initials TFF before the name of a molecule means that the drug was administrated by electronic transfer. The figure 1 illustrates the basic principles of this method.

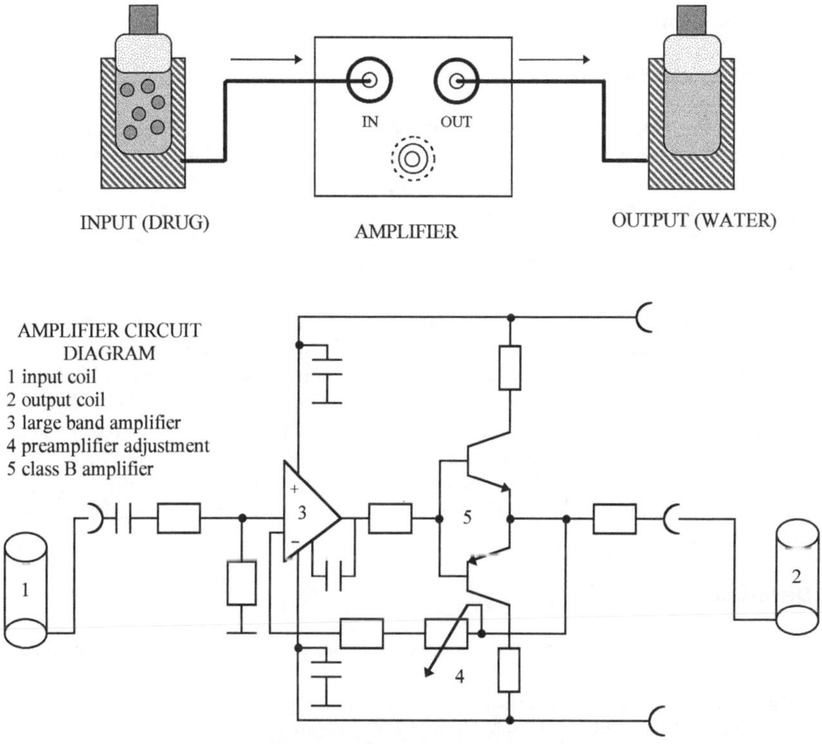

Figure 1. Above: set up for the TFF. Below: circuitry of MORA III 200.

3. Clinical experiences

On May 4th 1990 we realized the very first Pharmacological Frequency Transfer (TFF) of tetracycline to a cat dying of *Haemobartonella*. The cat rapidly and totally recovered from the disease (and is still healthy today). As a consequence of this success, treatments on humans were started. Since then we have collected a high number of cases treated with TFF of drugs belonging to different classes. Until now we have obtained our best therapeutic responses with TFF of antibiotics, anti-inflammatories, analgesics, benzodiazepin, bronchodilators, progestin, oestroprogestin associations, antihistaminics, antidepressants, hormones and opiates [2].

We wish to point out that in two cases the problem was heroin addiction. In both cases, the abstinential fits have been treated successfully with intravenous injections of TFF of heroin in physiological solution. It must be noted that in some cases, the TFF of drugs was also administrated with success to patients that were not acquainted of the nature of the original drug.

In March 1995 we have presented a collection of 89 cases (74% of positive response) treated with anti-inflammatories and analgesics TFF at the *6th Interscience World Conference on Inflammation, Geneva, Switzerland.* In total are more than 150 different cases that were treated and cured by TFF.

4. Laboratory experiments

From the series of positive results, we started in 1992 various investigations with experimental laboratory models in order to obtain more objective data and to exclude as many *placebo* effects as possible. In all our experiments the TFF of a drug was always tested along with a control, which was a TFF without drug in entry. This excludes the possibility of any interference of electromagnetic signals coming from the apparatus. The water used in both cases is always the same. The only variable is the presence-absence of the drug.

The models were animals and vegetals. From 1993 to 1994 we have worked with P.C. Endler (Zoologisches Institut, Universität, Graz, Austria) on an animal model with TFF Thyroxin on *Rana temporaria larvae*. The multicentric research, repeated blindly on 936 animals in late Spring 1993 and 1994 (Torino - Graz), proved that TFF Thyroxin had a significant influence on the tadpoles metamorphosis speed (15%; $p<0.01$). Since 1994 we are working on vegetal models with TFF of herbicides. The TFF Glyphosate significantly inhibits the growth of *Ervum lens* sprouts (31-63%; $p<0.001$) inducing sufferance and reproducing the herbicide's effects (see the report by Vinattieri, Picard and Citro, this volume).

5. Discussion

The TFF is a new method that is opening a large avenue to several applications. However, its fundamentals remains to be fully clarified. In order to provide a basis as a starting point for further elaboration, we are making the following assumptions concerning the TFF working principles (see also the figure 2):

1) All substances continuously and spontaneously emit non molecular signals that are both specific and informed, *i.e.* able to reproduce the properties and effects of a substance despite its absence.

2) These signals can be intercepted, amplified and transmitted directly to a biological system which will receive them, or else, they can be recorded in a liquid vector which will memorize them.

3) The administration of this vector or the direct application of the signals to a biological system will induce the specific effect of the original substance, in this case the drug.

The Metamolecular Informed Signals (MIS) can have the following properties:

1) The MIS seem to be released by all drugs;

2) The MIS seem to have an electromagnetic nature;

3) The MIS are presently impossible to measure because of their extreme weakness;

4) The MIS can be transmitted by TFF;

5) The MIS, after transmission, travel along a non-molecular nature capable to induce the same effects than the drug itself;
6) The living cells recognize the MIS sent by TFF as the drug itself;
7) The MIS might be assimilated to a sort of general electromagnetic file containing all the information that defines a given substance.

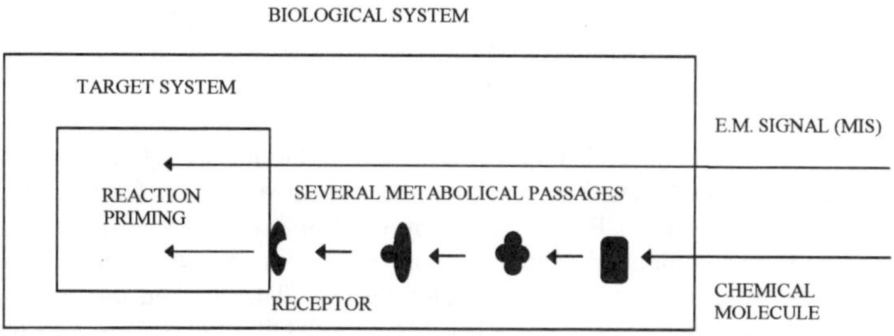

Figure 2. Hypothesis on the phenomenon of the receptive signal through TFF: the target system receives electromagnetic trigger signals and recognizes them even without molecular interaction.

Altogether, the observations tend to prove that living organisms were cured directly or indirectly by information picked up from drugs and transmitted by electronic devices to living systems, either humans, animals or simple vegetals. However, attempt to interact with isolated molecules did not show any result. In many cases, aqueous preparations were the principal vector. This validates the idea of clusters in which the information can be written, and this information seems to be red only by living organisms. As a consequence, our present effort is oriented toward the analysis of the electromagnetic signals, their transmission, and the study of living organisms and the effect of various RF frequencies on them.

6. References

[1] Bellavite, P. and A. Signorini, *Fondamenti teorici e sperimentali della medicina omeopatica*, IPSA, Palermo, 1992.
[2] Citro, M., "TFF, dal farmaco alla frequenza", *Vivibios* 2(3):66-72, 1992.
[3] Citro, M., "Meta Molecular Informed Signal theorie und TFF", in *Struktur und Funktion des Wassers im Organismus*, O. Bergsman ed., pp. 72-77, Facultas Universitätsverlag GmbH, Wien, Austria, 1994.
[4] Endler, P.C. and J. Schulte eds, *Ultra-High Dilution. Physiology and Physics*, Kluwer Dordrecht, 1994.
[5] Milde, K., "Medikamenten Informationsübertragung" in *Wassers*, I. Engler ed., pp. 121-123, Sommer Verlag, Teningen, 1989.
[6] Morell, F., "Neue Wege der Medikamententestung", in *Diagnose und Therapieverfahren im Ultrafeinen Bioenergiebereich,* H. Brügemann ed., pp. 115-148, Haug Verlag, Heidelberg, 1984.
[7] Voll, R., *20 Jahre Elektroakupunktur*, Medical Literature Verlagsgesellschaft, Uelzen, 1976.

NANOPARTICLES IN THIN LIQUID FILM. POSSIBLE INTERACTIONS WITH WATER CLUSTERS

G. PICARD
Istituto di Ricerca "A. Sorti", IDRAS, via Passalacqua, 19, Torino, Italia

1. Introduction

A new method called Pharmacological Frequency Transfer or TFF (Trasferimento Farmacologico Frequenziale) was conceived by M. Citro in 1990 [1]. This method consists essentially in reading by electronic circuitry an electromagnetic signal emitted from a compound, amplifying and sending it to another material where the message can be stored, either in solid or liquid vectors. This method proved to have a therapeutical effect on hundred of cases [2]. Moreover, TFF effects were also observed on animals and vegetals [3]. Essentially, the administration of an activated liquid to organisms procured the same therapeutical effects than the original compound. However, TFF works on life processes in a yet not well understood way. Indeed, it should be admitted that life itself is still an enigma.

Reductionists are trying to bring living processes to a series of biophysical and biochemical reactions in which no place for self-determination is given. However, alternative philosophies are now emerging. For example, in a recent work, the importance of electric and magnetic fields was demonstrated [10]. The bigger is the living organization, the smaller is the required signal to influence the whole chain of reaction. The invisible fields, magnetic or electric, can be indeed playing a guidance role in the life processes.

Life is showing us how to built organisms with water. As a matter of fact, water is the main constituent of cells, and biochemistry occurs in water. Water could just be considered as a useful solvent, giving a support to biochemically active molecules, either in the membranes or in the bulk. However, why water couldn't also be an active player and not only a passive support? How could this other role be observed? How could the information be stored in such a fluid material, without being canceled out in the next nanosecond? The mechanism of storage is still a matter of discussion, as no physical measurement has given a clear picture of the situation yet. However, until now, some hypotheses are being tested actively. One of the most attractive model is based on the presence of water clusters and clathrates. They are supposed to exist because of electrostatic interactions that keep the water dipoles either flickering or solidly attached. These water formations are suspected to have either dielectric properties suitable for the storage of information or specific physico-chemical properties. Their large dielectric constant should slow down the electromagnetic waves and be able to entrap long electromagnetic wavelengths. As a consequence, it could be possible to write in water, more precisely in water clusters.

The present document reports two recent and original experiments that are indicating that water clusters may be influencing the arrangement of particles on solid substrates. These two experiments were conducted independently, although they had a common goal, which was arranging proteins in 2D crystals. In both cases, the two experiments were tested with solid particles, *i.e.* polystyrene spheres. In one case, the arranging of particles was practically limited in smallness to 50 nm. In the other case, the method seemed limitless in smallness. The main difference was the thin liquid film thickness. This subject is discussed and some conclusions are drawn, indicating that special structures in water can be considered as a valid hypothesis for explaining the two different behaviors.

2. Materials and Methods

The polystyrene particles were from Interfacial Dynamics Corporation (IDC), Portland, OR. The buffers and NaCl (99.99%) were from Merck. Water was distilled (Aquatron BS I) and demineralized (Elgastat UHQ II) before use. Its surface tension was higher than 72 mNm^{-1} and its conductivity equal to 10^{18} $M\Omega.cm^{-1}$. One of the main parameter for proteins at the air-water interface is the time of exposure. Proteins are expected to have a lifetime in the order of the second before unfolding [4]. As a consequence, rapidly arranging proteins in monolayer at the air-water interface would be sufficient to avoid unfolding. In a recent and major project to make two-dimensional protein arrays, a set-up was created to coat a solid surface based on a water stream driven by evaporation (see figure 1A) [6,9].

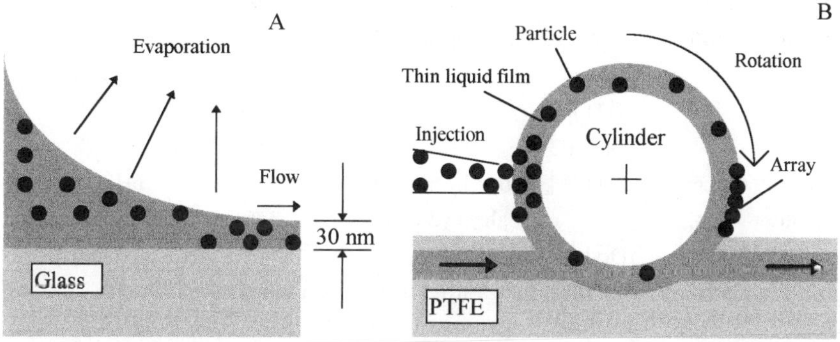

Figure 1: two methods for packing fine particle in monolayers. At the Figure 1A, the evaporation of a thin liquid film brings water molecules from the bulk to the triple contact line, driving suspended fine particles to the solid substrate. At the Figure 1B, the DTLF (Dynamic Thin Liquid Film) method is a rotating glass cylinder over PTFE (Teflon) covered by a 5-μm thin liquid film, packing adsorbed fine particles at the air-water interface.

In this case, the proteins (12-nm diameter holoferritin) were pushed toward the triple air-water-solid contact line and then pressed together by the incoming flux of water and proteins. The water film thickness in the final part of the operation was about 30-nm thick. In spite of a sufficient speed of proteins at the triple contact line, all attempts to arrange proteins in two-dimensional arrays did not work. However, the application of this principle with microspheres from 1 to 0.3 μm performed well. At smaller sizes, the difficulties increased steadily down to 50-nm diameter, the lower limit ever reached by this method.

Another project was also aiming to the same goal, *i.e.* to make protein monolayers with an optimum arrangement. The main strategy was the same: to drive proteins quickly on the air-water interface by a water stream. The water stream, created by the rotation of a cylinder under 5 μm of water, did succeed to well-pack proteins in monolayers (see figure 1B) [8]. This new method, named by now the dynamic thin liquid film (DTLF) method, has the virtue not to constrain the air-water and water-glass interfaces to interact. It succeeded also in making 53-nm particle monolayer preparations [7].

3. Results

The following table summarizes the monolayer preparations using the evaporation and DTLF methods. The last column comments on the feasibility of the preparations. The nature of the particles depends on the surface coating. Carboxyl and sulfate were added by IDC (Interfacial Dynamics Co.) in order to stabilize the particles in pure water. The CML particles, *i.e.* Carboxyl Modified Latex, were highly hydrophilic because of the surface high electric charge density.

D (μm)	Particle type	Method used	Hydro-affinity of the particle	Surface charge (μC/cm²)	Observations on feasibility
6.240	Sulfate	DTLF	Rather hydrophobic	7	pH 4.0; easy
1.170	Carboxyl	"	"	19	"
1.010	CML	"	Highly hydrophilic	923	1.0 M salt; easy
"	"	Drying	"	"	Easy
0.833	"	DTLF	"	1135	"
"	"	Drying	"	"	"
0.614	Carboxyl	DTLF	Rather hydrophobic	16	pH 4.0; easy
0.250	CML	"	Highly hydrophilic	226	1.0 M salt; easy
"	"	Drying	"	"	Uneasy
0.220	Carboxyl	DTLF	Rather hydrophobic	Unknown	"
0.127	"	"	"	4	pH 4.0; easy
"	"	Drying	"	"	Became difficult
0.053	"	DTLF	"	0.2	pH 4.0; easy
"	"	Drying	"	"	Nearly impossible
0.012	Holoferritin	"	Highly hydrophilic	Unknown	Impossible
0.005	Cyt P450	DTLF	"	"	pH 7.0; easy

It is clear that the behaviors of the nanoparticles in the 5-μm or the 30-nm thin liquid films were different. If the DTLF method is considered independently, the results indicate that the nature of the particles played the most important role. On the other hand, considering the evaporation method separately, the size of the particles was the determinant factor. In both cases, the efficiency depended on the water layer. In the DTLF method, electric charges on the outer particle surface prevented them from adsorbing at the air-water interface. Only the reduction of surface charges by either acidity or salts led to adsorption. In the case of the evaporation method, the dependence on the particle size is pointing toward another direction. The evaporation method required a water film thinner than the particle diameter, and during its evaporation, since this process is pumping continuously new water, particles were driven into the monolayer.

Globally, for particles bigger than about 200 nm, there was no difference between the two methods. On the contrary, going from 200-nm down to 54-nm particles, the evaporation method ran into increasing complications, as instabilities were observed. Meanwhile, the DTLF method was practically insensitive to the smallness and hydro-affinity of particles. The results with DTLF method are indicating that the strategy of arraying by surface lateral motion does work. This divergence in efficiency is bringing into light some anomalies that exist in water.

4. Discussion

Comparing the two methods, the only relevant difference is the thickness of the water film. In the DTLF method, the particles were moving free above 5 μm of streaming water. In this dynamic flow, water is considered as unstructured. In the evaporation method, a water film only 30-nm thick was supporting lateral motions of proteins. As it is well known that water binds to hydrophilic surfaces, it may be suspected that the failure in the attempts to array proteins with the evaporation method was due to special arrangements of the water molecules on surfaces. However, these layers are well structured, atomically thin and strong enough not to cause noticeable perturbations [6]. As a consequence, it seems that unexpected formations are playing an active role to impede the lateral motions of proteins.

Considering the increasing difficulty of the evaporation method down to 50-nm latexes, and the failure with the 12-nm proteins, one reasonable assumption could be the existence of impurities in the water. The impurities could have a size comparable to the thin liquid film, and provoke its rupture at the critical moment of assembling. This hypothesis must however take into account the always extreme precautions that prevailed, in particular in distillation and double filtration of water, one for ions and the other for organic compounds, and cleaning the glassware with acids and rinsing with the same pure water. Washing the particles did not change the results.

Another domain of thought can be the instabilities of the thin liquid film, 30-nm thick, under evaporation. It is not excluded that such thinness combined with the evaporation process provoked local temperature and pH lateral variations, that could lead to the rupture of the thin liquid film. This is the accepted hypothesis based on the present physico-chemistry.

However I suggest, among alternative explanations, the possibility that 10-nm water formations can always be present in the bulk. Water clusters may be a reality due to the fact that highly hydrophilic bodies are often suspended in water. A possibility can be that glass particles could be released from the glassware surfaces entering in the distillation, storage and experimentation itself. According to recent propositions by other groups [11], clusters of this size can be reached. The presence of clusters having about 10 nm would make the thin film not liquid anymore, but somewhat crystalline. The evaporation would become rather chaotic and would provoke unequal surface tensions, leading to film ruptures. It was also suggested that the water clusters can adsorb onto hydrophilic surfaces. Then, accumulation of a few clusters on either highly hydrophilic particles or glass surfaces should lead to perturbations in the lateral motions of ultra-fine particles.

5. References

[1] Citro M., R. Sacchi, C. Vinattieri. "L'arcano concerto che smuove una sottile energia" in *Medicina Naturale* 4(4):18-23, Tecniche Nuove ed., Milano, 1994.

[2] Citro M., C.W. Smith, A. Scott-Morley, W. Pongratz, P.C. Endler, "Transfer of information from molecules by means of electronic amplification - preliminary results", in *Ultra-High Dilution. Physiology and Physics,* Endler P.C. and J. Schulte eds., pp. 209-214, Kluwer, Dordrecht, 1994.

[3] Citro M., P.C. Endler, W. Pongratz, C. Vinattieri, C.W. Smith, J. Schulte, "Hormone effects by electronic transmission", *FASEB J.* 9:A392, 1995.

[4] Donaldson, T.L., E.F. Boonstra, J.M. Hammond, "Kinetics of protein denaturation at gas-liquid interfaces", *Colloid Interface Sci.* 74:441-450, 1980.

[5] Myers, D., *Surfaces, Interfaces and Colloids: principles and applications*, VCH Publishers Inc., New York, 1990.

[6] Nagayama, K..."Fabrication of two-dimensional colloidal arrays", *Phase Transition* 45:185-203, 1993.

[7] Picard, G., D. Alliata, "Making fine particle monolayers with DTLF method", in preparation.

[8] Picard, G., D. Alliata, I. Nevernov, C. Nicolini, "A novel method to make protein monolayers", sent for publication.

[9] Picard, G., S. Takeda, M. Yamaki, H. Yoshimura, S. Ebina, K. Nagayama, "Imaging of two-dimensional assembly of labeled apoferritin on mercury by Förster energy transfer microscopy", in *Sixth International Conference on Organized Molecular Films*, Trois-Rivières, Canada, 1993.

[10] Popp F.A., U. Warnke, H.L. König, W. Peschka, *Electromagnetic Bio-Information*, Urban and Schwarzenberg, München, 1989.

[11] Schulte, J. "Conservation of structures in aqueous ultra high dilutions", in *Ultra-High Dilution: Physiology and Physics*, Endler P.C. and J. Schulte eds., pp. 105-115, Kluwer, Dordrecht, 1994.

PHARMACOLOGICAL FREQUENCY TRANSFER ON ANIMAL AND VEGETAL MODELS

C. VINATTIERI, G. PICARD AND M. CITRO
Istituto di Ricerca "A. Sorti", IDRAS, via Passalacqua, 19, Torino, Italia

1. Introduction

A new method named Pharmacological Frequency Transfer or TFF (Trasferimento Farmacologico Frequenziale) was conceived by M. Citro in 1990 [2]. This method proved to have curative effects in hundreds of cases [1]. The diversity (over 150 different cases) and efficiency (over 70%) of this method have been observed for five years on patients (ages from 20 to 75). Experimentally, a drug is placed at the entry of an electronic device and, after amplification, an electric signal is sent directly to a patient, or stored in a liquid vector for further administration. A hypothesis was formulated to explain the curative effects. It is assumed that the drug or other chemical substances emit specific signals. The signals are assumed to be electromagnetic in nature and called MIS (Metamolecular Informed Signals).

This method is currently utilized and still giving excellent results. However, the intrinsical problem related to clinical investigation is the impossibility to make a systematic study based on parameter control. Also, in spite of the close survey of the clinical cases and the accurate control of the effects on patients, the *placebo* effects can be always evoked to explain the success of TFF, due to the complexity of human-environment relationship, even if hundreds of positive cases were actually recorded in favor of a truly TFF effect on patients.

In order to eliminate the *placebo* effect, while remaining in the field of TFF effects on living organisms, a series of experiments was launched in 1992. The complexity of the living systems was at two different levels: on animal and vegetal models. In the case of animal models, the experiments were conducted according to a protocol already established. In particular, the hormone thyroxin was used to test the effect of TFF on the growth of an amphibian, *Rana temporaria*, based on different steps in its evolution. The thyroxin is expected to accelerate the development of the amphibian. We are also working on vegetable models since 1994, in occurrence lentils (*Ervum lens*), treated with TFF of glyphosate, a total herbicide. The parameter chosen was the length of the plant during its growth from the fourth day up to 14 days later.

This report presents a compilation of the results with models that were recorded for over two years. A statistical analysis is presented. Together with clinical cases, the full spectrum of living form complexities has been covered by TFF. A discussion follows, and some propositions on further experiments are suggested, in order to better understand the TFF effect, and the interaction between aqueous preparations, electromagnetic waves and living matter.

2. Materials and Methods

We have used an amplifier (MORA III 200, Med Tronik, Friesenheim, Germany) with a selectable gain from 1 to 40 and a post-amplifier x12 (total gain from 1 to 480). The device is a large band amplifier with minimum frequency distortion fed by accumulators. It intercepts all together entry background noise and signals coming from the drug, amplifies the resonance frequencies and transfers them with a selectable periodism; we used a periodism of 7 s amplification with pauses of 3 s for a total duration of 15 min. The input is connected to a brass support where the drug to be transferred frequentially is placed, and the output is connected either to identical supports with bottles containing various preparations (*e.g.* aqueous, ethanolic or physiological solutions), or to brass electrodes placed in contact with the patient. The information is transmitted along insulated coaxial cables.

In all our experiments the TFF of a drug, called TFF thyroxin or TFF glyphosate, was always tested along with a control, which was a TFF without drug in entry, called TFF water. This excludes the possibility of any random interference of electromagnetic signals coming from the apparatus. The water used in both cases was always the same. All containers were in the same room, for an equal temperature, humidity and light environment. The only variable was the presence-absence of the drug.

TFF on an animal model

Thyroxin plays an important steering role in the metamorphosis of amphibian. For this reason, we used TFF of this hormone on *Rana temporaria*. For the experiments we chose only two-legged tadpoles which had just started to develop their hind legs. The tadpoles were monitored until they entered their four-legged stage (F_a). Furthermore, the animals were monitored until they had entered the stage with reduced tail (F_b). Having reached this stage, the animals were transferred in a natural environment. Six basins contained 18 animals each for a total of 108 animals. An amount of 8 ml of TFF thyroxin (sodium pentahydrate, Sigma $1{:}10^{-3}$) was transferred in each of the three basins, and 8 ml of control (TFF water) were added blindly to the three other basins (each basin contained 8 l of water). A gentle stirring every 8 hours maintained homogeneity in the various basins. The corresponding amount of liquid (8 ml) was always pipetted out of the basins.

Temperature was kept at about 18 °C. Indirect light was used. The tadpoles were fed with cooked green lettuce. The first experiment took place in Graz (Zoologisches Institut, Universität, Graz, Austria) in late spring 1993. The sets of experiments were coded at the University of Graz. Reproducibility was obtained in May/June 1994 with a multicentric research at Torino (184 aa) and Graz (540 aa).

TFF on a vegetal model

Glyphosate is a total herbicide that acts on the phenol production of a plant. In particular, it is expected to be effective on the growth rate of lentils. As a consequence, the impact should be visible at the macroscopic scale by a growth rate inferior to the reference one.

Plastic containers were half-filled with microbiologically controlled hydrophobic cotton. The seeds were placed on the cotton and, twice a day, 4 ml of TFF preparations or reference water was put in each container. After 4 days, the seeds have germinated and the sprouts were long enough for the first measurements. During 10 days, the sprout heights were measured once a day and the data put on graphics and analyzed for statistics.

3. Results

On an animal model

The figure 1 shows the difference in the development of the amphibians in the four-legged stage (F_a) and the reduced-tail stage (F_b) [3]. Thyroxin TFF caused a dropping of F_a and F_b by 10% and 30%, respectively. The difference referred to the sum of all the experiments (tot. 936 aa) is around 15% for both F_a and F_b with a statistic significance of $p<0.001$ in χ^2 test from the 10th interval. In few words, the amphibians treated with TFF thyroxin compared with normal treatment with TFF water entered later in the four-legged and the reduced tail phases.

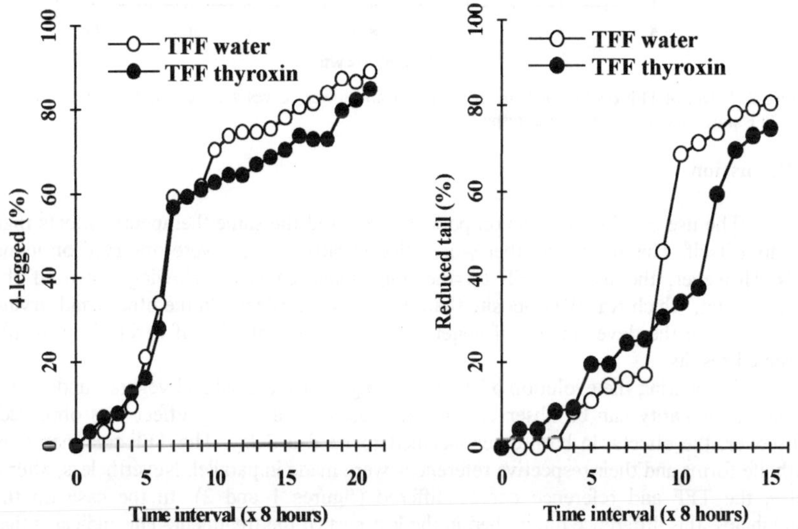

Figure 1 Effect of TFF on *Rana temporaria*. The left graph concerns the entering in the four-legged phase. The right part shows the second phase of development, *i.e.* the entering in the reduced-tail phase. The white circles are with TFF water and the black circles with TFF thyroxin.

On a vegetal model

The effects of TFF glyphosate on the growth of *Ervum lens* were clearly visible. On one row were the lentils treated with TFF of glyphosate, and on the other row the lentils with normal water. The difference was easily visible. The data are reported at the figure 2, where are the height of the sprouts day after day. It is obvious that during the first 6 days, there were no significant difference. However from the sixth day, a difference appeared, that increased in time. The lentils treated with TFF of glyphosate slowed down their growth rate, while those in normal conditions continued to grow at a steady rate.

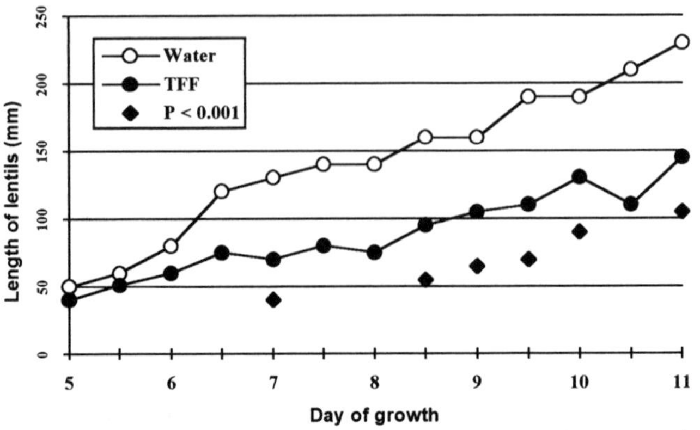

Figure 2 Effect of TFF on *Ervum lens*. In the upper and lower curves are the length of lentils treated with TFF water and TFF glyphosate, respectively.

4. Discussion

The use of TFF of drugs on patients procured the same therapeutic effects than the drug itself. This means in other words that curative effects were observed on living cells. However, the use of TFF on the amphibians caused a slowing down of the development, which was the opposite than the expected effect. On the other hand, trying to slow down the development of vegetables by the use of TFF of herbicide gave the expected results.

Comparing the evolution of the two living forms, animal and vegetal, under TFF effects, a similarity can be observed. In both cases, an inhibitory effect was observed. Moreover, the effects, in both cases, seemed to need a delay. The TFF treatments on both life forms and their respective references were made in parallel. Nevertheless, after a delay, the TFF and reference curves differed (figures 1 and 2). In the case on the amphibians, this difference diminished in the last part of the treatment. This indicates that the TFF treatment may have touched a particular mechanism in the development of the animal, but due to the complexity of the living forms, a compensation was found.

The reversal of the effect of thyroxin on the amphibians can be explained. The TFF being in its early stage of development, one of the present problem is the equivalent dose. In the case of thyroxin, it is known that, at high concentration, the effect is not anymore favorable to the development of the amphibians. At the opposite, thyroxin becomes toxic. As a consequence, too much TFF of thyroxin can slow down the development of the amphibians. It is clear that from this point, an influence of TFF was observed, but its qualitative and quantitative effects remain to be controlled.

Since this year we have increased the control on the various parameters that can influence the growth of vegetables. A Faraday cage was constructed and is now operational. It is capable to block ambient e.m. pollution coming from AM and FM radio frequencies, television stations, cellular phones and satellites. Vegetals are presently growing in the cage at a constant temperature and equal humidity and light conditions. Parallely, we are planning for the near future to clarify the role of RF frequencies that are responsible for the changes on animal and vegetal metabolisms, and to synthesize electromagnetic spectra for multifunctional purposes.

As a global conclusion, it is clear that electromagnetic waves are influencing the development of living organisms. It is known that electric and magnetic fields are playing an important role in the metabolism of cells. Thus, acting on the electric properties of such an important liquid vector like water must have consequences on the development of cells. The electronic amplificator did send signals either directly or via the liquid vector to living cells. The cells reacted in a manner that is still unclear to these signals, but it can be ascertained that they did not remained unaffected. The direction and the amplitude of the TFF impact on living forms still remain to be studied with greater care in the future, because the application range of TFF on living forms is very wide.

5. References

[1] Citro, M., R. Sacchi, C. Vinattieri, "L'arcano concerto che smuove una sottile energia" in *Medicina Naturale* 4(4):18-23, Tecniche Nuove ed., Milano, 1994.

[2] Citro, M., C.W. Smith, A. Scott-Morley, W. Pongratz, P.C. Endler, "Transfer of information from molecules by means of electronic amplification - preliminary results", in *Ultra-High Dilution. Physiology and Physics,* Endler P.C. and J. Schulte eds., pp. 209-214, Kluwer, Dordrecht, 1994.

[3] Citro, M., P.C. Endler, W. Pongratz, C. Vinattieri, C.W. Smith, J. Schulte, "Hormone effects by electronic transmission", *FASEB J.* 9:A392, 1995.

THE QUALITATIVE THEORY: AN EXPLANATION OF HIGH DILUTION EFFECTS, A JOURNEY THROUGH GREEK ONTOLOGY FROM PARMENIDES TO ARISTOTLE

PIETER MICHEL CONSTANT
Lammenschansweg 71, 2313 DK Leiden
The Netherlands

Facts of nature can only be explained within a scientific structure. The structure of modem science has two characteristics, a mathematical and an analytical one. In ancient philosophy we find these characteristics in Plato's Timaeus. However, Hippocrates wams us for the analytical approach and tells us to start our medical examination from the individual Aristotle adopted this advice in his ontology. With the aid of Aristotle's hylomorphism we can explain the action of highly diluted substances on living cells.

When we consider the influence of highly diluted substances on cells, three questions arise: first, the question of the character of the living, second, that of the character of the agent and third, that of the character of the interaction. What is life? In the various definitions we see key words like: *structure* and *high degree of complexity.*

The complexity of the living turns out to be an organisation of parts which are studied with the help of mathematical models. From the character of the parts and their function within the complex we deduce the qualities of the whole organism. In this view, the life function, the metabolism, is a (chemical) interchange of microforms within a certain structure. All processes are reduced to connections and separations of microforms, next to each other in one enclosing structure. Also the nonliving consists of these elementary parts, though within a simple (CH_3CH_2OH) or a very simple structure (O_2, etc.). The reaction upon this kind of stimulus is seen as a reaction of a high-complexity construction to a low-complexity construction.

A second connection is possible too. At higher temperatures ice becomes water. The base, that used to be solid, becomes liquid; the one quality has been exchanged for the other, but the basic structure (the molecules) is the same. These two data, quality and basic structure, form a stratification in which the layers are in the same place (synlocant). We also have the relation between unity and multitude: the one irreducible quality goes together and unites with a multitude of parts. When Aristotle defines the living, in De Anima, II,1[1], we also see this distinction. Summarized, this definition is: the soul is the substance, form or actuality of a natural organized body. Following our subject (the cell) we may restrict ourselves to the nutritive soul, this has 'movement in the sense of nutrition, decay and growth' (II,2,413a24). The example Aristotle gives of the nutritive soul is the plant: 'we think of plants also as living, for they are observed to possess in themselves an originative power through which they increase or decrease in all spatial directions' (II,2,413a25). This originative power is a real, consistent part of the plant or cell, which is fully connected with the structure and has the same extent. It governs the functions of the living. This principle of governing is not something alongside the material, but something synlocus, not as a helmsman on a boat, but everywhere parallel with the structure of the living matter. So we find two streams of understanding the living, which both have their roots in the antiquity. Let us make a trip to the Ancient Greek world of philosophy to find these roots.

Our first stop is Elea. In the 6th-5th century B.C. Parmenides lived here, who is known for his division of reality into two parts. To the first part belongs being, which is one, changeless (Coxon, p. 68[2]) and without becoming or perishing. To the second part belongs nature. To Parmenides nature is not real, but exists only in human beliefs. Men 'name two Forms and they chose opposites in body [..], on the one hand aetherial fire of flame, [..] on the other hand [..] unintelligent night [..]'(Coxon, p. 76-78). Parmenides deprives nature from its unity and leaves it in pieces. Since Parmenides we can call philosophy the science that is in search of the unity in physical nature. Physics becomes an assembly of elements in which the unity of the (living) construction will always be artificial. We travel further to the south of Italy, to Agregente on Sicily. In his 'About Nature' Empedocles (5th cent. B.C.) describes four roots (earth, water, air, fire), basic elements of all the existing (Diels 17[3]). Separated these elements are invisible, but compounds of them show visible things. These roots are qualitative elements: contrary to Plato's view (see below) water (liquid) really has the qualities of liquid, just like fire has the qualities of fire. The third place we stop is Crotone at the heel of Italy. Here lived Pythagoras and his disciples. Pythagoras is said to have laid the basis of Pythagorean cosmology (Guthrie I p. 220[4]). New in his vision is that Pythagoras made a connection not only between algebra and geometry, but also between these and body and soul: the number determined the mathematical figure, the parts of the body and the soul itself. Our fourth stop is Athens, where Plato used to live. Plato is known for his World of Ideas. He was not only an author but also the director of a school he founded. In this school highly gifted mathematicians like Theaetetus and physicists like Aristotle participated. They directed Plato's view towards nature, which resulted in the Timaeus dialogue (p. 73-79)[5] on the structure of the cosmos. Timaeus gives two kinds of triangles as basis of the cosmos, an isosceles, and a scalene whose angles are 90, 60 and 30°. Four isosceles form a square, six a cube (ibid.), which is the most stable figure and forms the element earth. Out of the scalenes the other three elements (water, air, fire) are formed. Out of six scalenes Timaeus forms an equilateral triangle. Four of these equilateral triangles make a fire-pyramid (ibid.), eight make the air-octahedron (ibid.) and twenty make the water-icosahedron (ibid.). Since these three figures have the same basic triangle and are less stable than the earth-cube, they can mingle. For instance: 1[fire] (4▵) + 2[air](2x8▵) = 1[water](20▵), the first chemical formula[6]. The earth-cube cannot mingle with any of the other figures. So we find: 1. The structure of the cosmos out of mathematical elements; 2. three elements can mingle; 3. quality is reduced to quantity; 4. the function of the living, how is it possible that elements that strive after their own place in the cosmos, come to function within one structure as a living object or a thing? Plato resolves this problem with his World of Ideas. Ideas are pure 'entities' in a separate world. He distinguishes Ideas like Justice, Plant, House, etc. and assumes the existence of a Demiurg that mingles the material cosmos (the triangles) with parts of the Ideas. Out of this mixture among other things, living objects are formed. A plant, a cell can only grow by sharing the Ideas of Life, Growth, Beauty, etc. Finally the question about the unity within things in the cosmos. The material elements out of which the platonic cosmos is built, can be reduced to two facts: the isosceles and the scalenes. Mixing the parts of the Ideas with these materials gives a thing, culture (house) or nature (plant, man) as a compound of elements from both groups. There is no unity to be found in this construction. The participation in the Idea 'One' is like a bandage around a package. It may be beautiful, but, as everything else, man remains a complexity and forms no unity.

Modern sciences follow Plato's theories, they also consider matter to be built out of basic elements within mathematical structures that mingle ($2H_2 + O_2 \rightarrow 2H_2O$): quality is reduced to quantity. Modern sciences, however, do not adopt the World of Ideas and so they lose the steering, purposive principle.

Now there turn out to be mechanisms that cannot be explained by these quantitative theories, for instance the effectiveness of highly diluted substances. So there has to be another way to explain the (living) nature. Hippocrates, Heraclitus and Aristotle furnish such an explanation. In the Hippocratic Collection we can read a reflection that puts an end to analyses. The author of the Ancient Medicine[7] (as Hippocrates is generally considered to be) turns against Empedocles and those who reduce man to their elements and he offers a new solution: 'one can have a clear knowledge about nature only from medicine' (A.M. 20,11). The author, writing about diet, makes his point clear with the aid of among other things the influence of food on man. In this study we learn to see the quality of man by looking at his utterances. Some people find their strength in for instance cheese or alcohol, others, however, get ill from these. In the description of the examination which, according to the author, one has to perform, a new, special aspect turns up. Not the separated parts become ill, but the harmony of the whole is disturbed. The author emphasizes: 'we must know what the pain is, the reasons for it, and which constituent of man is harmfully affected'(A.M.20.25). The basis of this vision we find with Heraclitus. Heraclitus made crystal-clear pronouncements that have stimulated philosophical thinking. We shall restrict us to his utterances about the concept of unity. Anywhere in the cosmos, in nature, in man and in social structures he sees unity (Diels 50). The unity of the cosmos is fire 'all things interchange for fire' (Diels 90). Fire is the element out of which all things arise. In nature we find unity in sayings as 'As they step into the same river, other and still other waters flow upon them' (Diels 12, Kahn L.52[8]). The unity is here that what rests in the alteration, namely the river, while the water is the changeable element. So not everything in Heraclitus' view is changeable, an utterance like: 'One cannot step twice into the same river' can mean 'it is always a different river', but also 'the river is constantly changing'. In the following fragment we see that he means the last: 'nor can one touch a living substance twice being in the same state' (Diels 91). So there is something that always rests in the alteration (the living substance) 'it rests by changing' (Kahn L II). The question is: 'what is it that changes in its rest?', and: 'what is the substance to which the changing of the condition happens?' 'What kind of unity is it, in which alterations can take place, that cannot be explained by constructing elements (Plato)?' It is Aristotle who gives the answer to this question. He continues the line Heraclitus - Hippocrates and distinguishes the substratum (the resting) and the characteristics (the changing) (Physica I,7[9]). In the coming to be and changing of things and characteristics we discern three simple principles, for instance unmusical, musical and man (ibid.). In the changing are the substratum (the resting, 'the man remains man') and the contraries: something that does not survive through the process (unmusical) and something that arises (the acquired art of music) (ibid.). We see this in every change: formless bronze becomes a statue, cold water becomes hot and evaporates. None of these paradoxes can exist without the substratum, 'the substratum is matter for them both' (Gen et Corr. II,1,31[10]). So one whole has essentially different principles: the subject itself and its characteristics. Hereby we have two kinds of subjects: the living (plants, etc.) and the nonliving (fire, etc.). The influence of highly diluted substances on cells is explained with the help of aristotelian views.

Highly diluted substance is nonliving material and is obtained by trituration (blending and shaking). After 24 times of trituration not a single molecule of the original matter is left. What is the character of a fluid like alcohol that it can take in characteristics by induction, without influencing its own molecular structure? This becomes clear when we look at the parts matter and form, of which the substance is composed. The matter is what stays the same in the change; the form is what comes and goes in the change. When the substance ice melts, it becomes water. The substance which was solid before has become a liquid; in this change 'something' stays the same: only part of the substance changes. It is the field of the structure, the molecular layer, which is the matter: this did not change but remained. Next to this unchangeable field of structure, we distinguish the field of the quality, in which the changing takes place: this is the form. Everywhere the field of the quality, the form, coincides with the field of the structure; they are 'synlocant'. Though the character of the structure is a necessary condition for the character of the form, it is no sufficient condition; so the field of the quality can change independently from the field of the structure. We have matter and form. The form is a stratum which has its own characteristics, particularly a tension. We can observe this tension with our tactile sense: wood is solid, rubber is elastic, liquid is fluid. This tension is so characteristic for the field of the quality, the form, that we can call this the field of tension[11].

A field of tension has a vibration like a string (we call it vibration colour) that differs according to the character of its substance (alcohol, sulphur, etc.). The outspoken vibration colour (of sulphur, etc.) stamps itself in the apparently indifferent field of tension of the alcohol. The living as the nonliving has a substratum and a form. The form of the living is the soul, the substratum the matter and form of the nonliving. As we saw the soul is an originative power (De Anima), which we may characterize as a controlling principle. This principle is a field in the living (synlocus with the other fields). It is this field that controls the vegetative processes. So for every living thing, and also for cells, we can distinguish the three strata of the living: the matter, the qualities with their field of tension and the soul, which is the controlling field. The relation between nonliving and living can be expressed in the following figure.

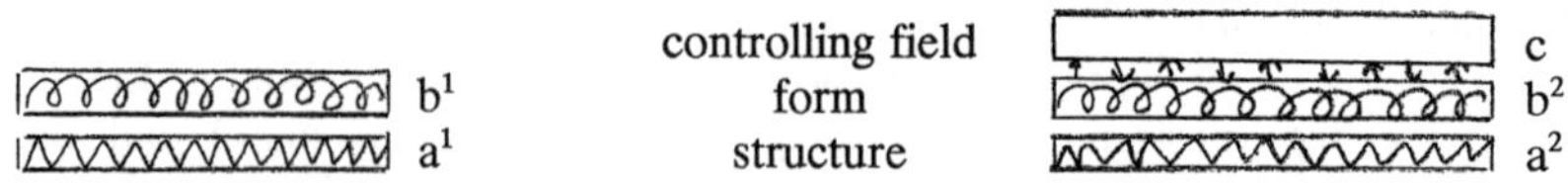

In this figure we see the highly diluted substance as the triturated field of tension (b^1), which connects with and is influenced by the field of tension of the living (b^2), which, in its turn, influences the controlling field (c). Since we only know the controlling field (c) indirectly, it (c) is not shaded.

Modern sciences have discovered the characteristics of the matter, its structure (molecular, etc.). This fits perfectly well in the aristotelian scheme if we use the definition of the soul for the form of the nonliving: the form is the field of tension for a natural structurated (molecular) body.

References

[1] Aristotle 'De Anima', transl. W.D. Ross *et al.*, 1963, Oxford University Press.

[2] A.H. Coxon 'The fragments of Parmenides', Assen/Maastricht, 1986, Van Gorcum.

[3] Hermann Diels 'Die Fragmente der Vorsokratiker', Zürich-Hildesheim 1992, Weidmann.

[4] W.K.C. Guthrie 'The earlier Presocratics and the Pythagoreans', Cambridge 1980, Cambridge University Press.

[5] Plato 'Timaeus and Critias', London 1977, Penguin Books Ltd.

[6] Platon 'Timée/Critias', trad. Luc Brisson, Paris, 1992, Flammarion, p. 301.

[7] Hippocrates 'Ancient Medicine', transl. W.H.S. Jones, Cambridge 1984, Harvard University Press.

[8] Charles H. Kahn 'The art and thoughts of Heraclites', London-New York 1981, Cambridge University Press.

[9] Aristotle 'Physica', transl. R.P. Hardie *et al.*, 1966, Oxford University Press.

[10] Aristotle 'De generatione et corruptione', transl. H.H. Joachim, London 1966, Oxford University Press.

[11] Pieter Michel Constant 'De kwalitatieve theorie ter verklaring van de homoeopathische wetten' Tijdschrift voor Intergrale Geneeskunde 1994; 9, 6: 275-282.

THE HISTORY OF NON-CONVENTIONAL MEDICINES AND HIGH DILUTION RESEARCH

PAOLO MAROTTA
International Center for the Study of the History of Medicine
IRFEST, Université Louis Pasteur, Stasbourg, France/Istituto Italiano per gli Studi Filosofici, Napoli, Italia

In the history of medicine an important role is played by those medical practices which are usually known as non-conventional medicines. Non conventional medicines (also called "alternative medicines" or "traditional medicines") make up a crowded group of different medical and therapeutic practices particularly distinct from so-called official or conventional medical science.

Non-conventional medicines consist in an alteration of current practices, theories and thechnology. They can be seen as an integration of techniques that have endured through the centuries with the advancements of contemporary biomedical research. A process is going on which is overturning gradually the present medical paradigm, enhancing the importance of human individuality against any biological reductionism [14]. In this frame it is possible to explain the new importance and success that different types of non-conventional medicines are attaining today.

Today new indications coming from the researches in different fields of science are facilitating a rethinking process of the fundamental paradigms in biology and, subsequently, in modern medicine. Academic medicine has performed incredible wonders, but the time is ripe for investigations in non-conventional ways of preserving and enhancing health. Not only physical but also mental well-being must become the goal for the fight against illness. New possibilities for health-care are offered to suffering people from non conventional medicines.

It is important to remember that among non-conventional medicines there are different types of medical systems. To give a definition of what a medical system is, it is possible to say that it consists in the attempt to shape medicine on the ground of a unifying principle in analogy with the traditional philosophical models of methaphysics. There are different medical systems which have had an important role in the history of medicine. Among these it is possible to quote the animism of G. B. Stahl (1600-1734), the *medicina rationalis systematica* of Frederich Hoffmann (1660-1742), the animal magnetism of Franz Messmer (1734-1815) and finally homeopathy founded by Samuel Hahnemann (1755-1843).

Homeopathy is a medical system focused on a holistic vision of man, stressing the complexity of the human organism. It consists in a therapeutical methodology which clinically applies the law of similitude: that is to say, treating an illness with a substance that, if tested in a healty person, produces symptoms similar to those displayed by an ill person. Current medical opinion takes the view that symptoms are a direct manifestation of the illness and must be eliminated. Homeopathy, in contrast, sees the symptoms as a reaction against the illness and it seeks to stimulate and not suppress this reaction. In order to achieve this goal, it uses as therapeutical remedies different substances at very diluted dosages, the so-called "infinitesimal dosages", including ultra high or ultramolecular dilutions at wich, according to Avogadro's law, it is extremely unlikely that any molecule of the initial substance persists [6]. Without delving inside the different aspects of homeopathy, its vision represents a criticism of positivism and cartesianism and it seems to be incredibly modern: it is a modern attempt at studying the ill person as a whole, underlining the importance of inter-relations among the different physiopathological aspects of illness. It can be said that all the process of diagnosis and therapy of the homeopathic medicine is based on exact empirical evidence and that its succesfull application becomes a confirmation of the foundations of homeopathy itself [17]. Before Claude Bernard (1813-1878), the French physiologist who first adopted the experimental method to solve problems of clinical medicine (publishing his principal work [14] only in 1865), Samuel Hahnemann is the first medical doctor who introduced the experimental method on a human scale [15].

From a historical point of view, homeopathy has its beginning in the period of the German romanticism and idealism, a very particular period of the history of scientific and philosophical thought, characterized by a fertile connection among natural and human science, art, philosophy, literature and often also theology. Shubert, Kerner and Heinroth are not only medical doctors, but also elegant poets; Carus from Dresda, apart from being a medical doctor, is also an esteemed painter; Schlegel (1772-1829) is an acute writer and an accurate researcher in the fields of medicine and mathematics; Hoffmann (1660-1742) and Brentano (1778-1842) are at the same time poets, painters and composers; Novalis (1772-1801) is involved in all these fields; finally Schelling (1775-1854) and Hegel (1770-1831) developed their *Naturphilosophie* on the base of deep and accurate scientific knowledge. This particularly rich moment in the history of thought seems to find a new particular role in to-day's scientific and philosophic renewal, also involving the modern debate in the field of medicine. Today the romantic holistic vision is again obtaining success, according to the integration between different fields of scientific research, like physics and biophysics, chemistry and biochemistry, molecular biology, genetics, cognitive science and psychology. In this frame non-conventional medicines in general can be seen as a fundamental tool in the contemporary methodological shift in the approach to the problematic relationship between nature and culture; in particular homeopathy represents one of the most interesting non-conventional medical systems having success today. New possibilities of dialogue must be able to find space in the scientific debate, recovering the gap between science, philosophy and art. It can be seen that there are four different cultures facing each other without being able to find a cultural mediation: the culture of the science of nature, the culture of the science of the spirit, the culture of the arts and finally the culture of the human behaviour [20]. Today there is an attempt to break down the boundaries between these different cultures: non-conventional medicines are a moment in this process, suggesting the use of diagnostic and therapeutic techniques addressed both to body and mind, and homeopathic medicine is part of this process. In this way homeopathy has a significant role in the dialectic process of the history of science and medicine. A lot of different disciplines are finding new importance in the medical debate, particularly the field of philosophy: medical activity is strictly connected to art because it can be seen as a synthesis between science, technology and clinical experience [12; tr. it. 1994, p. 108]. New philosophical concepts, on one hand, and the progress of scientific research, on the other, are intensely stimulating each other to introduce new categories of thought and paradigms. Epistemology (*Erkenntnistheorie*) of natural science offers systematic chances for interdisciplinary discussion. And the field of medicine is directly involved in this rethinking process because it applies methods both of natural science and humanities [18].

As far as it is concerned with the official recognition of the therapeutic efficacy of homeopathy, high and ultra-high dilution investigation may play an important role in giving possible scientific explanation of this particular type of non-conventional medicine. High and ultra-high dilutions are standardized aqueous or aqueous-alcoholic solutions which have been diluted in such a way that the concentration of solute becomes of the order of Avogadro's number, or even far below it [9, p. 263]. There are several pharmacological experiments which demonstrate the biological activity of the homeopathic preparations. There are many interesting collections of different scientific works on this topic [3]. It is also possible to recall some of the most significant works: immuno-pharmacological action of preparations of thymus and thymic hormone in infinitesimal dose [1]; biological effects of therapeutic substances in concentrations 10^{-13}-10^{-17} [7]; action of hahnemannian dilutions of mercuric chloride on animal cells [5]; carcinogenesis modulation using ultra-high diluted carcinogens [8]; experiments on dose-dependent effects [16]; control of theratogenic effects of high dilutions of purine-derived products in mouse [19]; control of the effects of high diluted and shaken hormonal substances [10]. But what is of real importance is the fact that the explanation of the method of action of ultrahigh dilutions and of homeopathic remedies can be able to suggest a different approach to the study of living phenomena. Clinical experiences in homeopathy show that therapy with shaken high dilutions corresponding to homeopathic remedies can be seen as a piece of biological information interacting with living systems, able to modulate the cell-defense reaction against patogen agents [2, pp. 49-54]. The research on the mechanisms of information storage in ultra high dilutions will be central to any understanding of their action, in order to generate new scientific concepts with consequences extending far beyond medicine and biology [11].

It must be said also that new hope and interest are being placed in non conventional or complementary medicine, because of the expensiveness of conventional medical therapy. Governments are now funding different research programmes in order to test the healing possibilities of non-conventional medicine. Particularly, very interesting indications can come from researche in the field of biophysics, in which the research on ultra hgigh dilutions can be included. Modern medicine has the important duty of verifying the possibilities of action of non-conventional therapeutic means.

«Medicine, despite all the words, is still an empirical process in which one does what works well» [13, pp. 3-4].

References

[1] Bastide M., Daurat V., Doucet-Jaboeuf M., Pèlegrin A., Dorfman P. (1987), Immunomodulator activity of very low doses of thymulin in mice, *Int. J. Immunother.* 3: 191-197.

[2] Bastide M., Lagache A. (1992), *The Paradigm of Signifiers - Le Paradigme du Sens*, Atelier Alpha Bleue, Paris.

[3] Bellavite P., Signorini A. (1992), *Fondamenti teorici e sperimentali della medicina omeopatica*, Nuova Ipsa, Palermo.

[4] Bernard C. (1973), *Introduzione allo studio della medicina sperimentale*, Ital. transl., Feltrinelli, Milano.

[5] Boiron J., Abecassis J., Cotte J., Bernard A.M. (1991), L'étude de l'action de dilutions hahnémanniennes de chlorure mercurique sur l'index mitotique de cultures de cellules animales, *Ann. Homéopath. Fr.* 23: 43-44.

[6] Bornoroni C. (1993), Teaching for homoeopathy, in: C. Bornoroni, ed., *Omeomed '92. First International Congress: The homeopathic medicine in Europe 1993. Physical - Chemical and Clinical Research,* Proceedings Book, pp. 273-277, Editrice Compositori, Bologna.

[7] Burlakova E.B., Konradov A.A., Khudyakov I.V. (1990), Effect of chemical agents in ultralow doses on biological agents, *J. Nonlinear Biol.* 1: 77-91.

[8] De Gerlashe J., Lans M. (1991), Modulation of experimental rat liver carcinogenesis by ultra low doses of the carcinogens, in: C. Doutremepuich, ed., *Ultra Low Doses*, p. 17, Taylor and Francis, London.

[9] Endler P.C., Schulte J., eds (1994), *Ultra high dilutions.* Physiology and Physics, Kluwer, Dordrecht, Boston, London.

[10] Endler P.C., Pongratz W., Kasterberger G., Wiegant F.A.C., Schulte J. (1994), The effect of highly diluted agitated thiroxine on the climbing activity of frogs, *Veterinary and Human Toxicology* 36, 1: 56-59.

[11] Fisher P. (1993), Prefatory Word, in: Endler P.C., Schulte J., eds (1994), *Ultra high dilutions.* Physiology and Physics, pp. VII-VIII, Kluwer, Dordrecht, Boston, London.

[12] Gadamer H.G. (1993), *Über die Verborgenheheit der Gesundheit*, Suhrkamp Verlag, Frankfurt am Main, [Ital. trasl. 1994, *Dove si nasconde la salute*, Raffaello Cortina, Milano].

[13] Lewontin R.C. (1993), *Biologia come ideologia.* La dottrina del DNA, Ital. transl., Bollati Boringhieri, Torino.

[14] Marotta P. (1996), Il ruolo dei sistemi medici nella storia del pensiero scientifico. Il caso della medicina omeopatica di Samuele Hahnemann, in: Marotta P., Shea R., eds, *New Perspectives on Holism and Non-Conventional Medicines,* Watson, Nantucket, in press.

[15] Marotta P., Taddei-Ferretti C. (1994), I principi della medicina omeopatica, in: Allocati F., Musio C., Taddei-Ferretti C., eds, *Cibernetica Biologica, Atti del Seminar Club di Cibernetica Biologica*, Istituto di Cibernetica, CNR, Arco Felice (Napoli), ciclo 1991-1992, pp. 332-356, Rosemberg & Sellier, Torino.

[16] Oberbaum M., Cambar J. (1994), Hormesis: Dose dependent reverse effects of low and very low doses, in: Endler P.C., Schulte J., eds, *Ultra high dilutions.* Physiology and Physics, pp. 5-18, Kluwer, Dordrecht, Boston, London.

[17] Righetti M. (1994), Characteristics and selected results of research on homoeopathy, in: Endler P.C., Schulte J., eds, *Ultra high dilutions.* Physiology and Physics, pp. 223-227, Kluwer, Dordrecht, Boston, London.

[18] Stieve H. (1995), Limits of natural science: brain research and computers, *Zeitschrift für Naturforschung* 50 c: 317-336.

[19] Taddei-Ferretti C., Cotugno A. (1995), Treatment of teratogenicity induced in mice by caffeine or adenine, *Progr. Biochem. Biotechnol.* 4, in press.

[20] von Henghelardt D. (1993), *La scienza nel periodo romantico*, in *Annali dell'Istituto Universitario Orientale di Napoli*, Sezione Germanica, Nuova Serie III, 1-3, pp. 15-44, Napoli.

CLOSING DISCUSSION

HOW CAN HIGH DILUTIONS BE EFFICACIOUS FROM A SCIENTIFIC POINT OF VIEW?

moderated by

KARL W. KRATKY
Institute for Experimental Physics, University of Vienna, Boltzmanng. 5, Vienna, Austria

The single contributions to the general discussion are here summarized.

Karl W. KRAKTY comes back to the different models that have been presented, describing them in details, and points out that one has to see how they fit together, or whereas there are some problems or inconsistencies. The main problem is how it is possible to obtain High Dilutions without any molecules and which is the process of "imprinting" that allows the information transfer. Later, at the end of each intervention to the discussion, he puts critically in evidence the main points that have been treated.

Madeleine BASTIDE invites to afford the problem from the pure position of science.

Cyril W. SMITH talks about the biophysical aspects of the High Dilution effects in cells; the important thing is that living systems may be regarded as microscopic systems containing water that has a high rate order. He gives a survey of different types of physical effects that involve oscillation processes at different frequency ranges. He makes also a critical review of the methodologies that may be useful to detect the various types of fluctuation phenomena in living systems.

Aurelia CRISTEA observes that the researches on High Dilutions may be grouped according to three main directions. The first one deals with the reproducibility of the preparations for the study of High Dilutions in different conditions. The second direction deals with the researches that try to demonstrate the existence of the effects of High Dilutions, as well as with the priority effects on the sensitive individuals. The third direction is towards understanding what is the informational substrate of the action of High Dilutions.

Peter Christian ENDLER stresses the responsibility of all types of scientist for an interdisciplinary work, especially in biophysics, and the need of very precise experimental data.

Menachem OBERBAUM stresses again the strong need of interdisciplinary work and observes that such characteristics is much more necessary in the study of High Dilutions than in other fields of researches. He makes a historical survey of the researches on High Dilutions: twenty years ago no scientist would have worked in this field and only people at the fringe of science had some interest in the field; after, only some medical doctors began clinical experimentations; later, the biologists began researches on living models, and in the last years also the physicists approached the field with their own methodologies. At present, it seems that the main researches in High Dilutions are brought by the physicists, and it might even be that we will find one day that the underlined mechanism responsible of the effects of High Dilutions is another kind of energy that we are not detecting at present. He reminds also that today an interdisciplinary approach to the problem of High Dilutions is carried out by some Organizations, such as the "Groupe International de Recherche sur l'Infinitesimale"

(GIRI), that are interested to the development of researches and discussions on High Dilutions from different points of view.

Aurelia CRISTEA observes that the problem of the information flow from a biological signal to a receptor, occurring not only as interaction of substances, but also as a different type of information transfer, is a problem for biology and pharmacology both for what it concerns High Dilutions, and also for the drugs that act in full concentration.

Madeleine BASTIDE thanks Cloe Taddei-Ferretti, who had the idea to organize the meeting, during which all have learned a lot of things, staying all together day by day, moving slowly towards an integrated comprehension of the studied phenomena. She puts the attention of everybody to the existence of two political structures: the COST B4 European Program on alternative therapies and the European Commission DG XII. A Research Group of such Commission deals with High Dilutions and their members have deliberative power; it is articulated in six Subgroups (1. Critical literature review, metanalysis and audit standards; 2. Dictionary and terminology; 3. Research methodology; 4. Constraint analysis; 5. Methodology for the study of low concentrations in biological systems; 6. Protocol review). She invites to send the program of the School to prof. Garattini, who is the co-ordinator of the fifth Subgroup of which she is a member, as well as the contents of the book of proceedings, that prove that there is a lot of people working seriously in the field of High Dilutions.

Karl W. KRAKTY thanks all the organizers, lecturers and participants.

Cloe TADDEI-FERRETTI reminds that when she first met the High Dilution field in 1972, there were on one side medical doctors dealing with their patients without caring to know the real underlying mechanisms for which High Dilutions have terapeutic effects, and on the other side the official research structure that did not care to afford researches on the field due to the Avogadro number problem. She was convinced since then that these two groups had to meet, exchange informations, confront different positions and that scientific investigations had to be done (as the group of research where she was working did, although at that time it could not do it in an official way). The problems are at three levels: to make interdisciplinary researches, as was stressed by Menachem Oberbaum; to change the mind by becoming convinced that what a researcher has to do is exactly to investigate what is not yet known, as in the case of the Avogadro number problem; to consider the economic and political reasons for which the High Dilution phenomena are very often officially ignored, and intervene at the right public structures, as Madeleine Bastide suggested. She thanks the Istituto Italiano per gli Studi Filosofici, that promoted the School and gave support for such meeting of high scientific level, as well as the lecturers, participants and co-organizers, who all contributed to create a very cordial and stimulating climate.

PARTICIPANTS

Members of the Scientific Committee, Lecturers and Members of the Organizing Committee are indicated respectively by three, two and one asterisks

**Anagnostatos G.S.
Institute of Nuclear Physics
National Center for Scientific Research
"Demokritos"
GR-15310 Attikis, Aghia Paraskevi
POB 60228
Greece
tel: 0030 1 6513111/6518770
fax: 0030 1 6511215
e-mail: anagnos@cyclades.nrcps.ariadne-t.gr

Andrioli Giuseppe
Istituto di Chimica e Microscopia Clinica
Università di Verona
Policlinico- Via delle Menegone
I-37134 Verona
Italy
tel: 0039 45 8202978
fax: 0039 45 8202978

**Arad Dorit
Department of Molecular Microbiology and Biotechnology
Wise Faculty for Life Sciences
Tel Aviv University
IL-Tel Aviv
Israel
tel: 00972 3 6408723/5
fax: 00972 3 6409407
e-mail: dorit@argaman.tau.ac.il

***,**Bastide Madeleine
Laboratoire d'Immunologie et Parasitologie
Faculté de Pharmacie
Université de Montpellier, 1
15, avenue Ch. Flahaut
F-34060 Montpellier Cedex 1
France
tel: 0033 467 635202
fax: 0033 467 411617

***Berezin Alexander A.
Department of Engineering Physics
McMaster University
Hamilton, L8S 4L7 Ontario
Canada
tel: 001 905 5259140/24546
fax: 001 905 5278409
e-mail: berezin@mcmail.cis.mcmaster.ca

**Burlakova E.B.
Institute of Chemical Physics
Chemical and Biological Kinetics
Russian Academy of Sciences
4, Kosygin str.
CSI-117334 Moscow, Russia
fax: 007 095 9382156
e-mail: chembio@glas.apc.org
spector@glas.apc.org

**Cambar Jean
Group d'Étude de Physiologie et Physiopathologie Rénales, GEPPR
Faculté de Pharmacie
Université de Bordeaux II
3 Place Victoire
F-33076 Bordeaux, France
or:
91, rue Leyteire
F-33076 Bordeaux Cedex, France
tel: 0033 57 571245
fax: 0033 56 571246

Certosino Marisa
Farmacia Istituto Omeopatico
Piazza Dante, 41/A
I-80135 Napoli, Italy
tel/fax: 0039 81 5499331

Chillemi Santi
Istituto di Biofisica, CNR
Università di Pisa
Via San Zeno, 26
I-56127 Pisa, Italy
tel: 0039 50 513111
fax: 0039 50 553501

Citro Massimo
Istituto di Ricerca "Alberto Sorti"
Via Passalacqua, 19
I-10122 Torino
Italy
tel/fax: 0039 11 535252

Constant Pieter Michel
Lammenschansweg, 71
NL-2313 DK Leiden
The Netherlands
tel: 0031 712 23250
fax: 0031 71 5223250

*Cotugno Antonio
Istituto di Cibernetica, CNR
Via Toiano, 6
I-80072 Arco Felice (Napoli)
Italy
tel: 0039 81 8534113
fax: 0039 81 5267654
e-mail: neurosist@imagm.na.cnr.it

**Cristea Aurelia Nicoleta
Department of Pharmacology
Facultea de Farmacie
Universitatea de Medicina si Farmacie "Carol Davila"
Str. Traian Vuia, 6
R-70138 Bucuresti
Romania
tel: 0040 1 2107557/58
fax: 0040 1 2112730

**Endler Peter-Christian
Technical Büro of Biophysics
Research Site for Low Energy Bio-Information
Wittenbauerstrasse 137
A-8042 Graz
Austria
tel and fax: 0043 316 423813
or
Zoologisches Institut Universität
A-8010 Graz
Austria

Fossati Andreina
Via Beniamino Cesi, 10
I-80127 Napoli
Italy
tel: 0039 81 5569077
5780278

Guglielmotti Vittorio
Istituto di Cibernetica, CNR
Via Toiano, 6
I-80072 Arco Felice (Napoli)
Italy
tel: 0039 81 8534114/117
fax: 0039 81 5267654
Haidvogl Max
Ludwig Boltzmann Institut
Schillerstrasse, 11
A-8010 Graz
Austria
tel/fax: 0043 316 384515

Halm René-Philippe
GIRI
18, rue des Boers
F-06100 Nice
France
tel: 0033 493 984626
fax: 0033 493 989816
e-mail: rp-halm@worldnet.net

Handschuh Jaqueline
Faculté de Pharmacie
4, av. de l'Observatoire
F-75270 Cedex 06 Paris
France
tel: 0033 1 43291208
fax: 0033 1 43290592

***Jacobs Jennifer
23200 Edmonds Way, #A
Edmonds
WA 98026
USA
or:
Department of Ephidemiology
University of Washington
Seattle
98195WA
USA
or:
School of Public Health
Seattle
98195 WA
USA
tel: 001 206 5425595/5425598
fax: 001 206 3680843

**Kratky Karl Wilhelm
Institute for Experimental Physics
University of Vienna
Boltzmanngasse, 5
A-1090 Vienna
Austria
tel: 0043 1 4087008/16
fax: 0043 1 3102338
e-mail: kratky@acpx.exp.univie.ac.at

Lussignoli Sabrina
Istituto di Chimica e Microscopia Clinica
Università di Verona
Policlinico- Via delle Menegone
I-37134 Verona
Italy
tel/fax: 0039 45 8202978

*Marotta Paolo
International Center for the Study of the History of Medicine
IRFEST, Université Louis Pasteur, Strasbourg, France/Istituto Italiano per gli Studi filosofici, Napoli, Italy
Via Gennaro Serra, 75
I-80132 Napoli
Italy
tel: 0039 81 5890120/230
fax: 0039 81 7645058

Martino Maurizio
Viale Europa, 33
I-80053 Castellammare di Stabia (Napoli)
Italy
tel: 0093 81 8703080

Materni Massimo
Via Arrigo Boito, 14
I-24125 Bergamo
Italy
tel: 0039 35 233465

Minichini Gaetana
Centro Hahnemann
Via Martucci, 62
I-80121 Napoli, Italy
tel: 0039 81 7613766

Mundo Anna Maria
Via Pessina, 56
I-80135 Napoli
Italy
tel: 0039 81 5641233

Muranov Konstantin
Institute of Chemical Physics
Kosygin Street 4
CSI-117977 Moscow
Russia
tel: 007 095 9397441
fax: 007 095 9382156
e-mail: chembio@glas.apc.org

Musio Carlo
Istituto di Cibernetica, CNR
Via Toiano, 6
I-80072 Arco Felice (Napoli)
Italy
tel: 0039 81 8534131
fax: 0039 81 5267654
e-mail: camus@imagm.na.cnr.it

Neugebauer Anna
Istituto Internazionale di Genetica e Biofisica, CNR
Via Marconi, 10
I-80125 Napoli
Italy
tel: 0039 81 7257268
fax: 0039 81 5963123
e-mail: neugebauer@iigbna.iigb.na.cnr.it

***,**Oberbaum Menachem
Hagefen Street 13
IL-48254 Ramat Hasharon
Israel
tel: 00972 3 5497512/5562
fax: 00972 3 5497521
or: Ruth Ben Ari Institute for Clinical Immunology, Kaplan Hospital
IL-Rehovot
Israel

*Piantedosi Salvatore
Istituto di Cibernetica, CNR
Via Toiano, 6
I-80072 Arco Felice (Napoli), Italy
tel: 0039 81 8534212
fax: 0039 81 5267654
e-mail: sp@imagm.na.cnr.it

Picard Gilles
Istituto di Ricerca "Alberto Sorti"
Via Passalacqua, 19
I-10122 Torino, Italy
tel/fax: 0039 11 535252

Poitevin Bernard
Laboratoire Boiron
20 rue de la Libération
F-69110 Sainte-Foy-Les-Lyon
France
tel: 0033 723 24056
fax: 0033 4785 96916

**Popp Fritz-Albert
International Institute of Biophysics
Biophoton Research
Technology Center and
Technology-Park II
Sauerwiesen, 6
D-67661 Kaiserslauten
Germany
tel: 0049 6301 71310/31/55
fax 0049 6301 703119/713220

Querinjean Pierre
Avenue des Fauvettes, 15
B-1341 Ottignies-LLN
Belgium
tel: 0032 10 613741
fax: 0032 10 615704

Radil Tomas
Institute of Physiology
Czech Academy of Sciences
Videnska 1083
14220 Praha 4
Czech Republic
tel: 0042 2 4752409
fax: 0042 2 4719517

Reber Annie
Université de Rouen
Faculté des Sciences
Laboratoire de Neurosciences
F-76821 Mont-Saint-Aignan cédex
France
tel: 0033 2351 46724
fax: 0033 2351 46724
e-mail: annie.reber@univ.rouen.fr

Sabelli Italo
Corso Vittorio Emanuele, 112
I-80122 Napoli
Italy
tel: 0039 81 680104

*Santillo Silvia
Istituto di Cibernetica, CNR
Via Toiano, 6
I-80072 Arco Felice (Napoli)
Italy
tel: 0039 81 8534113
fax: 0039 81 5267654
e-mail: silvia@imagm.na.cnr.it

Santoianni Marisa
Via Enrico de Nicola, 7
I-80059 Torre del Greco
Italy
tel:0039 81 8825733
0037 845981

Sapio Marina
Corso Vittorio Emanuele, 112
I-80122 Napoli
Italy
tel: 0039 81 680104

Schmidt Herta
Körblerg., 5
A-8010 Graz
Austria
tel: 0043 316 326356/34 65 3183
fax: 0043 316 384515

Silvestri Giovanna
Via Tasso, 194
I-80121 Napoli, Italy
tel: 0039 81 660983

Sinha Babu Santi Prasnad
Department of Zoology
Vsiva-Bharati University
IN-731235 Santiniketan
West Bengal, India
tel: 0091 3463 52751
fax: 0091 3463 52968/672
e-mail:root@vbharat.ernet.in

**Smith Cyril W.
Department of Electric and Electronic
Engineering
University of Salford
M5 4WT Salford, UK
tel: 00441 61 7894768/7455000
fax: 00441 61 7455999/7308
or:
36, Westminister Road
Ellesmere Park, Eccles
M30 9EA Manchester, UK

Sportiello Sergio
Università degli Studi di Napoli "Federico II"
Facoltà di Medicina e Chirurgia
Servizio Speciale di Psicoterapia
Via Sergio Pansini, 5
I-80131 Napoli, Italy
tel: 0039 81 7462646/47
fax: 0039 81 7462372

Stieve Hennig
Institut für Biologie II
RWTH Aachen
Kopernikusstrasse 16
D-5100 Aachen
Germany
tel: 0049 241 804835
fax: 0049 241 8888133
e-mail: ld010na@dacth11
stieve@RWTH-Aachen.DE
zoologie@RWTH-Aachen.DE

***,**Sukul Nirmal Chandra
Faculty of Science
Vsiva-Bharati University
IN-731235 Santiniketan
West Bengal
India
tel: 0091 3463 52978
fax: 0091 3463 52672/3346 82612/33 247 4229
e-mail: root@vbharat.ernet.in
nirmal@vbharat.ernet.in

**,*Taddei-Ferretti Cloe
Istituto di Cibernetica, CNR
Via Toiano, 6
I-80072 Arco Felice (Napoli)
Italy
tel: 0039 81 8534113
fax: 0039 81 5267654
e-mail: cloe@imagm.na.cnr.it

***,**Van Wijk Roeland
Department of Molecular and Cellular Biology
Utrecht University
Padualaan 8
NL-3584 CH Utrecht
The Netherlands
tel: 0031 30 533350
fax: 0031 30 513655
e-mail: R.VANWIJK@BIOL.RUU.NL
Vinattieri Cristina
Istituto di Ricerca "Alberto Sorti"
Via Passalacqua, 19
I-10122 Torino
Italy
tel/fax: 0039 11 535252

Wiegant Fred A.C.
Department of Molecular Cell Biology
Utrecht University
Padualaan, 8
NL-3508 TB Utrecht
The Netherlands
tel: 0031 30 533972/184
fax: 0031 30 513655